The Centromere

The Centromere

K. H. Andy Choo

Murdoch Institute, Royal Children's Hospital, Melbourne, Australia

Oxford   New York   Tokyo
OXFORD UNIVERSITY PRESS
1997

Oxford University Press, Great Clarendon Street, Oxford OX2 6DP
Oxford New York
Athens Auckland Bangkok Bogota Bombay
Buenos Aires Calcutta Cape Town Dar es Salaam
Delhi Florence Hong Kong Istanbul Karachi
Kuala Lumpur Madras Madrid Melbourne
Mexico City Nairobi Paris Singapore
Taipei Tokyo Toronto Warsaw
and associated companies in
Berlin Ibadan

Oxford is a trade mark of Oxford University Press

Published in the United States
by Oxford University Press Inc., New York

Oxford University Press, 1997

A catalogue record for this book is available from the British Library

Library of Congress Cataloging in Publication Data
(Data available)

ISBN 0 19 857781 8 (Hbk)
0 19 857780 X (Pbk)

Typeset by Hewer Text Composition Services, Edinburgh
Printed in Great Britain by Redwood Books, Trowbridge, Wiltshire

To my parents and my wife

Preface

There can be very little doubt that the study of the centromere is an important and rapidly growing area of research. It has been known for quite some time now that the centromere is the critical structure that regulates the proper pairing and segregation of chromosomes during mitotic and meiotic cell divisions. Through the recognition of an increasing battery of proteins that are, at one time or another during the cell division cycle, associated with the centromere, more recent work has added a multitude of other possible vital cellular functions to this structure. In the past few years, the molecular analysis of model systems based on lower organisms is also coming to fruition and much of the information is rapidly being utilized in the study of the centromeres of higher organisms. With many new results accumulated, and an increasing interest from workers in related major areas of research, such as the study of the cell cycle, chromosome structure and function, molecular and clinical cytogenetics, genome sequencing, and gene therapy, the field has become massive. Although a number of excellent reviews have been published in journals, these are inevitably focused on a small part of the total subject. As such, information on the centromere has remained scattered. A book aimed at reviewing the whole subject would therefore both be timely as well as provide the impetus to spur the field forward.

As the aim of this book is to review the centromere field, the treatment of the subjects is necessarily broad. The book is written with both nonspecialists and specialists in mind, a task not often straightforward. For nonspecialists, the book is supported by many easy to read diagrams and explanatory segments that should provide a thorough background and understanding on all the major areas of centromere research. For specialists, all the subject matters have been comprehensively reviewed, with much of the existing complex and scattered information clearly summarized in diagramatic or tabulated form wherever possible. It is hoped that such a subject treatment will provide valuable scholastic material for graduate students, clinical and molecular cytogeneticists, scientists new to the centromere field, as well as scientists who have worked in this and closely related areas for many years.

Melbourne   K.H.A.C.
July 1996

Acknowledgements

I would like to thank Oxford University Press for the opportunity to write this book. I am especially grateful to my colleagues at the Murdoch Institute and all members of my laboratory staff for their support, to Paul Kalitsis for his contribution to Chapter 5, Ivan Francis for tireless computer advice, Michele Winsor for graphic work, Fiona Keltie for secretarial help, and Sue Shaw for library searches. Most of all, I wish to thank my wife Mei Ling and my sons Howard and Raymond for their love and unfailing understanding of the need to divert many of my home hours and much of my attention away from them throughout the gestation of this book.

Contents

1

Introduction

Fundamental to all the life processes is the ability of the cells to divide faithfully. This division can take the form of the equipartitioning of the genetic materials of a mother cell between two daughter cells (mitosis), or the accurate reduction of chromosome number (meiosis). That 'the cell has no other mode of origin than by division of a pre-existing cell', a realization made by Wilson (1925) and his contemporaries decades ago, clearly expounds the pivotal role played by these two divisional processes. In exercising this extraordinary ability to divide, the cells are equipped with innumerable mechanisms to achieve the high degree of accuracy that is needed. That this is so is evident from the observed low error rates of approximately 10^{-5} per generation for mitotic chromosome segregation and 10^{-3} per generation for meiotic chromosome segregation (Hawley 1988; Brown *et al.* 1991). How the cells accomplish these unusual and understandably complex feat is presently far from being fully understood. It is against this backdrop of realized importance and expected complexity that many keen biologists have risen to the challenge. In this pursuit, the centromere, a highly specialized structure of the chromosome, has long been recognized as having a centre stage role in the processes of cell division. Through many years of dedicated investigation, much has now been learnt about the centromere, and it is the purpose of this book to review all the recent advances in our understanding of this complex and highly intriging structure.

The book begins in this chapter with a brief overview of a number of the key features of the mitotic and meiotic processes, highlighting in particular aspects that involve the centromere directly. Chapter 2 takes an in-depth look at the centromere of a single cell eukaryote, the budding yeast *Saccharomyces cerevisiae*. This organism has a relatively small centromere DNA region that is highly amenable to genetic and biochemical manipulation, and much information has been gained through its study. Chapter 3 examines the centromere of another single cell eukaryote, the fission yeast *Schizosaccharomyces pombe*, and provides a good comparison between the centromeres of two simple organisms. The remaining six chapters of the book concentrate mainly on the centromeres of the higher eukaryotes. This begins in Chapter 4 with a description of the structural and various general properties of the higher eukaryotic centromere. The next two chapters then examine the two basic constituents of the centromere: its DNA (Chapter 5) and its proteins (Chapter 6). Chapter 7 continues the study of the higher eukaryotic centromere and reviews all the known clinical anomalies that are associated with the human centromere. The book concludes in Chapter 8 with a discussion of some practical applications that have emerged through the study of the centromere.

Universal stages of mitosis and meiosis

Mitosis has classically been divided into interphase (within which are the G_1 or Gap 1 phase, S or DNA-synthesis phase, and G_2 or Gap 2 phase), prophase, prometaphase, metaphase, anaphases A and B, telophase, and cytokinesis (Fig. 1.1). The first division of meiosis is generally composed of similar stages and events, with the obvious exception of an extended prophase that includes a period in which homologous chromosomes pair and undergo recombination. This recombination event allows most homologous chromosomes to remain attached to each other by virtue of chiasmata. These attachments constrain the homologues to adopt and maintain orientations to opposite spindle poles during prometaphase and metaphase, thus ensuring reductional division at anaphase I.

While many detailed aspects of the various mitotic and meiotic phases may differ between organisms, the accurate separation of chromosomes during these two divisional processes is accomplished by three universally shared stages (reviewed by Koshland 1994; Fuller 1995). First, special structures necessary to mediate microtubule-dependent movements have to be assembled. These structures include the spindle and the centromere. Second, the sister chromatids (in mitosis) or paternal and maternal homologues (in meiosis), assisted by correct pairing, have to attach to microtubules extending from opposite poles of the spindle. Third, the oppositely oriented sister chromatids or homologues, which tend to align midway between the two spindle poles following microtubule attachment, have to become unpaired and segregate from each other toward their respective poles.

The spindle and microtubules

The spindle is a macromolecule that attaches to, organizes, and directs the movements of chromosomes during mitosis and meiosis. This macromolecule, which is most prominently composed of microtubules, is a highly dynamic structure, forming anew during the cell cycle when it is time to divide and disassembling when the division is over (Salmon 1989a, 1989b; Karsenti 1991; Wadsworth 1993). Microtubules, on the other hand, are assembled from heterodimeric tubulin subunits and display both structural and kinetic polarity. This polarity is reflected in the designation of the microtubule ends: the *minus end*, where tubulin subunits are added and lost more slowly, and the *plus end*, where tubulin subunits are rapidly added and lost (Cassimeris *et al.* 1987). Spindle microtubules are oriented with their minus ends at the microtubule-organizing centres (or spindle poles) and their plus ends pointing toward the centre of the cell or the cell cortex (Fig. 1.1).

The centromere

The centromere is a highly differentiated structure of the chromosome that fulfils a multitude of essential mitotic and meiotic functions. (i) *Sister chromatid pairing*. The juxtapositioning of replicated sister chromatids is thought to result from an underlying physical attachment at the centromere. At the metaphase/anaphase transition, some signals act on the centromere to release the cohesion between the sister chromatids. (ii) *Mitotic and meiotic spindle attachment*. The centromere is a major chromosomal site onto which the mitotic and meiotic spindle attaches. Attachment occurs at the specialized disc-shaped kinetochore, which is a proteinaceous structure found at the periphery of the centromere. Such attachments can involve from 1 to over 100 microtubules per chromosome. (iii) *Chromosome movement*. The centromere harbours a number of molecular motors that move chromosomes during mitosis and meiosis. This has been shown by movement of chromosomes relative to spindle microtubules both *in vivo* and *in vitro*. (iv) *Checkpoint control*. The centromere is directly involved in cell cycle checkpoint control. For example, in most cells, centromeres of mitotic chromosomes that have not yet attained a stable bipolar orientation on the spindle would send a signal to delay the onset of metaphase/anaphase transition. (v) *Marshalling of passenger proteins*. A relatively large number of proteins have now been shown to concentrate at centromeres during the early stages of the mitotic cell division cycle. These proteins move with the chromosomes to the metaphase plate and transfer from the chromosomes to various mitotic organelles to serve a wide range of roles.

Microtubule capture and bipolar attachment of centromeres

Sister chromatids segregate from each other because their centromeres are bound to microtubules from opposite spindle poles. This orientation of sister chromatids is achieved prior to chromosome segregation by a process that ensures a bipolar attachment of the sister chromatids (reviewed by Koshland 1994). Pairing between sister chromatids and the proper linking of the kinetochores in opposite directions (Fig. 1.2A) clearly play a key role in this process. Bipolar attachment begins with the stable association of a microtubule with one of the paired sister centromeres to achieve a monopolar attachment (Rieder and Salmon 1994). Following monopolar attachment, the sister chromatid pair undergoes slow oscillations toward and away from the spindle poles by the loss and gain of tubulin subunits on the microtubule, until eventually a microtubule is captured by the unoccupied centromere on the other sister chromatid. If the microtubules bound to the two sister

Introduction

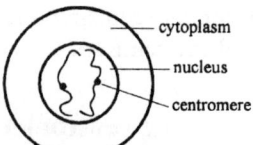

Interphase Gap 1

G₁ phase

Chromosome replication

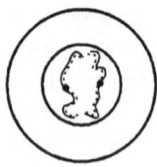

S phase

Interphase Gap 2

G₂ phase

Centrosome separation;
Chromosome
condensation;
Kinetochore assembly

Prophase

Nuclear membrane
breakdown;
Kinetochore microtubule
capture;
Chromosome
congression onto
metaphase plate

Prometaphase

Spindle stabilization;
Chromosome alignment
at metaphase plate

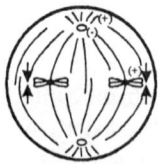

Metaphase

Sister chromatid
dissociation;
Chromosome-to-pole
movement

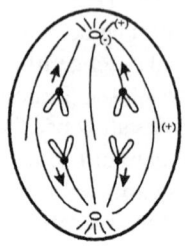

Anaphase A

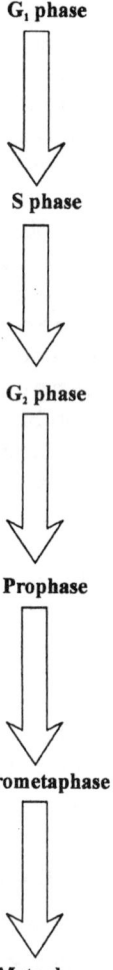

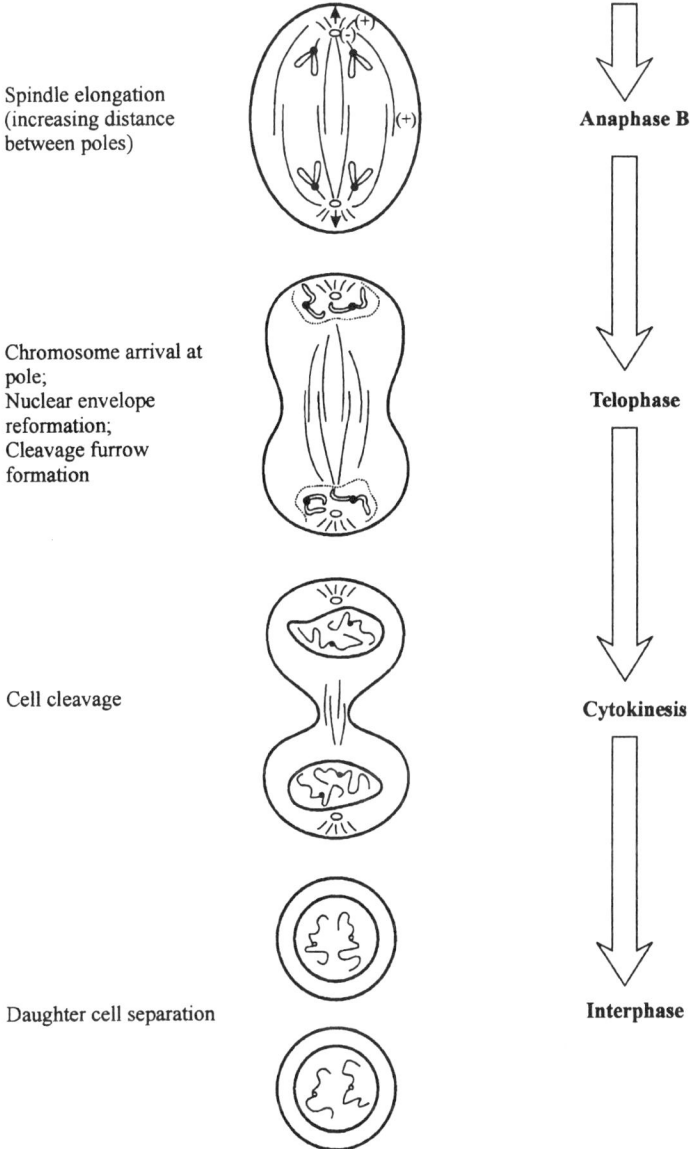

Spindle elongation
(increasing distance
between poles)

Anaphase B

Chromosome arrival at
pole;
Nuclear envelope
reformation;
Cleavage furrow
formation

Telophase

Cell cleavage

Cytokinesis

Daughter cell separation

Interphase

Fig. 1.1 Stages and significant events of mitosis, showing in particular the roles of the centrosome, the spindle, and the centromere in accomplishing chromosome segregation. (+) and (−) denote the plus and minus ends of spindle microtubules, respectively. Arrows indicate the direction of the force generators that are required to power the various movements in mitosis.

chromatids emanate from opposite poles, the centromeres remain stably bound to them, thus establishing a bipolar attachment. If the microtubules bound to the two sister chromatids emanate from the same pole, a mal-oriented attachment results and the centromere on one of the sister chromatids eventually dissociates from its microtubule. This process is repeated until the unoccupied centromere captures a microtubule from the correct spindle pole. Once bipolar attachment is achieved, tension between the two oppositely linked kinetochores is necessary for the subsequent oscillatory movements to achieve congression at the metaphase plate (Goldstein 1980; McNeill and Berns 1981; Kerrebrock *et al.* 1992; Skibbens *et al.* 1993; Cassimeris *et al.* 1994). During this movement, when one kinetochore is stretched and moving towards its pole, the other is passively following or actively moving away from its pole (Skibbens *et al.* 1993). The chromatids are then maintained at the metaphase plate by this 'cooperative switching' of forces on each sister kinetochore until the commencement of anaphase.

Unlike in mitosis and meiosis II, the mechanics of chromosome association and movement in meiosis I are somewhat different because of the unique requirement of homologous chromosomes to pair and recombine to form a bivalent in most organisms. Here, the replicated homologous chromosomes are brought into alignment by the synaptonemal complex, a structure present transiently from zygotene until diplotene of prophase I. Reciprocal recombination leads to the formation of chiasmata that act to physically link the two homologous chromosomes of a bivalent after the dissolution of the synaptonemal complex (Fig. 1.2B). Chiasmata are thought to enable the bivalent to orient and form stable attachments to both meiosis I poles, thus functioning analogously to sister kinetochore cohesion in mitosis and meiosis II. The bivalent attains stable bipolar attachment because of tension resulting from spindle forces from the two poles counteracted by attachment of the homologues to each other at the chiasma (Darlington 1932; Nicklas and Koch 1969; Miyazaki and Orr-Weaver 1992; Rockmill and Roeder 1994). The migration of sister chromatids of each homologue as a paired unit to the pole is achieved by the two sister kinetochores orienting in the same direction to attach to the same pole (Fig. 1.2B).

'Wait anaphase' checkpoint at the centromere

Centromere separation marks the onset of anaphase. If aneuploidy is to be avoided, it is essential that this separation occurs relatively synchronously on all sister chromatids. This is achieved by the cells exacting a 'wait anaphase' checkpoint control to delay anaphase onset until metaphase congression is completed with all chromosomes attaining bipolar attachment to microtubules and properly aligned at the spindle midzone (reviewed by Pluta *et al.* 1995). Evidence suggests that tension generated at the kinetochore is an important component of this checkpoint control. Presumably, the application of tension switches off the 'wait anaphase' signal-generating mechanism at the kinetochore and allows the cells to proceed through anaphase (Li and Nicklas 1995). However, other studies have indicated that, rather than kinetochore tension, it is the unattached kinetochores that transmit the 'wait anaphase' signal (Rieder *et al.* 1995).

Chromosome movement

The direction and magnitude of chromosome motion are ultimately determined by the net force acting on the chromosome. The kinetochore is a major site on the chromosome upon which forces act to effect mitotic and meiotic chromosome movement (reviewed by Ault and Rieder 1994). Both plus end-directed and minus end-directed microtubule motors are known to be present at the kinetochore (Rieder and Alexander 1990; Hyman and Mitchison 1991; Rieder 1991; Skibbens *et al.* 1993; Wadsworth 1993). A motor is classified as plus end-directed if its activity moves a chromosome from the minus end to the plus end of a microtubule, or as minus end-directed if its activity moves a chromosome from the plus end to the minus end of a microtubule. In addition to motor activities residing at the kinetochore, motor molecules based in the microtubules or on chromosome arms (Afshar *et al.* 1995; Murphy and Karpen 1995; Vernos *et al.* 1995; Wang and Adler 1995; reviewed by Fuller 1995) that are capable of transducing directed forces to move chromosomes have been described (reviewed by Barton and Goldstein 1996). The different motors can work alone or in combination, and in many cases they cooperate to create oppositely acting forces the sum of which actually produce the necessary movements (reviewed by Fuller 1995; Barton and Goldstein 1996). Furthermore, in powering these movements, it has long been recognized that microtubule polymerization and depolymerization are two potential sources of force (Inoue and Sato 1967; Salmon 1989a, 1989b; Cassimeris *et al.* 1994; Rieder and Salmon 1994; Desai and Mitchison 1995).

(A) Mitosis

Metaphase

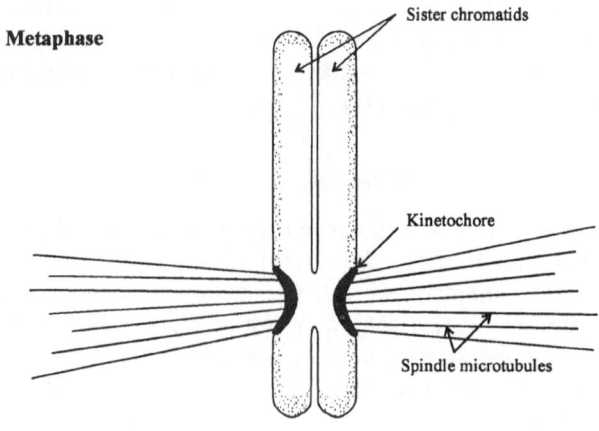

Anaphase

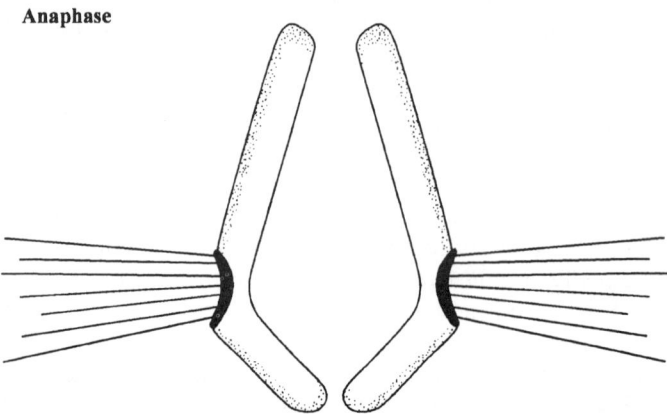

Fig. 1.2 Centromere pairing and separation. (A) At mitotic metaphase, newly replicated sister chromatids are joined at both the centromere and along the arms. The kinetochore discs are facing away from each other so that they can attach to spindle microtubules to effect movement to opposite poles. At the onset of anaphase, the sister chromatids separate completely. (B) At meiotic metaphase I, the newly replicated sister chromatids (S1/S1 or S2/S2) of the two homologues are joined at both the centromere and along the arms. The kinetochore discs of each pair of sister chromatids are facing in the same direction so that they attach to the same pole. The sister chromatid arm pairing is thought to be necessary to stabilize the chiasma. At the onset of anaphase I, the association of the arms is lost as the chiasma resolves and homologues segregate. However, association of the centromere is maintained until the onset of anaphase II.

(B) Meiosis

Metaphase I

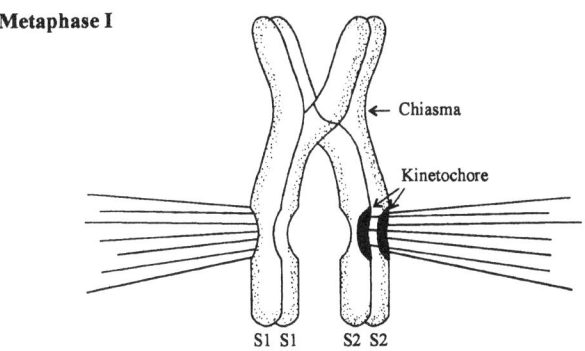

Chiasma

Kinetochore

S1 S1   S2 S2

Anaphase I

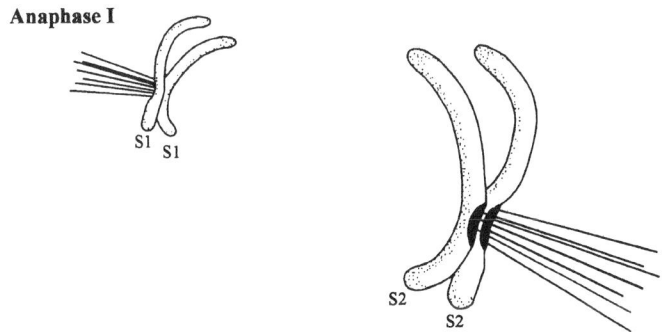

S1 S1

S2

S2

Anaphase II

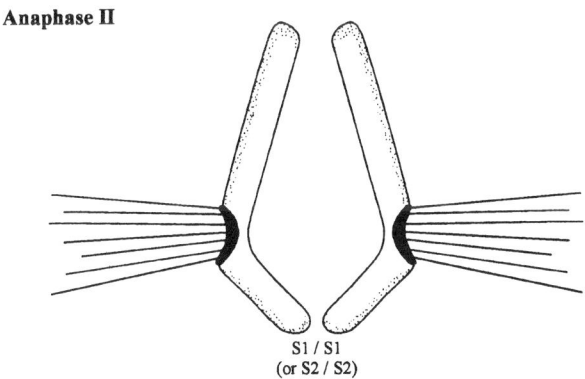

S1 / S1
(or S2 / S2)

References

Afshar, K., Barton, N. R., Hawley, R. S., and Goldstein, L. S. B. (1995). DNA binding and meiotic chromosomal localization of the *Drosophila* nod kinesin-like protein. Cell *81*, 129–138.

Ault, J. G. and Rieder, C. L. (1994). Centrosome and kinetochore movement during mitosis. Cell Biol *6*, 41–49.

Barton, N. R. and Goldstein, L. S. B. (1996). Going mobile: microtubule motors and chromosome segregation. Proc. Natl Acad. Sci. USA *93*, 1735–1742.

Brown, M., Garvik, B., Hartwell, L., Kadyk, L., Seeley, T., and Weinert, T. (1991). Fidelity of mitotic chromosome transmission. Cold Spring Harbor Symp. Quant. Biol. *56*, 359–365.

Cassimeris, L. U., Walker, R. A., Pryer, N. K., and Salmon, E. D. (1987). Dynamic instability of microtubules. BioEssays *7*, 149–154.

Cassimeris, L., Rieder, C. L., and Salmon, E. D. (1994). Microtubule assembly and kinetochore directional instability of vertebrate monopolar spindles: implications for the chromosome congression. J. Cell Sci. *107*, 285–297.

Darlington, C. D. (1932). Recent Advances in Cytology. Churchill, London.

Desai, A. and Mitchison, T. J. (1995). A new role for motor proteins as couplers to depolymerizing microtubules. J. Cell Biol. *128*, 1–4.

Fuller, M. T. (1995). Riding the polar winds: chromosomes motor down east. Cell *81*, 5–8.

Goldstein, L. (1980). Mechanisms of chromosome orientation revealed by two meiotic mutants in *Drosophila melanogaster*. Chromosoma *78*, 79–111.

Hawley, R. (1988). Genetic Recombinations, eds. Kucherlapati, R and Smith, G. R. Am. Soc. Microbiol., Washington, DC, 497–527.

Hyman, A. A. and Mitchison, T. J. (1991). Two different microtubule-based motor activities with opposite polarities in kinetochores. Nature (Lond.) *351*, 206–211.

Inoue, S. and Sato, H. (1967). Cell motility by labile association of molecules. The nature of mitotic spindle fibers and their role in chromosome movement. J. Gen. Physiol. *50*, 259–292.

Karsenti, E. (1991). Mitotic spindle morphogenesis in animal cells. Semin. Cell Biol. *2*, 251–260.

Kerrebrock, A. W., Miyazaki, W. Y., Birnby, D., and Orr-Weaver, T. L. (1992). The *Drosophila mei-S332* gene promotes sister-chromatid cohesion in meiosis following kinetochore differentiation. Genetics *130*, 827–841.

Koshland, D. (1994). Mitosis: back to the basics. Cell *77*, 951–954.

Li, X. and Nicklas, R. B. (1995). Mitotic forces control a cell-cycle checkpoint. Nature *373*, 630–632.

McNeill, P. A. and Berns, M. W. (1981). Chromosome behavior after laser micro-irradiation of a single kinetochore in mitotic PtK2 cells. J. Cell Biol. *88*, 543–553.

Miyazaki, W. Y. and Orr-Weaver, T. L. (1992). Sister-chromatid misbehavior in *Drosophila ord* mutants. Genetics *132*, 1047–1061.

Murphy, T. D. and Karpen, G. H. (1995). Localization of centromere function in a *Drosophila* minichromosome. Cell *82*, 599–609.

Nicklas, R. B. and Koch, C. A. (1969). Chromosome micromanipulation III. Spindle fiber tension and the reorientation of mal-oriented chromosomes. J. Cell Biol. *43*, 40–50.

Pluta, A., Mackay, A., Ainsztein, A., Goldberg, I., and Earnshaw, W. (1995). The centromere: hub of chromosomal activities. Science *270*, 1591–1594.

Rieder, C. L. (1991). Mitosis: towards a molecular understanding of chromosome behavior. Curr. Opin. Cell Biol. *3*, 59–66.

Rieder, C. L. and Alexander, S. P. (1990). Kinetochores are transported poleword along a single astral microtubule during chromosome attachment to the spindle in newt lung cells. J. Cell Biol. *110*, 81–95.

Rieder, C. L. and Salmon, E. D. (1994). Motile kinetochores and polar ejection forces dictate position on the vertebrate mitotic spindle. J. Cell Biol. *124*, 223–233.

Rieder, C. L., Cole, R. W., Khodjakov, A., and Sluder, G. (1995). The checkpoint delaying anaphase in response to chromosome monoorientation is mediated by an inhibitory signal produced by unattached kinetochores. J. Cell Biol. *130*, 941–948.

Rockmill, B. and Roeder, G. S. (1994). The yeast *medl* mutant undergoes both meiotic homolog nondisjunction and precocious separation of sister chromatids. Genetics *136*, 65–74.

Salmon, E. D. (1989a). Mitosis: Molecules and Mechanisms, eds. Hyams, J. S. and Brinkley, B. R. Academic Press, London, 119–181.

Salmon, E. D. (1989b). Cell Movement: Kinesin, Dynein, and Microtubule Dynamics. Liss, New York, Vol. 2, 431–440.

Skibbens, R. V., Skeen, V. P., and Salmon, E. D. (1993). Directional instability of kinetochore motility during chromosome congression and segregation in mitotic newt lung cells; a push-pull mechanism. J. Cell Biol. *122*, 859–875.

Vernos, I., Raats, J., Hirano, T., Heasman, J., Karsenti, E., and Wylie, C. (1995). Xklp1, a chromosomal *Xenopus* kinesin-like protein essential for spindle organization and chromosome positioning. Cell *81*, 117–127.

Wadsworth, P. (1993). Mitosis: spindle assembly and chromosome motion. Curr. Opin. Cell Biol. *5*, 123–128.

Wang, S. Z. and Adler, R. (1995). Chromokinesin: a DNA-binding, kinesin-like nuclear protein. J. Cell Biol. *128*, 761–768.

Wilson, E. (1925). The Cell in Development and Inheritance, 2nd Edn. Macmillan, New York.

2

Centromere of budding yeast
S. cerevisiae

The budding yeast *Saccharomyces cerevisiae*, commonly used by brewers and bakers, is an amenable system that is easily manipulated genetically, molecularly, and biochemically, and has proved to be a powerful experimental tool for the study of the eukaryotic centromere. This organism contains in the haploid state 16 Mb of DNA per cell, or approximately 0.5% of the amount found in the haploid human genome of 3000 Mb. The DNA of *S. cerevisiae* is contained within 16 chromosomes, ranging in size from about 260 kb to 3 Mb (Carle and Olson 1985). The entire genome of this organism has now been completely sequenced (see *Saccharomyces* Genome Database).

Life cycle of *S. cerevisiae*

S. cerevisiae is an oblately spheroid or ovoid shaped cell of some 3 μm in diameter which, under optimal nutritional conditions, divides mitotically every 90 minutes. Mitotic division is achieved with the cell forming a bud-like appendage (hence the name budding yeast) into which a full complement of newly replicated chromosomes moves (Fig. 2.1) (Watson *et al.* 1987). The bud is then pinched off to yield a parent cell and a smaller daughter cell. The daughter cell will grow to the size of its parent cell and then, in turn, will form a new bud. Like all yeasts, *S. cerevisiae* can exist in a haploid or diploid state, with the haploid cells being either of two mating types, designated **a** and α. Fusion of **a** and α cells yields **a**/α diploid cells which, under good nutritional conditions, will grow and divide mitotically, maintaining the diploid state. However, when starved, the diploid cells undergo meiosis to yield four progeny haploid cells, which become encapsulated as spores within a thick-walled, sac-like ascus structure. (This structure, in which the four primary spore products of a given meiosis are contained as a 'tetrad' within a single ascus that separates them from the products of other meiotic divisions, makes yeasts an excellent tool for studying meiotic events.) Subsequently, upon sporulation and rupture of the ascus, the spores are released. These spores then germinate and undergo new rounds of haploid existence.

In comparison with a typical eukaryotic mitotic cell cycle, the budding yeast has normal G_1 and S phases but does not appear to have a G_2 phase; the microtubule-based spindle begins to form early in the cycle, during S phase, and the cell proceeds directly into the M phase. In contrast to higher eukaryotes, there is no visible chromosome condensation during mitosis. Furthermore, the nuclear envelope does not break down, and the micro-tubules of the mitotic spindle are formed inside the nucleus and attached to spindle pole bodies at the nuclear periphery (see Fig. 3.4).

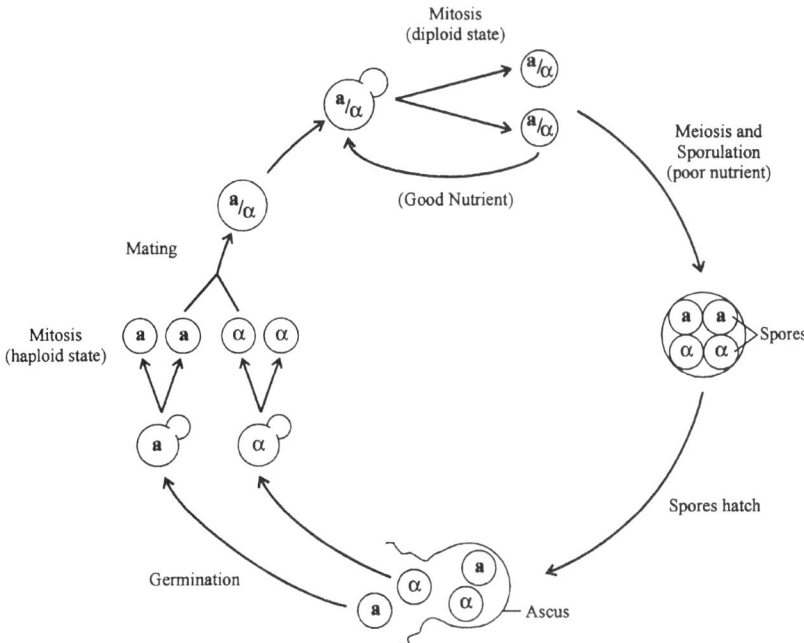

Fig. 2.1 Life cycle of *S. cerevisiae*. **a** and α are cells of the two mating types. Diploid cells are formed by mating between **a** and α haploid cells. Under good nutritional conditions, haploid and diploid yeast cells will maintain their respective genomic states and reproduce by mitosis. In a nutritionally deficient environment, diploid cells will divide meiotically.

Centromere assays

A unique advantage of using *S. cerevisiae* for centromere study is the availability of well-established functional assays for isolated structural centromeric components. The three most widely used assay systems are the plasmid assay, homologous recombination assay, and YAC assay.

Plasmid assay

Yeast cells are easily transformed by exogenous DNA, and numerous 'shuttle' plasmid vectors have been constructed that can replicate in either *Escherichia coli* or *S. cerevisiae*. Yeast plasmids constructed *in vitro* that contain a functional replicator (autonomously replicating sequence *ARS*, or the yeast 2 μm plasmid origin of replication) are capable of efficient extrachromosomal replication but are very unstable during mitotic cell divisions (Fig. 2.2A). Due to preferential segregation to the parent cell (Murray and Szostak 1983), the plasmid is rapidly lost from the population of cells under nonselective conditions of growth and the small percentage of cells that retain the plasmid accumulate high copy numbers (20–50 plasmids per cell). Introduction of a functional centromere (*CEN*) to an *ARS* plasmid provides an active mechanism for the segregation of the resulting aneuploid minichromosome (reviewed by Szostak and Blackburn 1984) (Fig. 2.2B). A marker that provides yeast colony colour can also be incorporated into the plasmid to simplify assessment of the stability of the minichromosome (Hieter *et al.* 1985a; Koshland *et al.* 1985). Stable maintenance of a single copy of the minichromosome in daughter cells (1:1 segregation) through successive cell divisions marks the inclusion of a mitotically active *CEN* DNA element in the *CEN* plasmid. However, the rate of mitotic chromosome loss for these small circular *CEN/ARS* plasmids is approximately 10^{-2} per generation (Fitzgerald-Hayes *et al.* 1982; Stinchcomb *et al.* 1982; Murray and Szostak 1983), which is three orders of magnitude greater than the rate of loss for authentic yeast chromosomes (Hartwell *et al.* 1982). This higher rate of *CEN/ARS* minichromosome loss could be accounted for by aberrant 1:0 segregation (i.e. chromosome loss; due, for example, to failure of the plasmid to replicate during the preceding cell cycle, or the destruction or physical loss of one of the two replicated copies of the plasmid) or 2:0 segregation (i.e. chromosome nondisjunction; due, for example, to failure of sister chromatids to disjoin or engage in the segregation apparatus) (Hieter *et al.* 1985a) (Fig. 2.2B).

Meiotic function of the *CEN* DNA is normally assayed by the conventional technique of tetrad analysis. A haploid yeast strain containing the *CEN* plasmid or minichromosome is crossed with a haploid of the opposite mating type. Since both the mating parents are mutationally defective for a wild-type gene carried by the plasmid, this genetic marker can be used to score the four haploid progeny (tetrad) for the presence of the plasmid. The sister spores resulting from the second meiotic division are identified by the presence in the cross of a heterozygous marker known to be tightly linked to the centromere on any of the host chromosomes. When diploids containing *CEN* plasmids are analysed by this procedure, the genetic marker on the plasmid segregates in a 2+:2− manner in the majority of the tetrads, i.e. two haploid progeny with the minichromosome and two without, indicating the presence in most of the diploid cells of a single copy of the minichromosome that has undergone an aneuploid segregation pattern typifying normal occurrence of both meiosis I and II (Fig. 2.2C). Precocious sister chromatid separation in meiosis I due, for example, to failure of proper sister chromatid attachment, will similarly result in 2+:2− tetrads, but segregation in this case

between these observations remains to be resolved, given the greater reliability of the YAC system for meiotic assay, it is probable that CDEI has at least some functional role in meiosis II segregation of the yeast chromosomes.

CDEII

Deletions or insertions that alter the length of the AT-rich CDEII domain decrease mitotic chromosome stability by 10- to 1000-fold, and deletions within CDEII also result in some random segregation during the first meiotic division (reviewed by Gaudet and Fitzgerald-Hayes 1990). Missegregation during meiosis is also seen when a 31 bp deletion of CDEII is placed on a YAC where, in meiosis I, precocious sister chromatid separation is increased about 30-fold and nondisjunction 5-fold, while in meiosis II, nondisjunction is increased 30-fold compared to wild-type (Hegemann and Fleig 1993). Centromere function in CDEII deletion mutants can be almost completely restored by insertion of an equivalent length of a random AT-only DNA sequence (Gaudet and Fitzgerald-Hayes 1990; Murphy *et al.* 1991). Thus, high AT content and length, perhaps influencing the conformation of this DNA element, rather than specific nucleotide sequence, are critical in the role of CDEII for faithful mitotic and meiotic chromosome transmission.

CDEIII

Mutational analyses have shown that CDEIII is absolutely essential for centromere function. Certain single base pair changes within this domain and deletions of several nucleotides at the boundaries of CDEIII completely inactivate the centromere. Extensive mutational analysis (McGrew *et al.* 1986; Cumberledge and Carbon 1987; Gaudet and Fitzgerald-Hayes 1987; Ng and Carbon 1987; Jehn *et al.* 1991; Niedenthal *et al.* 1991) reveals the CCG sequence located at the centre of bilateral symmetry to be absolutely essential for centromere function. A change in any of these three nucleotides inactivates the *S. cerevisiae* centromere. No other single base pair exchange within *CEN* nor partial deletion of CDEII or complete deletion of CDEI has such a drastic effect on centromere function. Interestingly, analysis of symmetrical mutations in the CDEIII core sequence reveals a greater overall contribution to mitotic centromere function by the right half-site of the palindrome (Hegemann and Fleig 1993). Thus, in both CDEIII and CDEI, the half-sites facing CDEII are less important than the half-sites pointing away from the rest of the *CEN* DNA.

CDEIII is also essential for meiosis, as point mutations in the central core element (TTCCGAA) strongly impair meiotic chromosome transmission. However, exceptions have been reported for mutants at the two outer nucleotides (i.e. C and G) of the highly conserved CCG sequence located at the centre of bilateral symmetry, which have a less pronounced effect on meiotic segregation (Gaudet and Fitzerald-Hayes 1989, 1990).

It is noteworthy that despite the apparent bilateral symmetry of sequences surrounding the central C nucleotide within the consensus CDEIII sequence, a closer examination of individual *CEN* DNA (Fig. 2.3) reveals that only the central 5 bp (CCGAA) sequence is fully conserved in all 16 centromeres. In most of the *CEN* DNA, the symmetry outside of the 5 bp region is either partial or completely absent. It is therefore unclear what, if any, the relevance of bilateral symmetry outside of this region is.

Centromere proteins

A protein (CBF1) that binds to CDEI, and a quadripartite protein complex (CBF3) that binds to CDEIII, have been identified. In addition, several other proteins (CSE4p, Kar3p, CSE1, CSE2, CBF5p, and Mif2) have been reported to have a possible interactive involvement with either the *CEN* DNA domains and/or the *CEN*-associated proteins. The properties of these proteins are summarized in Table 2.1 and Fig. 2.4, and discussed below.

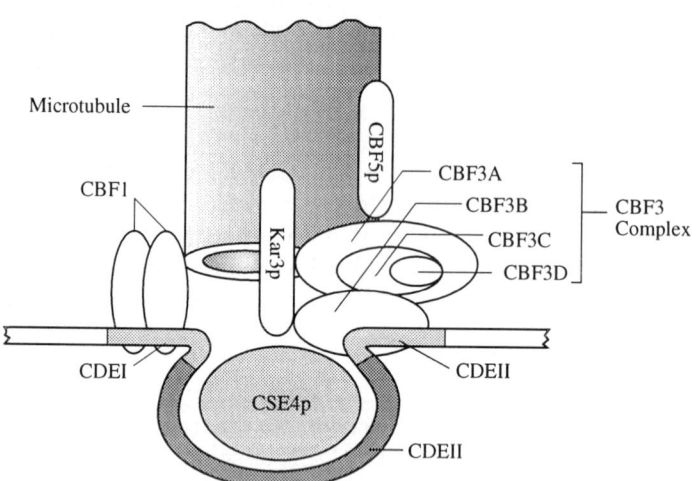

Fig. 2.4 Structural model of the *S. cerevisiae* kinetochore. CBF1 is depicted to bind CDEI as a dimeric structure (Mellor *et al.* 1990; Dowell *et al.* 1992) and interacts with microtubules (Cai and Davis 1990). The CBF3 complex contains one unit each of CBF3A, CBF3B, CBF3C, and CBF3D with CBF3B binding directly onto CDEIII (at the highly conserved CCG triplet within CDEIII) (Lechner 1994). In the presence of CDEIII-bound CBF3, Kar3p becomes active in microtubule binding and minus end-directed movement (Middleton and Carbon 1994). Another protein that interacts with CBF3 and binds microtubules is CBF5p (Jiang *et al.* 1993b). The *CEN* DNA is shown wrapped around a nucleosomal core that may contain the histone H3-like protein, CSE4p (Pluta *et al.* 1995; Stoler *et al.* 1995). Three other proteins not included in the diagram, Mif2 (Meluh and Koshland 1995), CSE1 and CSE2 (Xiao *et al.* 1993), have also been implicated to have a possible interactive role with CDEII, CBF1, components of the CBF3 complex, and/or the microtubule.

Table 2.1 Protein components of the *S. cerevisiae* centromere

| Protein component | Interacts with | | MW (kDa) | | Proposed functions | Putative human |
Designation[1]	DNA	Protein	Gel[2]	ORF[3]		homologue
CBF1/CBF1p/CPF1/Cep1p/CP1[4]	CDEI	self–self; Microtubule?	58	39	Microtubule binding; transcription	n.a.
CBF3	CDEIII	Kar3p	240	n.a.	(see subunits below)	
CBF3A/cbf2p/ctf14p/Ndc10p	n.a.	CBF3B;CBF3C; Self–self	110	112	Microtubule binding; sister chromatid cohesion	n.a.
CBF3B/Cep3p		CDEIII CBF3A;CBF3C	64	71	*CEN* DNA binding	n.a.
CBF3C/ctf13p	n.a.	CBF3A;CBF3B	58	56	n.a.	CENP-B
CBF3D/Skp1p	n.a.	3BF3C	29	22	CBF3-*CEN* complex assembly; cell cycle regulation	p19Skp1
CSE4p	CDEII?	n.a.	n.a.	27	Centromere-specific chromatin	CENP-A
Kar3p	*CEN?*	CBF3; Microtubule	80	84	Motor movement	CENP-E
CSE1	CDEII?	Microtubule?	n.a.	109	n.a.	CAS
CSE2	CDEII?	Microtubule?	n.a.	17	n.a.	n.a.
CBF5p/p64'	n.a.	CBF3? Microtubule?	64	55	Microtubule binding	n.a.
Mif2	CDEII?	CBF1? CBF3?	n.a.	63	n.a.	CENP-C

n.a. = not available or applicable.
[1] Alternative designations used for the same component are separated by '/'. For ease of reference, the designation in bold will be employed throughout the text. The same designations in italics are used in the text to denote the corresponding genes that encode the proteins.
[2] Molecular weight as determined by SDS gel electrophoresis.
[3] Molecular weight as predicted from the open reading frame of the gene sequence.
[4] The CBP1 designation is no longer in use because it is now adopted for the cytochrome b processing genes (Cai and Davis 1990).

CDEI-binding protein, CBF1

CBF1 (or centromere-binding factor 1) is a relatively abundant polypeptide (greater than 500 copies per cell) that binds the CDEI DNA element directly (Bram and Kornberg 1987; Baker *et al.* 1989; Cai and Davis 1989; Jiang and Philippsen 1989). The protein is unusually stable and retains its specific DNA-binding activity after boiling or following elution and renaturation from SDS gels. Its predicted sequence (Baker and Masison 1990; Cai and Davis 1990; Mellor *et al.* 1990) indicates a region near the C-terminus with marked homology to a DNA-binding and protein dimerization motif known as the basic helix–loop–helix (bHLH) (Murre *et al.* 1989) identified in factors controlling tissue-specific transcription in higher eukaryotes.

In functional studies, haploid cells containing a disrupted *CBFI* gene with no detectable CBFI protein activity are viable, indicating that the gene is nonessential (Baker and Masison 1990). However, such cells grow more slowly than their isogenic counterparts, resulting in a 30–35% increase in cell doubling time. In addition, the cells exhibit a 5- to 10-fold increased rate of mitotic chromosome and *CEN* plasmid loss. Interestingly, *CBFI* gene disruption also converts the cells to a methionine auxotrophic phenotype (see below). The observation that loss of *CBF1* function results in hypersensitivity to low levels of the antimitotic, microtubule-disrupting drug thiabendazole (Cai and Davis 1989) suggests that one role of the CBF1 protein in chromosome segregation may be to improve the efficiency with which contact between the kinetochore and spindle microtubules is established or maintained (Foreman and Davis 1993).

CBF1 is also believed to be a transcription factor. Within the yeast genome, the DNA sequence of CDEI is found at several noncentromeric locations, in some instances in promoter regions of genes, including those for *Gal2*, nuclear encoded mitochondrial cytochrome *c* oxidoreductase subunits, *TRP1*, *MET25*, and the other six genes in the methionine biosynthesis pathway that are co-regulated with *MET25* (*MET2*, *MET3*, *MET8*, *SAM2*, *MET14*, and *MET16*) (Bram and Kornberg 1987; Dorsman *et al.* 1988; Baker *et al.* 1989; Thomas *et al.* 1989, 1990; Cherest *et al.* 1990; Mellor *et al.* 1991; Korch *et al.* 1991). For example, deletion of CDEI sites upstream of the *MET25* gene, which encodes *O*-acetylhomoserine sulfhydrylase, results in a reduction of enzyme activity by over 90% and a corresponding decrease in the steady-state levels of *MET25* mRNA (Thomas *et al.* 1989); thus, this severe impairment in the transcription of *MET25*, and possibly the other CDEI-containing genes in the methionine metabolism pathway, offers a good explanation for the observed methionine auxotrophy in cells containing a disrupted *CBFI* gene.

Several studies have examined the mechanism of the bimodal (centromere and transcription) role of CBF1. Analysis of *CBFI* alleles with mutations clustered in or just downstream from the bHLH domain has demonstrated a class that is more compromised for transcriptional activation and a class that is more compromised for chromosome loss and thiabendazole hypersensitivity. These results indicate that at least some aspects of the role of CBF1 protein in chromosome segregation and transcriptional activation are distinct. In contrast, increased chromosome loss and thiabendazole hypersensitivity are not separated in any of the alleles, suggesting that these phenotypes reflect the same mechanistic defect. The notion that the roles of CBF1 in transcription and centromere function are different is supported by the isolation of mutant *S. cerevisiae* strains that contain a defective *CBF1* gene (and show the expected defect in centromere function), but are methionine prototrophs (McKenzie *et al.* 1993). Mechanisms on how CBF1 may influence transcription at the CDEI site have also been described (McKenzie *et al.* 1993; Kent *et al.* 1994).

CBF1 binds to *CEN* CDEI DNA as a dimer (Mellor *et al.* 1990; Dowell *et al.* 1992). Binding involves the region C-terminal to the bHLH domain of CBF1 (Dowell *et al.* 1992). This region contains a potential long amphipathic helix with hydrophobic faces which could provide protein–protein interaction domains in an analogous way to the leucine repeat in the bZIP (Vinson *et al.* 1989; Dingwall and Laskey 1991) and bHLH-ZIP proteins (Hu *et al.* 1990; Blackwood and Eisenman 1991). Indirect evidence further suggests that the CDEI–CBF1 complex may interact with other DNA–protein complexes located at CDEII and/or CDEIII. For example, double point mutants in CDEI and CDEIII show a much higher mitotic chromosome loss rate than the sum of the individual chromosome loss rates of the single mutants, i.e. the effect of the double mutation is synergistic and not purely additive (Niedenthal *et al.* 1991). In another study, Wilmen *et al.* (1994) showed that the ability of CBF1 protein to bind a mutant CDEI element *in vitro* does not parallel the ability of that mutant to confer *in vivo CEN* activity. Based on the observation that binding of CBF1 to CDEI results in DNA bend angles ranging from 66° to 71°, protein-induced DNA bending is postulated as a possible mechanism that may facilitate interaction between the CDEI–CBF1 complex and other DNA–protein complexes (Niedenthal *et al.* 1993).

CDEII-binding proteins

The existence of CDEII-binding proteins is suggested by chromatin analysis showing a protected CDEII region. However, no specific CDEII-binding protein has so far been directly identified, although three proteins (CSE1, CSE2, and CSE4p; see below) have been implicated to interact with the CDEII domain.

CDEIII-binding proteins

CBF3 and its subunits: CBF3A, CBF3B, CBF3C, and CBF3D

CBF3 is a 240 kDa multisubunit protein complex that binds specifically to the CDEIII DNA (Lechner and Carbon 1991). This protein complex is absolutely essential for centromere function. It binds with high affinity to wild-type *CEN* DNA sequence, but not to a functionally inactive *CEN* DNA fragment containing a single base alteration in the central CCG nucleotides of CDEIII (Ng and Carbon 1987). The CBF3 protein is composed of four major subunits, CBF3A (110 kDa), CBF3B (64 kDa), CBF3C (58 kDa), and CBF3D (29kD), that are tightly associated in equimolar amounts. DNase I footprinting analysis of the CBF3–*CEN* complex shows that 56 bp of *CEN* DNA are protected. These are the entire CDEIII sequence plus 6 bp of CDEII to the left and 24 bp of adjacent DNA to the right of CDEIII (Lechner and Carbon 1991). The position of the right boundary of the protected region coincides with the right boundary of the nuclease-resistant centromere chromatin (see below). It is now clear that CBF3 and associated proteins form a kinetochore structure that performs essential centromere functions, including mediating attachment and movement of the *S. cerevisiae* chromosomes on the microtubules.

CBF3A

The gene that encodes CBF3A has been isolated by three independent routes. Jiang *et al.* (1993a) utilized partial amino acid sequence of purified CBF3A to prepare synthetic oligonucleotides and used these as probes to isolate the gene from yeast genomic libraries. Doheny *et al.* (1993) identified the *ctf14* mutation in this gene employing a genetic screen for chromosome transmission fidelity (*ctf*) mutants, whereas Goh and Kilmartin (1993) identified another mutation (*ndc10*, for **n**uclear **d**ivision **c**ycle) in the same gene by screening a library of temperature-sensitive mutants for spindle defects. The deduced amino acid sequence indicates that the protein is substantially hydrophilic, especially toward the C-terminal one-third of the molecule. Except for some partial homology with the consensus GTP-binding site in various G-proteins (Bourne *et al.* 1991), the sequence shows no significant homologies with other known proteins, including the microtubule-based motors (Yang *et al.* 1989; Kosik *et al.* 1990; Meluh and Rose 1990; Cyr *et al.* 1991; Gibbons *et al.* 1991; Wright *et al.* 1991; Hoyt *et al.* 1992; Lillie and Brown 1992; Roof *et al.* 1992). Interestingly, the consensus GTP-binding sequence found in CBF3A is also present in dynamin (Obar *et al.* 1990), a protein that *in vitro* induces microtubules to form hexagonally packed bundles and has microtubule-stimulated ATPase and GTPase activities (Shpetner and Vallee 1989, 1992).

Functionally, *CBF3A* is an essential gene for cell growth. The gene product has been localized to the spindle pole body region during the cell cycle and along some short, presumably metaphase, spindles. The phenotype of the gene is a detachment of chromosomes from the spindle while DNA replication, anaphase, and cytokinesis continue. Mutation in this gene leads to a pronounced chromosome segregation defect. In *ndc10* mutant strains, the chromosomes remain at one pole of the anaphase spindle and do not move to the bud junction at nonpermissive temperature. Overexpression of the *CBF3A* gene in *S. cerevisiae* results in the *in vitro* formation of a CBF3–*CEN* complex that carries as many as four copies of CBF3A to every one copy of the other CBF3 subunits, with complexes containing one, two, three, and four copies of the CBF3A subunit all being observed in electrophoretic mobility shift assay (Jiang and Carbon 1993). This oligomerization process probably occurs by self-association of CBF3A rather than by binding directly to the *CEN* DNA, since DNase I footprints of complexes containing one or two CBF3A subunits are not dramatically different (Lechner and Carbon 1991), and CBF3A does not bind to *CEN* DNA in the absence of the other CBF3 subunits. This observation suggests two possible *in vivo* functions for CBF3A (Jiang and Carbon 1993). First, oligomerization of CBF3A could result in the formation of a structure similar to the fibrous corona (an outer diffuse structure found on the kinetochores of many higher eukaryotes; see Chapter 4) (Ris and Witt 1981; Brinkley *et al.* 1989; Brinkley 1990) that could serve as an interface for interaction of the centromere with microtubules or the spindle pole body. This postulation is supported by results of *in vivo* immunolocalization experiments indicating the association of CBF3A in some manner with the mitotic spindle and with the spindle pole bodies (Goh and Kilmartin 1993; Jiang and Carbon 1993). Second, the apparent ability of monomeric CBF3A proteins to bind to both core CBF3 complex and to each other should allow linkage of *CEN* loci on two adjacent double-stranded DNA. This offers a molecular mechanism by which sister chromatids could be held together at the centromere throughout metaphase or during the entire first meiotic division.

CBF3B

Cloning of the *CBF3B* gene was achieved by screening a *S. cerevisiae* genomic library using a DNA probe prepared from CBF3B peptide sequence information (Lechner and Carbon 1991; Lechner 1994). The predicted polypeptide contains an N-terminal Zn_2Cys_6-type zinc finger domain, as well as a C-terminal acidic domain and a putative coiled coil dimerization domain that might enable protein–protein interaction with CBF3A/CD or other kinetochore proteins. Functionally, CBF3B is shown to be essential for cell growth. Mutations within the zinc finger domain result in cells that exhibit a G_2/M cell cycle delay and increased chromosome loss in each mitotic cell division. Therefore, CBF3B has a key role in chromosome segregation and the zinc finger domain executes an important part of this role.

Three lines of evidence suggest that CBF3B is the subunit within the CBF3 complex that is responsible for specific *CEN* DNA binding. First, anti-CBF3B antibodies prevent the formation of the CBF3–*CEN* complex. Interestingly, anti-CBF3B antibodies do not affect the electrophoretic mobility of the CBF3B–*CEN* complex when added to pre-assembled complexes, indicating that CBF3B might constitute the core of the CBF3 complex with only the DNA-binding domain of CBF3B accessible for antibody interaction. Second, the DNA-binding function of the Zn_2Cys_6 zinc finger domain in CBF3B has previously been demonstrated for transcriptional activators such as Gal4 (Keegan *et al.* 1986; Marmorstein *et al.* 1992). Third, CBF3B mutants are unable to support segregation of minichromosomes with mutations in the central part of CDEIII (Strunnikov *et al.* 1995). Thus, CBF3B is the only one of the CBF3 components that contains a known DNA-binding domain. X-ray crystallography data has revealed that Gal4 makes direct base contacts only to a highly conserved CCG triplet at each end of the Gal4 binding site (Marmorstein *et al.* 1992). CCG triplets are also present in the binding sites of three other transcriptional activators (LEU3, LAC9, and PPR1) (Marmorstein *et al.* 1992). Therefore, it could be speculated that CBF3B might interact directly with the highly conserved and functionally indispensable CCG triplet of CDEIII.

CBF3C

The gene encoding CBF3C has been isolated as another chromosome transmission fidelity mutant, *ctf13* (Doheny *et al.* 1993). Sequence determination (Lechner 1994) indicates that the protein contains a short acidic serine-rich region of about 30 amino acids that is approximately 40% identical to the first acidic block found in the mammalian centromere-binding protein, CENP-B (Chapter 6). There is also a possible phosphorylation site in the CBF3C protein.

In vitro and *in vivo* evidence demonstrate that the CBF3C protein is an essential component of the *S. cerevisiae* kinetochore (Doheny *et al.* 1993; Sorger *et al.* 1995). *In vitro*, the protein has been shown to be a component of the *CEN* DNA–protein complex and, in concert with CBF3B, interacts specifically with CDEIII core and flanking DNA. *In vivo*, a *CBF3C* mutation confers relaxation of a transcriptional block mediated by the kinetochore, stabilizes a test dicentric chromosome fragment, causes an increase in the mitotic rate of chromosome missegregation, and results in a terminal phenotype indicative of a defect in the G_2/M phase of the cell cycle.

CBF3D

CBF3D or Skp1p (for suppressor of kinetochore protein-1p) has been independently identified by Stemmann and Lechner (1966) as a protein that copurifies with a specific subunit of CBF3 and is required for reconstitution of CBF3 *in vitro* from purified components, and by Connelly and

Fig. 6.4) of highest similarity with CENP-C. Finally, the Mif2 protein has two features that suggest that it might act at the centromere by binding to the AT-rich CDEII region: an acidic domain and a proline-rich, 'AT hook' motif common to several chromatin proteins that bind AT-rich DNA [e.g. *Drosophila* D1 and mammalian HMGI(Y) proteins; reviewed by Churchill and Travers 1991] (Brown *et al.* 1993). Taken together, these data suggest that the Mif2 protein may have a direct role in centromere function by associating with CDEII and interacting with CBF1 and the CBF3 complex (Meluh and Koshland 1995).

Centromere chromatin

At a higher-order organizational level, nuclease-sensitivity studies indicate that the consensus *CEN* sequence resides within a 220–250 bp DNA segment composed of specially compacted chromatin that is resistant to nuclease cleavage and is flanked on both sides by nucleosomes containing hypersensitive cleavage sites (Bloom and Carbon 1982; Saunders *et al.* 1988; Cottarel *et al.* 1989). On some centromeres at least, a smaller (150–160 bp) protected region has been demonstrated, again with all three conserved centromere DNA elements included (Funk *et al.* 1989). The boundaries of the smaller *CEN* chromatin structure have been shown to start about 10 bp left of CDEI and end 20–30 bp right of CDEIII (Fig. 2.5). Mutations that destroy centromere function will destroy the integrity of this core region, so will conditions that dissociate chromatin structures, such as treatment with high salt.

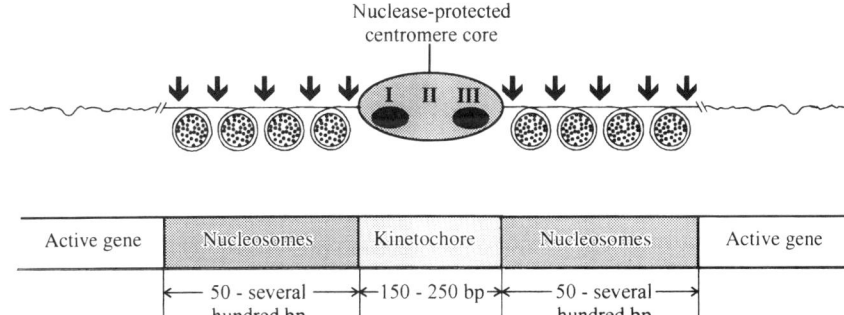

Fig. 2.5 Diagrammatic representation of the higher-order chromatin organization of *S. cerevisiae* kinetochore. I–III denote the positions of CDEI, CDEII, and CDEIII within the nuclease-protected centromere core region. Four nucleosome 'beads' (spotted circles) are shown flanking each side of the centromere. Nuclease-sensitive sites surrounding the kinetochore and nucleosome 'beads' are indicated by the arrows. Transcribed genes may exist as close as 50 to several hundred base pairs from the core kinetochore structure.

The pattern of nuclease accessibility or protection of the centromeric chromatin is unaltered in the different stages of the cell cycle, suggesting that this core chromatin structure is present throughout the entire division cycle (see Bloom *et al.* 1989; Hegemann and Fleig 1993). A possible exception may be during early S phase, when this structure may have to be relaxed to allow replication of the centromere DNA. Similar studies in the fission yeast *Schizosaccharomyces pombe* have shown that the distinct centromeric chromatin structure of this organism is also unchanged during its cell cycle (Chapter 3).

Centromere activity at different stages of the cell cycle

The presence of *CEN* DNA–protein complexes throughout all or the greater part of the yeast cell cycle raises the question of whether the centromeres are active throughout the cell cycle, or are only activated at the onset of mitosis. This has been investigated by *in vitro* measurement of the ability of yeast *CEN* plasmids, isolated as chromatin from different stages of the cell cycle, to bind microtubules. Cells arrested at the G_2/M boundary are shown to have a 7-fold increase in the ability to bind microtubules over that of cells arrested at G_1 (Kingsbury and Koshland 1991). These results indicate that microtubule-binding activity of the centromere is cell cycle regulated, perhaps through binding of additional factors to existing centromere complexes, or by modification of one or more components of the complexes.

Presence of active genes near the centromere

Unlike the situation in higher eukaryotic species where there is generally a paucity of trancribed genes near the heterochromatic centromere region, the centromeres of *S. cerevisiae* are surrounded by transcribed genes at their chromosomal locations (Fig. 2.5). Mapping studies on chromosome XI, for example, have located the *SPO15* and *MET14* genes within several hundred base pairs on each side of the *CEN* consensus sequence (Yeh *et al.* 1986). For *CEN6*, the nearest active gene is the *DEG1* gene, which is transcribed towards the centromere and where the 3' end of the RNA is about 50 bp away from CDEI (Carbone *et al.* 1991). The presence of the *DEG1* gene does not appear to interfere with centromere function, nor does the centromere chromatin structure have a major effect on its transcription. However, it has been reported that transcripts from genes located close to *CEN6* are less abundant than those from the genes in the central region of either of the chromosome VI arms (Yoshikawa and Isono 1991). It is unclear if the reduced transcription is due to some as yet unidentified transcriptional terminators flanking the *CEN* DNA. Alternatively, it is possible that only promoters that are relatively weak are tolerated in extreme close proximity to centromeres since transcription from strong promoters towards the centromere has been shown to result in termination of transcription and inactivation of the centromere

(Panzeri *et al.* 1983; Hill and Bloom 1987). Furthermore, the possibility exists that *CEN* proximal genes may not be transcribed during mitosis.

Time of *CEN* DNA replication

The observation that sister chromatids remain associated at the centromeres until anaphase dates back more than half a century ago (Darlington 1937). This observation suggested the possibility that the duplication of kineto-chores might occur immediately prior to chromosome separation during mitosis, and that replication of the centromere DNA itself might be delayed until mitosis, with the unreplicated centromere region maintaining the association of sister chromatids until anaphase (Tschumper and Carbon 1983; Murray and Szostak 1985). McCarroll and Fangman (1988) and Reynolds *et al.* (1989) tested these proposals in *S. cerevisiae* by determining the time at which centromeres replicate during the yeast cell cycle. The results indicated that replication of the centromere occurred in the first third of S phase, thereby discharging the speculation that a delay in the replication of centromeres until mitosis was responsible for sister chromatid adherence and proper chromosome segregation at anaphase.

The observation that *S. cerevisiae* centromeres replicate early in S phase appears to be at odds with findings in higher eukaryotes where cytological analyses of replication timing have shown that DNA of the centromeric regions replicates late in S phase. However, the resolution obtained in such studies is limited to relatively large regions of heterochromatic DNA involving at least several hundred kilobases. It is possible that the critical centromeric DNA element of higher eukaryotes also replicates early and that only the surrounding regions replicate late. Early replication of the centro-mere, at least in the demonstrated case of *S. cerevisiae*, raises the possibility that such an occurrence may play a role in centromere function. Perhaps replicated centromere DNA is required for initiating the assembly of the kinetochore, or for achieving proper orientation of sister chromatids on the developing spindle apparatus (McCarroll and Fangman 1988).

Replication fork pauses at the centromere

The two-dimensional agarose gel electrophoresis method that permits the identification of replication intermediates has been used to analyse the replication of the chromosomal copies of the *S. cerevisiae CEN* DNA and to determine the fate of replication forks that encounter the protein–DNA complex at the centromere (Greenfeder and Newlon 1992). The results indicate that all yeast centromeres cause replication forks to pause. In addition, analysis of replication of plasmids containing mutant centromere derivatives that vary in function as well as the ability to form the nuclease-resistant core structure indicates that the ability to cause replication forks to pause correlates with the ability to form the nuclease-resistant core structure and not with the presence or absence of a particular DNA sequence. These results demonstrate that the centromere is fully replicated during S phase, confirming the findings of McCarroll and Fangman (1988) and Reynolds *et al.* (1989) described above. In addition, they suggest that the centromere protein–DNA complex is present during S phase when replication forks encounter the centromere and therefore may be present throughout the cell cycle (see 'Centromere chromatin' above).

Centromeres of other budding yeasts

Like the *CEN* DNA of *S. cerevisiae*, the centromeres of budding yeasts *Saccharomyces uvarum*, *Yarrowia lipolytica*, and *Kluyveromyces lactis* are all relatively small. The DNA sequence and organization of the centromeres of *S. uvarum* (Huberman *et al.* 1986) and *K. lactis* (described below) are both closely similar to those of *S. cerevisiae*. However, while the centromeres of *S. uvarum* can function in *S. cerevisiae*, those of *K. lactis* cannot. On the other hand, although the size of the *Y. lipolytica* centromere is also similar to that of *S. cerevisiae*, its structure is quite different (Fournier *et al.* 1993), and its instability in *S. cerevisiae* suggests that it is not functionally recognized in *S. cerevisiae*.

Centromere of *Kluyveromyces lactis*

Phylogenically, *K. lactis* and *S. cerevisiae* are both budding yeasts and are more closely related to each other than they are to *S. pombe* (Hendriks *et al.* 1992). These two budding yeasts are intimately related in terms of the sequences and functional interchangeability of many of the genes analysed, yet display characteristic lifestyles and differences both at the level of individual metabolic pathways (Kuzhandaivelu *et al.* 1992; Mazzoni *et al.* 1992; Wésolowski-Louvel *et al.* 1992) and the structures of an increasing number of regulatory proteins (Goncalves *et al.* 1992; Na and Hampsey 1993; Zachariae and Breunig 1993; Mulder *et al.* 1994a). At the genomic level, although the total cellular DNA contents of *K. lactis* and *S. cerevisiae* are both about 14 Mb, *K. lactis* has a haploid number of 6, instead of 16.

Centromere DNA elements

Determination of the nucleotide sequence of five of the *K. lactis* centromeres reveals the presence of four conserved centromeric DNA elements (Heus *et al.* 1993a). Three of these elements have a counterpart in *S. cerevisiae* and are designated CDEI (*kl*), CDEII (*kl*), and CDEIII (*kl*) to distinguish them from CDEI (*sc*), CDEII (*sc*), and CDEIII (*sc*) of *S. cerevisiae*. The fourth DNA element, named CDE0 (*kl*), is not found in *S. cerevisiae*. The main features of the *K. lactis* centromeric DNA elements can be summarized as follows (Fig. 2.6).

(Hartwell and Weinert 1989; Murray 1992; Sheldrick and Carr 1993; Roeder 1995; Toyn *et al.* 1995). The three best studied checkpoints in *S. cerevisiae* are the responses to DNA damage, unreplicated DNA, and spindle defects (reviewed by Murray 1994, 1995).

Cell cycle checkpoints must have at least three components. The checkpoint that monitors the assembly of a mitotic structure must have a monitoring system that detects defects in that structure, a means of transmitting the information from this monitoring system to the cell cycle machinery, and a target that the signal acts on to prevent chromosome segregation and exit from mitosis. Checkpoint mutants have been isolated on the basis of their increased sensitivity to drugs (Hoyt *et al.* 1991; Li and Murray 1991; Enoch *et al.* 1992) and DNA damage induced by irradiation (Weinert and Hartwell 1988; Al-Khodairy and Carr 1992; Rowley *et al.* 1992), and their ability to overcome the cell cycle arrest caused by *cell division cycle* mutants (Weinert *et al.* 1994). It has been shown that several different lesions can activate the same checkpoint. In the spindle assembly checkpoint system, for example, the *mad1* and *mad2* mutants fail to arrest in response to lesions in the spindle pole body, in mutants that decrease microtubule stability, and in the presence of multiple copies of small centromeric plasmids.

Although less well studied, the *S. cerevisiae* kinetochore also appears to be subjected to cell cycle checkpoint surveillance. A *CEN* DNA mutation (comprising a 31 bp deletion within CDEII) carried on a single chromosome, while causing chromosome missegregation in only 1% of cell divisions, is shown to induce a dramatic cell cycle delay observed as retarded mitosis in the cells (Spencer and Hieter 1992). Analysis of strains containing this mutation indicates that most (and possibly all) cells experience delay in each cell cycle and that the delay is not due to increased chromosome copy number. Furthermore, a synchronous population of cells containing the mutation undergoes DNA synthesis on schedule with wild-type kinetics, but subsequently exhibits late chromosomal separation and concomitant late cell separation. It is speculated that this delay in cell cycle progression before the onset of anaphase, mediated perhaps by surveillance at a cell cycle checkpoint that monitors the completion of chromosomal attachment to the spindle, provides a mechanism for the stabilization of chromosomes with defective kinetochore structure (Spencer and Hieter 1992).

Other factors involved in chromosome segregation

Chromosome segregation is a complex process. In addition to the proteins described above that have been demonstrated to interact, directly or indirectly, with the *CEN* DNA elements, many other *trans*-acting proteins are expected to be involved in controlling the precise movement of chromosomes in *S. cerevisiae*. A growing number of such proteins have now been identified, generally through genetic screens for mutants with defective chromosome segregation phenotype. Mutant collections that have been isolated include *ctf* (chromosome transmission fidelity; Spencer *et al.* 1990), *chl* (chromosome loss; Kouprina *et al.* 1993), *cin* (chromosome instability; Hoyt *et al.* 1990), *cse* (chromosome segregation; Xiao *et al.* 1993), *mcm* (minichromosome maintenance; Maine *et al.* 1984), *ndc* (nuclear division cycle; Goh and Kilmartin 1993), *mif* (mitotic fidelity; Meeks-Wagner *et al.* 1986), and *cdc* (cell division cycle; Hartwell and Smith 1985; Palmer *et al.* 1990) mutants. Some of these collections are quite large. For example, the *ctf* collection consists of 136 independent mutants that exhibit increased loss of a nonessential chromosome and represents approximately 50 genes whose products are required for high fidelity chromosome transmission in the mitotic cell cycle (Spencer *et al.* 1990).

Secondary criteria can be, and are often, applied to identify those mutants that are defective in a particular structure or process. Some of the mutants, such as the *ctf13* and *ctf14* mutants described earlier, are expected to involve genes that directly encode kinetochore components and, as such, determination of centromere function may be used as an informative secondary criterion. In other cases, using *in vitro* motility assay or the detection of sensitivity to microtubule-destabilizing drugs, a number of proteins have been shown to be microtubule-based motors, or have microtubule-related functions. Examples of genes that encode such proteins are *CIN8* and *KIP1* (*CIN9*) (Hoyt *et al.* 1992; Roof *et al.* 1992; Saunders and Hoyt 1992), *SMY1* (Lillie and Brown 1992), *SRP1* (Yano *et al.* 1992), *BIK1* (Berlin *et al.* 1990), and *CIN1*, *CIN2*, and *CIN4* (Hoyt *et al.* 1990; Stearns *et al.* 1990). In addition, several other genes, such as *SPA1* (Snyder and Davis 1988) and *DBF8* (Houman and Holm 1994), have been shown to be important for chromosome segregation but whose modes of action have not been defined. It is without doubt that through the study of such extensive mutant collections, many new important genes of the cell division cycle of *S. cerevisiae* will continue to be discovered.

Conclusion

The centromere of *S. cerevisiae* has a number of unique features that make it a highly tractable system for investigation. It has a relatively small and simple DNA structure that is not associated with any of the hundreds to thousands of kilobases of repetitive DNA so universally found in the centromeres of higher eukaryotic chromosomes. Unlike the centromeres of other organisms which are known to bind from 2 to 120 microtubules (reviewed by Bloom 1993), each *S. cerevisiae* centromere needs only bind a single microtubule (Peterson and Ris 1976; Moens 1979; Clarke and Carbon 1985) and, as such, presumably requires a far less complex kinetochore structure to interact with the microtubules. The relative ease whereby genetic mutants carrying chromosome segregation defects can be isolated and characterized, the availability of good assay systems for centromere functions, and the recent discovery that a number of the centromeric components of *S. cerevisiae* have a counterpart in mammalian centromeres, should further define *S. cerevisiae* as an organism of choice for centromere studies. To date, relatively more information is known about the full spectrum of the DNA, protein, and functional components of the centromere of *S. cerevisiae* than for any other organism, and there is every reason to believe that this friendly baker's yeast will continue to serve us well in our future quests for an understanding of the centromere of both the lower and higher eukaryotes.

References

Al-Khodairy, F. and Carr, A. (1992). DNA repair mutants defining G2 checkpoint pathways in S. pombe. EMBO J. *11*, 1343–1350.

Bai, C., Sen, P., Hofmann, K., Ma, L., Goebl, M., Harper, J., and Elledge, S. (1996). SKP1 connects cell cycle regulators to the ubiquitin proteolysis machinery through a novel motif, the F-box. Cell 86, 263-274.

Baker, R., Fitzgerald-Hayes, M., and O'Brien, T. (1989). Purification of the yeast centromere-binding protein CP1 and a mutational analysis of its binding site. J. Biol. Chem. *264*, 10843–10850.

Baker, R. and Masison, D. (1990). Isolation of the gene encoding the *Saccharomyces cerevisiae* centromere-binding protein CP1. Mol. Cell. Biol. *10*, 2458–2467.

Basrai, M. A. and Hieter, P. (1995). Is there a unique form of chromatin at the *Saccharomyces cerevisiae* centromeres? BioEssays *17*, 669–672.

Berlin, V., Styles, C., and Fink, G. (1990). BIK1, a protein required for microtubule function during mating and mitosis in *Saccharomyces cerevisiae*, colocalizes with tubulin. J. Cell Biol. *111*, 2573–2586.

Blackwood, E. M. and Eisenman, R. N. (1991). Max: a helix-loop-helix zipper protein that forms a sequence-specific DNA-binding complex with Myc. Science *251*, 1211–1217.

Bloom, K. (1993). The centromere frontier: kinetochore components, microtubule-based motility, and the CEN-value paradox. Cell *73*, 621–624.

Bloom, K. and Carbon, J. (1982). Yeast centromere DNA is in a unique and highly ordered structure in chromosomes and small circular minichromosomes. Cell *29*, 305–317.

Bloom, K., Hill, A., Kenna, M., and Saunders, M. (1989). The structure of a primitive kinetochore. Trends Biochem. Sci. *14*, 223–227.

Bourne, H., Sanders, D., and McCormick, F. (1991). The GTPase super-family: conserved structure and molecular mechanism. Nature (Lond.) *349*, 117–127.

Bram, R. and Kornberg, R. (1987). Isolation of a *Saccharomyces cerevisiae* centromere DNA-binding protein, its human homolog, and its possible role as a transcription factor. Mol. Cell. Biol. *7*, 403–409.

Brinkley, B. (1990). Toward a structural and molecular definition of the kinetochore. Cell Motil. Cytoskeleton *16*, 104–109.

Brinkley, B., Valdivia, M., Tousson, A., and Balezon, R. (1989). The kinetochore: structure and molecular organization. In Mitosis: Molecules and Mechanisms, (eds. Hyams, J. S. and Brinkley, B. R.). Academic Press, London, 77.

Brown, M. (1995). Sequence similarities between the yeast chromosome segregation protein Mif2 and the mammalian centromere protein CENP-C. Gene *160*, 111–116.

Brown, M. T., Goetsch, L., and Hartwell, L. H. (1993). MIF2 is required for mitotic spindle integrity during anaphase spindle elongation in *Saccharomyces cerevisiae*. J. Cell Biol. *123*, 387–403.

Busch, S. and Sassone-Corsi, P. (1990). Dimers, leucine zippers and DNA-binding domains. Trends Genet. *6*, 36–40.

Cai, M. and Davis, R. (1989). Purification of a yeast centromere binding protein that is able to distinguish single basepair mutations in its recognition site. Mol. Cell Biol. *9*, 2544–2550.

Cai, M. and Davis, R. (1990). Yeast centromere binding protein CBF1, of the helix-loop-helix protein family, is required for chromosome stability and methionine prototrophy. Cell *61*, 437–446.

Carbon, J. and Clarke, L. (1984). Structural and functional analysis of a yeast centromere (CEN3). J. Cell Sci. *1*, 43–58.

Carbone, M., Solinas, M., Sora, S., and Panzeri, L. (1991). A gene tightly linked to CEN6 is important for growth of *Saccharomyces cerevisiae*. Curr. Genet. *19*, 1–8.

Carle, G. and Olson, M. (1985). An electrophoretic karyotype for yeast. Proc. Natl Acad. Sci. USA. *82*, 3756–3760.

Cherest, H., Thomas, D., and Surdin-Kerjan, Y. (1990). Nucleotide sequence of the MET8 gene of *Saccharomyces cerevisiae*. Nucleic Acids Res. *18*, 659.

Churchill, M. and Travers, A. (1991). Protein motifs that recognize structural features of DNA. Trends Biomed. Sci. *16*, 92–97.

Clarke, L. and Carbon, J. (1980). Isolation of a yeast centromere and construction of functional small circular chromosomes. Nature *187*, 504–509.

Clarke, L. and Carbon, J. (1983). Genomic substitutions of centromeres in *Saccharomyces cerevisiae*. Nature *305*, 23–28.

Clarke, L. and Carbon, J. (1985). The structure and function of yeast centromeres. Annu. Rev. Genet. *19*, 29–56.

Connelly, C., and Hieter, P. (1996). Budding yeast SKP1 encodes an evolutionarily conserved kinetochore protein required for cell cycle progression. Cell 86, 275-285.

Cottarel, G., Shero, J., Hieter, P., and Hegemann, J. (1989). A 125-base pair CEN6 DNA fragment is sufficient for complete meiotic and mitotic centromere functions in *Saccharomyces cerevisiae*. Mol. Cell Biol. *9*, 3342–3349.

Cumberledge, S. and Carbon, J. (1987). Mutational analysis of meiotic and mitotic centromere function in *Saccharomyces cerevisiae*. Genetics *117*, 203–212.

Cyr, J., Pfister, K., Bloom, G., Slaughter, C., and Brady, S. (1991). Molecular genetics of kinesin light chains: generation of isoforms by alternative splicing. Proc. Natl Acad. Sci. USA *88*, 10114–10118.

Darlington, C. (1937). Recent Advances in Cytology. Blakiston Co., Philadelphia.

Dingwall, C. and Laskey, R. (1991). Nuclear targeting sequences—a consensus? Trends Biochem. Sci. *16*, 478–481.

Doheny, K., Sorger, P., Hyman, A., Tugendreich, S., Spencer, F., and Hieter, P. (1993). Identification of essential components of the *Saccharomyces cerevisiae* kinetochore. Cell *73*, 761–774.

Dorsman, J., Heeswijk, W. v., and Grivell, L. (1988). Identification of two factors which bind to the upstream sequences of a number of nuclear genes coding for mitochondrial proteins and to genetic elements important for cell division in yeast. Nucleic Acids Res. *16*, 7287–7301.

Dowell, S., Tsang, J., and Mellor, J. (1992). The centromere and promoter factor 1 of yeast contains a dimerisation domain located carboxy-terminal to the bHLH domain. Nucleic Acids Res. *20*, 4229–4236.

Enoch, T., Carr, A., and Nurse, P. (1992). Fission yeast genes involved in coupling mitosis to DNA replication. Genes Dev. *6*, 2035–2046.

Fitzgerald-Hayes, M. (1987). Yeast centromeres. Yeast *3*, 187–200.

Fitzgerald-Hayes, M., Clarke, L., and Carbon, J. (1982). Nucleotide sequence comparisons and functional analysis of yeast centromere DNAs. Cell *29*, 235–244.

Fleig, U., Beinhauer, J., and Hegemann, J. (1995). Functional selection for the centromere DNA from yeast chromosome VIII. Nucleic Acids Res. *23*, 922–924.

Foreman, P. and Davis, R. (1993). Point mutations that separate the role of *Saccharomyces cerevisiae* Centromere Binding Factor 1 in chromosome segregation from its role in transcriptional activation. Genetics *135*, 287–296.

Fournier, P., Abbas, A., Chasles, M., Kudla, B., Ogrydziak, D., Yaver, D.*et al.* (1993). Colocalization of centromeric and replicative functions on autonomously replicating sequences isolated from the yeast *Yarrowia lipolytica*. Proc. Natl Acad. Sci. USA *90*, 4912–4916.

Funk, M., Hegemann, J., and Philippsen, P. (1989). Chromatin digestion with restriction endonucleases reveals 150–160 bp of protected DNA in the centromere of chromosome 14 in *Saccharomyces cerevisiae*. Mol. Gen. Genet. *219*, 153–160.

Gaudet, A. and Fitzgerald-Hayes, M. (1987). Alterations in the adenine-plus-thymine-rich region of CEN3 affect centromere function in *Saccharomyces cerevisiae*. Mol. Cell. Biol. *7*, 68–75.

Gaudet, A. and Fitzgerald-Hayes, M. (1989). Mutations in CEN3 cause aberrant chromosome segregation during meiosis in *Saccharomyces cerevisiae*. Genetics *121*, 477–489.

Gaudet, A. and Fitzgerald-Hayes, M. (1990). The function of centromeres in chromosome segregation. In The Eukaryotic Nucleus, eds. Strauss P.R. and Wilson, S. H. Telford, New Jersey, Vol. 2, 845–881.

Gibbons, I., Gibbons, B., Mocz, G., and Asai, D. (1991). Multiple nucleotide-binding sites in the sequence of dynein beta heavy chain. Nature *352*, 640–643.

Goh, P. and Kilmartin, J. (1993). NDC10: a gene involved in chromosome segregation in *Saccharomyces cerevisiae*. J. Cell Biol. *121*, 503–512.

Goldstein, L. S. (1993). Functional redundancy in mitotic force generation. J. Cell Biol. *120*, 1–3.

Goncalves, P., Maurer, K., Mager, W., and Planta, R. (1992). *Kluyveromyces* contains a functional ABF1-homologue. Nucleic Acids Res. *20*, 2211–2215.

Greenfeder, S. and Newlon, C. (1992). Replication forks pause at yeast centromeres. Mol. Cell. Biol. *12*, 4056–4066.

Hartwell, L., and Smith, D. (1985). Altered fidelity of mitotic chromosome transmission in cell cycle mutants of *S. cerevisiae*. Genetics *110*, 381–395.

Hartwell, L., and Weinert, T. (1989). Checkpoints: controls that ensure the order of cell cycle events. Science *246*, 629–634.

Hartwell, L., Dutcher, S., Wood, J., and Garvik, B. (1982). The fidelity of mitotic chromosome reproduction in *S. cerevisiae*. Recent Adv. Yeast Mol. Biol. *1*, 28.

Hegemann, J. H. and Fleig, U. N. (1993). The centromere of budding yeast. Bioessays *15*, 451–60.

Hegemann, J., Shero, J., Cottarel, G., Philipssen, P., and Hieter, P. (1988). Mutational analysis of the centromere DNA from chromosome VI of *Saccharomyces cerevisiae*. Mol. Cell Biol. *8*, 2523–2535.

Hendriks, L., Goris, A., Peer, Y. V. d., Neefs, J.-M., Vancanneyt, M., Kersters, K., *et al.* (1992). Phylogenetic relationships among ascomycetes and ascomycete-like yeasts as deduced from small ribosomal subunit RNA sequences. System. Appl. Microbiol. *15*, 98–104.

Heus, J., Zonneveld, B., Steensma, H., and Berg, J. V. d. (1990). Centromeric DNA of *Kluyveromyces lactis*. Curr. Genet. *18*, 517–522.

Heus, J., Zonneveld, B., Steensma, H., and Berg, J. v. d. (1993a). The consensus sequence of *Kluyveromyces lactis* centromeres shows homology to functional centromeric DNA from *Saccharomyces cerevisiae*. Mol. Gen. Genet. *236*, 355–362.

Heus, J., Bloom, K., Zonneveld, B., Steensma, H., and Berg, J. V. d. (1993b). Chromatin structures of *Kluyveromyces lactis* centromeres in *K. lactis* and *Saccharomyces cerevisiae*. Chromosoma *102*, 660–667.

Heus, J., Zonneveld, B., Steensma, H., and Berg, J. V. d. (1994). Mutational analysis of centromeric DNA elements of *Kluyveromyces lactis* and their role in determining the species specificity of the highly homologous centromeres from *K. lactis* and *Saccharomyces cerevisiae*. Mol. Gen. Genet. *243*, 325–333.

Hieter, P., Mann, C., Snyder, M., and Davis, R. (1985a). Mitotic stability of yeast chromosomes: a colony color assay that measures nondisjunction and chromosome loss. Cell *40*, 381–392.

Hieter, P., Pridmore, D., Hegemann, J., Thomas, M., Davis, R., and Philippsen, P. (1985b). Functional selection and analysis of yeast centromeric DNA. Cell *42*, 913–921.

Hill, A. and Bloom, K. (1987). Genetic manipulation of centromere function. Mol. Cell Biol. *7*, 2397–2405.

Houman, F. and Holm, C. (1994). DBF8, an essential gene required for efficient chromosome segregation in *Saccharomyces cerevisiae*. Mol. Cell. Biol. *14*, 6350–6360.

Hoyt, M., Stearns, T., and Botstein, D. (1990). Chromosome instability mutants of *Saccharomyces cerevisiae* that are defective in microtubule-mediated processes. Mol. Cell Biol. *10*, 223–234.

Hoyt, M., Trotis, L., and Roberts, B. (1991). *S. cerevisiae* genes required for cell cycle arrest in response to loss of microtubule function. Cell *66*, 507–517.

Hoyt, M. A., Hee, L., Loo, K., and Saunders, W. (1992). Two *Saccharomyces cerevisiae* kinesin-related gene products required for mitotic spindle assembly. J. Cell Biol. *118*, 109–120.

Hu, J., O'Shea, E., Kim, P., and Sauer, R. (1990). Sequence requirements for coiled-coils: analysis with lambda repressor–GCN4 leucine zipper fusion. Science *250*, 1400–1403.

Huberman, J., Pridmore, R., Jager, D., Zonneveld, B., and Philippsen, P. (1986). Centromeric DNA from *Saccharomyces uvarum* is functional in *Saccharomyces cerevisiae*. Chromosoma *94*, 162–8.

Huffaker, T., Thomas, J., and Botstein, D. (1988). Diverse effects of beta-tubulin mutations on microtubule formation and function. J. Cell Biol. *106*, 1997–2010.

Hyman, A., Middleton, K., Centola, M., Mitchison, T., and Carbon, J. (1992). Microtubule-motor activity of a yeast centromere-binding protein complex. Nature *359*, 533–536.

Jacobs, C., Adams, A., Szaniszlo, P., and Pringle, J. (1988). Functions of microtubules in the *Saccharomyces cerevisiae* cell cycle. J. Cell Biol. *107*, 1409–1426.

Jager, D. (1990). Investigations and constructions with and on functional elements of *Saccharomyces cerevisiae* chromosomes. PhD Thesis, University of Giessen, Giessen, Germany.

Jehn, B., Niedenthal, R., and Hegemann, J. (1991). *In vivo* analysis of the *Saccharomyces cerevisiae* centromere CDEIII sequence: requirements for mitotic chromosome segregation. Mol. Cell. Biol. *11*, 5212–5221.

Jiang, W. and Carbon, J. (1993). Molecular analysis of the budding yeast centromere/kinetochore. Cold Spring Harbor Symp. Quant. Biol. LVIII, 669–676.

Jiang, W. and Philippsen, P. (1989). Purification of a protein binding to the CDEI subregion of *Saccharomyces cerevisiae* centromere DNA. Mol. Cell Biol. *9*, 5585–5593.

Jiang, W., Lechner, J., and Carbon, J. (1993a). Isolation and characterization of a gene (*CBF2*) specifying a protein component of the budding yeast kinetochore. J. Cell Biol. *121*, 513–519.

Jiang, W., Middleton, K., Yoon, H.-J., Fouquet, C., and Carbon, J. (1993b). An essential yeast protein, CBF5p, binds *in vitro* to centromeres and microtubules. Mol. Cell. Biol. *13*, 4884–4893.

Keegan, L., Gill, G., and Ptashne, M. (1986). Separation of DNA binding from the transcription-activating function of a eukaryotic regulatory protein. Science *231*, 699–704.

Kent, N., Tsang, J., Crowther, D., and Mellor, J. (1994). Chromatin structure modulation in *Saccharomyces cerevisiae* by centromere and Promoter Factor 1. Mol. Cell. Biol. *14*, 5229–5241.

Kingsbury, J. and Koshland, D. (1991). Centromere-dependent binding of yeast minichromosomes to microtubules *in vitro*. Cell *66*, 483–495.

Korch, C., Mountain, H., and Bystrom, A. (1991). Cloning, nucleotide sequence and regulation of MET14, the gene encoding the APS kinase of *Saccharomyces cerevisiae*. Mol. Gen. Genet. *229*, 96–108.

Koshland, D., Kent, J., and Hartwell, L. (1985). Genetic analysis of the mitotic transmission of minichromosomes. Cell *40*, 393–403.

Kosik, K., Orecchio, L., Schnapp, B., Inouye, H., and Neve, R. (1990). The primary structure of the squid kinesin heavy chain. J. Biol Chem. *265*, 3278–3283.

Kouprina, N., Tsouladze, A., Koryabin, M., Hieter, P., Spencer, F., and Larionov, V. (1993). Identification and genetic mapping of *CHL* genes controlling mitotic chromosome transmission in yeast. Yeast *9*, 11–19.

Kuzhandaevelu, N., Jones, W., Martin, A., and Dickson, R. (1992). The signal for glucose repression of the lactose-galactose regulon is amplified through subtle modulations of transcription of the *Kluyveromyces lactis KIGAL4* activator gene. Mol. Cell. Biol. *12*, 1924–1931.

Landschulz, W., Johnson, P., and McKnight, S. (1988). The leucine zipper: a hypothetical structure common to a new class of DNA binding proteins. Science *240*, 1759–1764.

Langkopf, A., Hammarback, J., Mueller, R., Vallee, R., and Garner, C. (1992). Microtubule-associated proteins 1A and LC2. J. Biol. Chem. *267*, 16561–16566.

Lechner, J. (1994). A zinc finger protein, essential for chromosome segregation, constitutes a putative DNA binding subunit of the *Saccharomyces cerevisiae* kinetochore complex, Cbf3. EMBO J. *13*, 5203–5211.

Lechner, J. and Carbon, J. (1991). A 240 kD multisubunit protein complex (CBF3) is a major component of the budding yeast centromere. Cell *64*, 717–727.

Li, R. and Murray, A. W. (1991). Feedback control of mitosis in budding yeast. Cell *66*, 519–531.

Lillie, S. and Brown, S. (1992). Suppression of a myosin defect by a kinesin-related gene. Nature (Lond.) *356*, 358–361.

Maine, G., Sinha, P., and Tye, B.-K. (1984). Mutants of *S. cerevisiae* defective in the maintenance of minichromosomes. Genetics *106*, 365–385.

Marmorstein, R., Carey, M., Ptashne, M., and Harrison, S. (1992). DNA recognition by GAL4; structure of a protein–DNA complex. Nature *356*, 408–414.

Mazzoni, C., Saliola, M., and Falcone, C. (1992). Ethanol-induced and glucose-insensitive alcohol dehydrogenase activity in the yeast *Kluyveromyces lactis*. Mol. Microbiol. *6*, 2279–2286.

McCarroll, R. and Fangman, W. (1988). Time of replication of yeast centromeres and telomeres. Cell *54*, 505–513.

McGrew, J., Diehl, B., and Fitzgerald-Hayes, M. (1986). Single base-pair mutations in centromere element III cause aberrant chromosome segregation in *Saccharomyces cerevisiae*. Mol. Cell. Biol. *6*, 530–538.

McKenzie, E., Kent, N., Dowell, S., Moreno, F., Bird, L., and Mellor, J. (1993). The centromere and promoter factor 1, CPF1, of *Saccharomyces cerevisiae* modulates gene activity through a family of factors including SPT21, RPD1 (SIN3), RPD3 and CCR4. Mol. Gen. Genet. *240*, 374–386.

Meeks-Wagner, D., Wood, J., Garvik, B., and Hartwell, L. (1986). Isolation of two genes that affect mitotic chromosome transmission in *S. cerevisiae*. Cell *44*, 53–63.

Mellor, J., Jiang, W., Funk, M., Rathjen, J., Barnes, C., Hinz, T. *et al.* (1990). CPF1, a yeast protein which functions in centromeres and promoters. EMBO J. *9*, 4017–4026.

Mellor, J., Rathjen, J., Jiang, W., Barnes, C., and Dowell, S. (1991). DNA binding of CPF1 is required for optimal centromere function but not for maintaining methionine prototrophy in yeast. Nucleic Acids Res. *19*, 2961–2969.

Meluh, P. and Koshland, D. (1995). Evidence that the *MIF2* gene of *Saccharomyces cerevisiae* encodes a centromere protein with homology to the mammalian centromere protein CENP-C. Mol. Biol. *6*, 793–807.

Meluh, P. and Rose, M. (1990). *KAR3*, a kinesin-related gene required for yeast nuclear fusion. Cell *60*, 1029–1041.

Middleton, K. and Carbon, J. (1994). KAR3-encoded kinesin is a minus-end-directed motor that functions with centromere binding proteins (CBF3) on an *in vitro* yeast kinetochore. Proc. Natl Acad. Sci. USA *91*, 7212–7216.

Moens, P. (1979). Kinetochore microtubule numbers of different sized chromosomes. J. Cell Biol. *83*, 556–561.

Mulder, W., Scholten, I., Boer, R. d., and Grivell, L. (1994a). Sequence of the HAP3 transcription factor of *Kluyveromyces lactis* predicts the presence of a novel 4-cysteine zinc-finger motif. Mol. Gen. Genet. *245*, 96–106.

Mulder, W., Winkler, A., Scholten, I., Zonneveld, B., Winde, J. d., Steensma, H.*et al.* (1994b). Centromere promoter factors (CPF1) of the yeasts *Saccharomyces cerevisiae* and *Kluyveromyces lactis* are functionally exchangeable, despite low overall homology. Curr. Genet. *26*, 198–207.

Murphy, M., Fowlkes, D., and Fitzgerald-Hayes, M. (1991). Analysis of centromere function in *Saccharomyces cerevisiae* using synthetic centromere mutants. Chromosoma *101*, 189–197.

Murray, A. (1992). Creative blocks: cell cycle checkpoints and feedback controls. Nature *359*, 599–604.

Murray, A. (1994). Cell cycle checkpoints. Curr. Opin. Cell. Biol. *6*, 872–876.

Murray, A. (1995). The genetics of cell cycle checkpoints. Curr. Opin. Genet. Dev. *5*, 5–11.

Murray, A. and Hunt, T. (1993). The Cell Cycle: An Introduction. Oxford University Press, New York.

Murray, A. W. and Szostak, J. W. (1983). Construction of artificial chromosomes in yeast. Nature *305*, 189–193.

Murray, A. and Szostak, J. (1985). Chromosome segregation in mitosis and meiosis. Annu. Rev. Cell Biol. *1*, 289–315.

Murre, C., McCaw, P., and Baltimore, D. (1989). A new DNA binding and dimerization motif in immunoglobulin enhancer binding, daughterless, MyoD, and myc proteins. Cell *56*, 777–783.

Na, J. and Hampsey, M. (1993). The *Kluyveromyces* gene encoding the general transcription Factor-IIB—structural analysis and expression in *Saccharomyces cerevisiae*. Nucleic Acids Res. *21*, 3413–3417.

Newlon, C. (1988). Yeast chromosome replication and segregation. Microbiol. Rev. *52*, 568–601.

Ng, R. and Carbon, J. (1987). Mutational and *in vitro* protein-binding studies on centromere DNA from *Saccharomyces cerevisiae*. Mol. Cell. Biol. *7*, 4522–4534.

Niedenthal, R., Stoll, R., and Hegemann, J. (1991). *In vivo* characterization of the *Saccharomyces cerevisiae* centromere DNA element I, a binding site for the helix-loop-helix protein CPF1. Mol. Cell. Biol. *11*, 3543–3553.

Niedenthal, R., Sen-Gupta, M., Wilmen, A., and Hegemann, J. (1993). Cpf1 protein induced bending of yeast centromere DNA element I. Nucleic Acids Res. *21*, 4726–4733.

Noble, M., Lewis, S., and Cowan, N. (1989). The microtubule binding domain of microtubule-associated protein MAP1B contains a repeated sequence motif unrelated to that of MAP2 and Tau. J. Cell Biol. *109*, 3367–3376.

Obar, R., Collins, C., Hammarback, J., Shpetner, H., and Vallee, R. (1990). Molecular cloning of the microtubule-associated mechano-chemical enzyme dynamin reveals homology with a new family of GTP-binding proteins. Nature (Lond.) *347*, 256–261.

Palmer, R., Hogan, E., and Koshland, D. (1990). Mitotic transmission of artificial chromosomes in *cdc* mutants of the yeast, *Saccharomyces cerevisiae*. Genetics *125*, 763–774.

Panzeri, L., Groth-Clausen, I., Altenburger, W., and Philippsen, P. (1983). DNA in the centromere of chromosome VI. In The Molecular Biology of Yeast. Cold Spring Harbor Laboratory Press, Cold Spring Harbor, 69.

Peterson, J. and Ris, H. (1976). Electron-microscopic study of the spindle and chromosome movement in the yeast *Saccharomyces cerevisiae*. J. Cell Sci. *22*, 219–242.

Pluta, A., Mackay, A., Ainsztein, A., Goldberg, I., and Earnshaw, W. (1995). The centromere: hub of chromosomal activities. Science *270*, 1591–1594.

Ransone, L., Visvader, J., Sassone-Corsi, P., and Verma, I. (1989). Fos–Jun interaction: mutational analysis of the leucine zipper domain of both proteins. Genes Dev. *3*, 770–781.

Reynolds, A., McCarroll, R., Newlon, C., and Fangman, W. (1989). Time of replication of *ARS* elements along yeast chromosome III. Mol. Cell. Biol. *9*, 4488–4497.

Ris, H. and Witt, P. (1981). Structure of the mammalian kinetochore. Chromosoma *82*, 153–170.

Roeder, G. (1995). Sex and the single cell: meiosis in yeast. Proc. Natl Acad. Sci. USA *92*, 10450–10456.

Roof, D., Meluh, P., and Rose, M. (1992). Kinesin-related proteins required for assembly of the mitotic spindle. J. Cell Biol. *118*, 95–108.

Rose, M. and Fink, G. (1987). *KAR1*, a gene required for function of both intranuclear and extranuclear microtubules in yeast. Cell *48*, 1047–1060.

Rothstein (1983). One-step gene disruption in yeast. Methods Enzymol. *101*, 202–211.

Rowley, R., Subramani, S., and Young, P. (1992). Checkpoint controls in *Schizosaccharomyces pombe*: rad1. EMBO J. *11*, 1335–1342.

Saunders, M., Fitzgerald-Hayes, M., and Bloom, K. (1988). Chromatin structure of altered yeast centromeres. Proc. Natl Acad. Sci. USA *85*, 175–179.

Saunders, W. and Hoyt, M. (1992). Kinesin-related proteins required for structural integrity of the mitotic spindle. Cell *70*, 451–458.

Schatz, P., Solomon, F., and Botstein, D. (1988). Isolation and characterization of conditional-lethal mutations in the *TUB1* alpha-tubulin gene of the yeast *Saccharomyces cerevisiae*. Genetics *120*, 681–695.

Scherf, U., Pastan, I., Willingham, M., and Brinkmann, U. (1996). The human CAS protein which is homologous to the *CSE1* yeast chromosome segregation gene product is associated with microtubules and mitotic spindle. Proc. Natl Acad. Sci. USA *93*, 2670–2674.

Sears, D., Hegemann, J., and Hieter, P. (1992). Meiotic recombination and segregation of human-derived artificial chromosomes in *Saccharomyces cerevisiae*. Proc. Natl Acad. Sci. USA *89*, 5296–5300.

Sheldrick, K. and Carr, A. (1993). Feedback controls and G2 checkpoints—fission yeast as a model system. BioEssays *15*, 775–782.

Shpetner, H. and Valle, R. (1989). Identification of dynamin, a novel mechanochemical enzyme that mediates interactions between microtubules. Cell *59*, 421–432.

Shpetner, H. and Vallee, R. (1992). Dynamin is a GTPase stimulated to high levels of activity by microtubules. Nature (Lond.) *355*, 733–735.

Snyder, M. and Davis, R. (1988). SPA1: a gene important for chromosome segregation and other mitotic functions in *S. cerevisiae*. Cell *54*, 743–754.

Sorger, P., Doheny, K., Hieter, P., MKopski, K., Huffaker, T., and Hyman, A. (1995). Two genes requied for the binding of an essential *Saccharomyces cerevisiae* kinetochore complex to DNA. Proc. Natl Acad. Sci. USA *92*, 12026–12030.

Spencer, F. and Hieter, P. (1992). Centromere DNA mutations induce a mitotic delay in *Saccharomyces cerevisiae*. Proc. Natl Acad. Sci. USA *89*, 8908–8912.

Spencer, F., Gerring, S., Connelly, C., and Hieter, P. (1990). Mitotic chromosome segregation fidelity mutants in *Saccharomyces cerevisiae*. Genetics *124*, 237–249.

Sreekrishna, K., Webster, T., and Dickson, R. (1984). Transformation of *Kluyveromyces lactis* with the kanamycin (G418) resistance gene of Tn*903*. Gene *28*, 73–81.

Stearns, T., Hoyt, M., and Botstein, D. (1990). Yeast mutants sensitive to antimicrotubule drugs define three genes that affect microtubule function. Genetics *124*, 251–262.

Stemmann, O., and Lechner, J. (1996). The Saccharomyces cerevisiae kinetochore contains a cyclin-CDK complexing homologue, as identified by in vitro reconstitution. EMBO J 15, 3611-3620.

Stinchcomb, D., Mann, C., and Davis, R. (1982). Centromeric DNA from *Saccharomyces cerevisiae*. J. Mol. Biol. *158*, 157–179.

Stoler, S., Keith, K., Curnick, K., and Fitzgerald-Hayes, M. (1995). A mutation in *CSE4*, an essential gene encoding a novel chromatin-associated protein in yeast, causes chromosome nondisjunction and cell cycle arrest at mitosis. Genes Dev. *9*, 573–586.

Strunnikov, A., Kingsbury, J., and Koshland, D. (1995). *CEP3* encodes a centromere protein of *Saccharomyces cerevisiae*. J. Cell Biol. *128*, 749–760.

Szostak, J. and Blackburn, E. (1984). The molecular structure of centromeres and telomeres. Annu. Rev. Biochem. *53*, 163–194.

Thomas, D., Cherest, H., and Surdin-Kerjan, Y. (1989). Elements involved in S-adenosylmethionine-mediated regulation of the *Saccharomyces cerevisiae* MET25 gene. Mol. Cell. Biol. *9*, 3292–3298.

Thomas, D., Barbey, R., and Surdin-Kerjan, Y. (1990). Gene–enzyme relationship in the sulfate assimilation pathway of *Saccharomyces cerevisiae:* study of the 3'-phosphoadenylsulfate reductase structural gene. J. Biol. Chem. *265*, 15518–15524.

Toyn, J., Johnson, A., and Johnston, L. (1995). Segregation of unreplicated chromsomes in *Saccharomyces cerevisiae* reveals a novel G_1/M-phase checkpoint. Mol. Cell. Biol. *15*, 5312–5321.

Tschumper, G. and Carbon, J. (1983). Copy number control by a yeast centromere. Gene *23*, 221–232.

Turner, R. and Tjian, R. (1989). Leucine repeats and an adjacent DNA binding domain mediate the formation of functional cFos–cJun heterodimers. Science *243*, 1689–1694.

Vinson, C., Sigler, P., and McKnight, S. (1989). Scissors-grip model for DNA recognition by a family of leucine zipper proteins. Science *246*, 911–916.

Watson, J., Hopkins, N., Roberts, J., Steitz, J., and Weiner, A. (1987). Molecular Biology of the Gene,.4th Edn. Benjamin/Cummings Publishing Co. Menlo Park, California. Vol.1., 549–557.

Weinert, T. and Hartwell, L. (1988). The *RAD9* gene controls the cell cycle response to DNA damage in *Saccharomyces cerevisiae*. Science *241*, 317–322.

Weinert, T., Kiser, G., and Hartwell, L. (1994). Mitotic checkpoint genes in budding yeast and the dependence of mitosis on DNA replication and repair. Genes Dev. *8*, 652–665.

Wells, D. and McBride, C. (1989). A comprehensive compilation and alignment of histones and histone genes. Nucleic Acids Res. *17*, 311–346.

Wesolowski-Louvel, M., Goffrini, P., Ferrero, I., and Fukuhara, H. (1992). Glucose transport in the yeast *Kluyveromyces lactis*. I. Properties of an inducible low-affinity glucose transporter gene. Mol. Genet. *233*, 89–96.

Wilmen, A., Pick, H., Niedenthal, R., Sen-Gupta, M., and Hegemann, J. (1994). The yeast centromere CDE1/Cpf1 complex: differences between *in vitro* binding and *in vivo* function. Nucleic Acids Res. *22*, 2791–2800.

Wright, B., Henson, J., Wedaman, K., Willy, P., Morand, J., and Scholey, J. (1991). Subcellular localization and sequence of sea urchin kinesin heavy chain: evidence for its association with membranes in the mitotic apparatus and interphase cytoplasm. J. Cell Biol. *113*, 817–833.

Wustinger, K. and Spevak, W. (1993). Isolation of CENIX and CENXII from *Saccharomyces cerevisiae*. Nucleic Acids Res. *21*, 3321.

Xiao, Z., McGrew, J., Schroeder, A., and Fitzgerald-Hayes, M. (1993). CSE1 and CSE2, two new genes required for accurate mitotic chromosome segregation in *Saccharomyces cerevisiae*. Mol. Cell. Biol. *13*, 4691–4702.

Yang, J. T., Laymon, R. A., and Goldstein, L. S. B. (1989). A three domain structure of kinesin heavy chain revealed by DNA sequence and microtubule-binding analyses. Cell *56*, 879–889.

Yano, R., Oakes, M., Yamaghishi, M., Dodd, J., and Nomura, M. (1992). Cloning and characterization of *SRP1*, a suppressor of temperature-sensitive RNA polymerase I mutations, in *Saccharomyces cerevisiae*. Mol. Cell. Biol. *12*, 5640–5651.

Yeh, E., Carbon, J., and Bloom, K. (1986). Tightly centromere-linked gene (SPO15) essential for meiosis in the yeast *Saccharomyces cerevisiae*. Mol. Cell. Biol. *6*, 158–167.

Yoshikawa, A. and Isono, K. (1991). Construction of an ordered clone bank and systematic analysis of the whole transcripts of chromosome VI of *Saccharomyces cerevisiae*. Nucleic Acids Res. *19*, 1189–1198.

Zachariae, W. and Breunig, K. (1993). Expression of the transcriptional activator LAC9 (KIGAL4) in *Kluyveromyces lactis* is controlled by autoregulation. Mol. Cell Biol. *13*, 3058–3066.

Zhang, H., Kobayashi, R., Galaktionov, K., and Beach, D. (1995). p19Skp1 and p45Skp2 are essential elements of the cyclin A-CDK2 S-phase kinase. Cell 82, 915-925.

3

Centromere of fission yeast *S. pombe*

Like the budding yeast *S. cerevisiae*, the fission yeast *Schizosaccharomyces pombe* is a relatively simple and tractable system that is easily manipulated genetically and biochemically. It contains in its haploid state about 14 Mb of genomic DNA that is embodied in 3 chromosomes with sizes of approximately 5.7 Mb (chromosome I), 4.7 Mb (chromosome II) and 3.5 Mb (chromosome III) (Fan *et al.* 1988). In many ways, *S. pombe* appears to be more similar to higher eukaryotic cells than it is to the budding yeast. For example, an intron in a gene derived from a mammalian system (the gene encoding the SV40 small T antigen) is correctly spliced in *S. pombe* (Kaufer *et al.* 1985), and the human gene CDC2HS has been screened by genetic complementation in *S. pombe* when transformed by a human cDNA library (Lee and Nurse 1987). Unlike the relatively small and repeat DNA-free centromeres of *S. cerevisiae*, *S. pombe* centromeres are large and contain untranscribed repetitive DNA, as is the case with the centromeres of higher eukaryotes. During early mitosis, discrete kinetochore structures are formed in *S. pombe* that each associates with the ends of 2–4 microtubules (Ding *et al.* 1993). This is similar to the attachment of multiple spindle fibres per kinetochore in higher eukaryotes, and is again in contrast to the single microtubule attachment observed for *S. cerevisiae* centromeres (Peterson and Ris 1976). These characteristics therefore make the centromere of *S. pombe* a good model system for higher eukaryotes.

Life cycle of *S. pombe*

The life cycle of *S. pombe* is quite similar to that of *S. cerevisiae*, with a few differences. The main difference is that *S. pombe* cells divide by fission rather than budding. The cells do this by growing in size until an equatorial cell division process produces two equally sized daughter cells. In *S. pombe*, the two haploid mating types are called h^+ and h^-. Unlike *S. cerevisiae*, *S. pombe* cells typically proliferate as haploid cells. These cells conjugate or mate in response to starvation to form diploid cells that promptly go through meiosis and sporulation to regenerate haploid cells (Fig. 3.1A). However, diploid strains can be obtained (Egel 1973) and shifted to nitrogen-free medium to induce an azygotic meiosis (Egel and Egel-Mitani 1974; Bahler *et al.* 1991, 1993) (Fig. 3.1B). Compared with *S. cerevisiae*, the mitotic cell cycle of the fission yeast is more typical of that of higher eukaryotes in that all four G_1, S, G_2, and M phases are present, and that the chromosomes undergo of these, are present also in *cen1* and *cen3*. As mentioned before, the B repeat shows sequence homology with the B' repeat of *cen1* in sharing common tRNA genes.

cen3

The Sp233 *cen3* region consists of a 5.4 kb central core (cc3) flanked on each side by inverted repeats K, L', J" and M (Steiner *et al.* 1993; Baum *et al.* 1994). As with *cen2*, the arrangement of these repeats is highly asymmetrical, with 11 tandem K repeats present on one side of cc3 and one K repeat present on the other side (Fig. 3.2C). One copy of the L', J" and M repeats is present on each side of the central core. Immediately flanking the central core is a 5.2 kb inverted repeat (black box; Fig. 3.2C) that is specific to *cen3*. Unlike the cc2 DNA, which is uniquely found on *cen2*, cc3 shares approximately 4 kb of homologous sequence with cc1 (Murakami *et al.* 1991; Takahashi *et al.* 1992). The K repeats of cen3 are similar to the K repeats in *cen2* with respect to orientation and structure. The L' repeat in *cen3* is in the same orientation as the L repeat in *cen2*, except that L' is a truncated ($\sim$ 3 kb) derivative of the 5 kb L repeat. J" is a truncated version of the J repeat found at the tail ends of the left-side K repeat and right-side K repeat arrays. As with the J' repeat of *cen1*, J" is part of the head of the 3.9 kb J repeat and contains a duplicated 300 bp region of the K repeat head (Baum *et al.* 1994). It is, however, not clear if the sequences and boundaries of J' and J" are identical. The M repeats, which are largely *cen3*-specific, are believed to contain tRNA genes and are thus somewhat analogous to the B' and B repeats of *cen1* and *cen2* (Steiner *et al.* 1993). In addition to the M repeats, tRNA has been detected within the central core-associated repeats and the central core itself of *cen3* (Takahashi *et al.* 1991, 1992).

Comparison between *cen1*, *cen2*, and *cen3*

In comparing the DNA structures of the three centromeres of Sp223, it is clear that they have a substantial amount of sequence elements in common, yet each centromere possesses some distinct characteristics. All three centromeres contain a central core sequence that is flanked by inverted structures composed of the basic K, L, J, and B repeats, or derivatives of these repeats. The number of copies, the detailed organization of these repeats, and the overall sizes of *cen1, cen2*, and *cen3* (estimated to be 38 kb, 65 kb and 97 kb, respectively; Steiner *et al.* 1993) are, however, quite different on the three chromosomes. The central core of *cen2* is unique, whereas those of *cen1* and *cen3* share considerable sequence homology. Immediately flanking each of the central cores is an inverted region that is unique for each centromere. Furthermore, since these centromeres have not been entirely sequenced, the possible presence of small amounts of other unrelated DNA within the centromeres cannot be excluded.

cen DNA variability in different strains of *S. pombe*

In addition to Sp223, *S. pombe* centromeres from a number of other laboratory strains (972h⁻, 975h⁺, SBPD400, and SBP120390) have been characterized (Chikashige *et al.* 1989; Hahnenberger *et al.* 1989, 1991; Clarke and Baum 1990; Murakami *et al.* 1991; Polizzi and Clarke 1991; Steiner *et al.* 1993). Strains 972h⁻ and 975h⁺, derived from the original Osterwalder Swiss *S. pombe* strain *liquefaciens* (Leupold 1950), are the standard heterothallic, prototrophic strains used in most *S. pombe* research. Many *S. pombe* auxotrophic strains, including Sp223 and SBP120390, are derivatives of 972h⁻ and 975h⁺. Although the *cen* structures of the different *S. pombe* strains have mainly been established by restriction analyses and thus lacked fine detail (e.g. in the exact number of copies of the K repeat in *cen3*, or in sequence information), useful comparisons can still be made. The results have indicated that *cen1* is well conserved between strains. In contrast, both *cen2* and *cen3* show a high degree of variability in different strains (Steiner *et al.* 1993) (Fig. 3.3). With *cen2*, the main variation is in the number of the K-L-B tandem repeats on the left-side of the central core sequence. With *cen3*, the differences observed include two orientations of the cc3 DNA with respect to flanking sequences, variable numbers of K repeats, and different extent of the asymmetrical arrangement of tandem K repeats around the central core region.

In addition to laboratory strains, Steiner *et al.* (1993) studied the centromeres of eight *S. pombe* wild-type strains isolated from various food and beverage sources and locations worldwide: IGC 2769 (Portugal), NCYC 132 (England), CSIR Y-457 (Switzerland), 0202 and 0209 (Germany), 0342 and EF4 (Japan), and EF1 (Czechoslovakia). The results, together with those obtained with the laboratory strains, indicate extensive polymorphic variation in the structural organization and overall sizes of the centromeres among the different strains. Furthermore, they demonstrate that, beyond the apparent variations, a conserved minimal structure is present in all three centromeres (Fig. 3.3). This conserved structure consists of the central core DNA element plus flanking regions extending up to a portion (as is the case for *cen1*) or the end of the most proximal K repeat unit. The importance of conserving this minimal structure probably underlies the homogeneity seen with *cen1* in the different *S. pombe* strains, since this may represent a minimal centromere in its natural state and will, therefore, not tolerate much variability between strains.

Fig. 3.3 Comparison of *cen* structures in five different laboratory strains of *S. pombe* (Steiner *et al.* 1993; Baum *et al.* 1994). Refer to Fig. 3.2 legend for general explanation. Open arrows indicate the orientation of the cc3 sequence relative to flanking sequences in the different strains. The boundaries and approximate sizes of the minimal structures that are conserved amongst the centromeres of the different laboratory strains (plus 8 other wild-type strains analysed but not shown here; Steiner *et al.* 1993) are indicated above the figures.

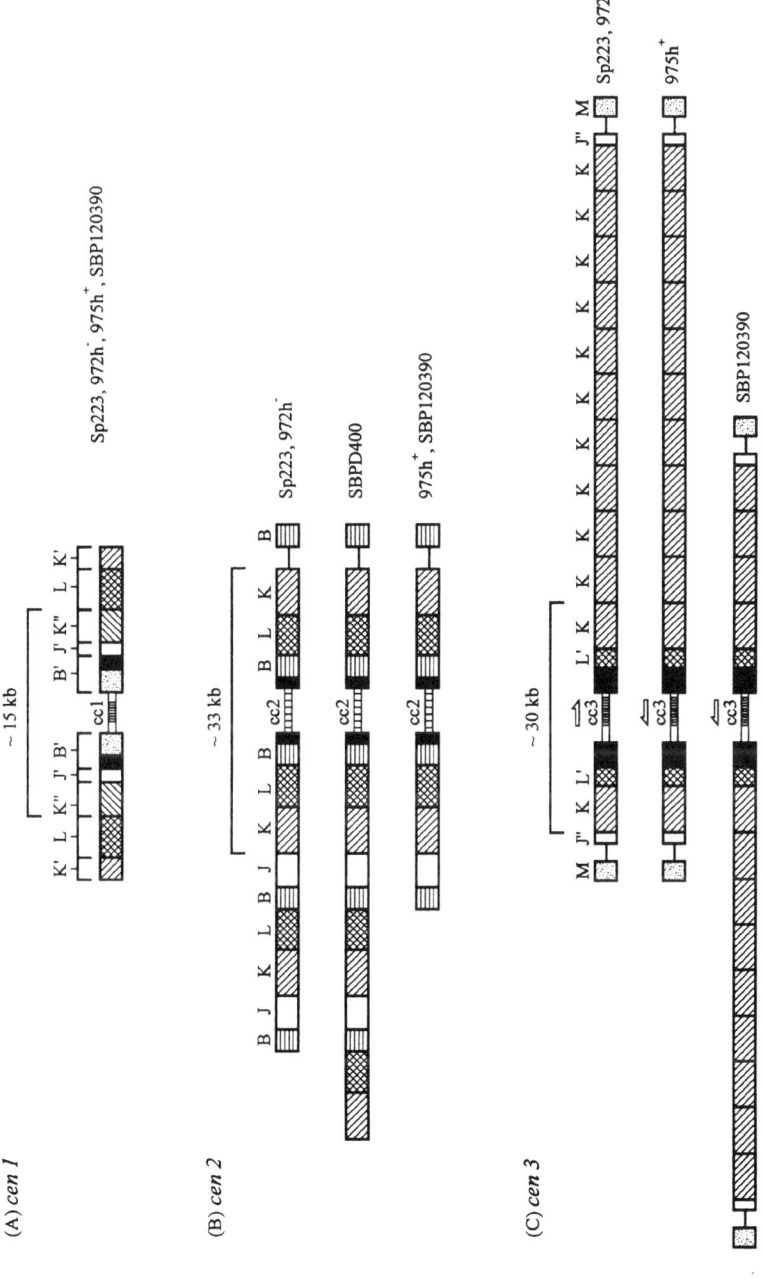

(A) *cen 1*

(B) *cen 2*

(C) *cen 3*

The majority of the differences depicted in Fig. 3.3 could be explained by simple recombinational events, either in mitosis, or in meiosis. For example, during mitotic division, recombinational looping out of repeated units could result in the loss of a K-L-B unit in *cen2*, or recombinational flip across homologous regions of the inverted repeat could be responsible for the two orientations of the central core observed for *cen3* (Steiner *et al.* 1993). Alternatively, misalignment or slippage within the repeats during pairing of homologous chromosomes in meiosis, followed by unequal crossover exchange (Smith 1976), could account for some of the observed alterations. Although the frequencies of meiotic recombination in *S. pombe* centromere regions are known to be low (Clarke *et al.* 1986; Nakaseko *et al.* 1986; Chikashige *et al.* 1989), its occasional occurrence, if accompanied by some selective advantage, could lead to the clonal expansion of a strain carrying a modified *cen* region. However, the exact mechanisms and underlying reasons for the evolution of sequences (such as cc2, or the DNA directly flanking the central cores) that are unique to each of the three centromeres remain unknown.

Nucleotide sequence conservation

The nucleotide sequences of the three central core DNAs from selected *S. pombe* strains, and representatives of each class of the centromeric repeats, have been determined (Clarke *et al.* 1986; Nakaseko *et al.* 1986, 1987; Chikashige *et al.* 1989; Kuhn *et al.* 1991; Murakami *et al.* 1991; Takahashi *et al.* 1991, 1992). The analyses demonstrate that the bulk of the 4 kb cc1 sequence is contained within the longer cc3, that the cc2 sequence is unique and shows no obvious sequence homology with cc1 or cc3, and that no part of any of the *cen* DNA, central cores or repeats contains an obvious open reading frame. Although individual copies of a particular type of repeat are highly homologous to one another, they are never 100% identical. Further- more, no good homologies are found between *S. pombe* centromeric sequences and those of *S. cerevisiae* or any other eukaryotic repeats. Interestingly, the 4.4 kb *S. pombe* K" repeat contains a relatively large number (22 copies; Takahashi *et al.* 1992) of a TGGAAA motif that resembles part of the 17 bp human CENP-B box sequence motif [CT(T/A)(C/T)G(T/G)TGGAAA(C/A)GG(G/A)A].

A striking finding when comparing a substantial segment ($\sim$ 30 kb) of nucleotide sequences within the conserved minimal structures (Fig. 3.3) of *cen1*, *cen2*, and *cen3* is that the two sides of the inverted repeats, which are separated by 4–7 kb of central core sequences, are strictly identical (Takahashi *et al.* 1992). Variations in different *S. pombe* strains occur at corresponding positions on the left and right sides, suggesting that strong constraints exist to maintain perfect sequence identity between the inverted repeats in the central domain. Since the rate of change in the sequences is as frequent as in other parts of the *S. pombe* genome, the results indicate that mutations are tolerated in the conserved minimal structures but only in a concerted fashion, possibly through a 'gene conversion' mechanism.

Functions of *cen* DNA components

The entire functional region of *cen1*, *cen2*, and *cen3* of *S. pombe* Sp223 can be isolated as single fragments of approximately 65, 100, and 150 kb, respectively (Hahnenberger *et al.* 1989). When any of these fragments is introduced into *S. pombe* on an artificial minichromosome carrying an *S. pombe* DNA replicator and genetic marker, the minichromosome behaves as an independent, fourth genetic linkage group that segregates stably and faithfully through mitosis and meiosis (Hahnenberger *et al.* 1989; Clarke and Baum 1990; Clarke *et al.* 1993). For example, a *cen2*-containing minichromosome is lost at a frequency of less than 1 in every 1000 mitotic cell divisions and segregates properly in meiosis in a $2+:2-$ manner and predominantly to sister spores in more than 90% of tetrads examined (Clarke and Baum 1990). The major findings of functional studies using these and other derivative minichromosomes are summarized below (Clarke and Baum 1990; Hahnenberger *et al.* 1991; Takahashi *et al.* 1992; Clarke *et al.* 1993; Baum *et al.* 1994).

Central core DNA is essential for *cen* function

The central core (cc) DNA is essential for centromere function but is insufficient on its own. Replacement of the central core sequences by an equivalent amount of heterologous DNA results in a minichromosome that is totally inactive in mitotic and meiotic centromere activities. Minichromosomes containing portions or all of the central core but only limited inverted repeat sequences also lack centromere function entirely. The central core sequences, together with at least the K-type repeats (see below), are necessary to establish a functional *S. pombe* centromere. Furthermore, the central core region appears to exhibit functional redundancy because deletion of various 1–3-kb portions from this region does not significantly reduce centromere function, whereas removal of the entire region abolishes this function (Baum *et al.* 1994). The lack of conserved sequence of any appreciable length between cc2 and cc1/cc3 suggests that the functionally important central core sequences may be quite small and/or degenerate.

The above deletion experiments, together with the observed differences in the lengths of native central core DNA for the three centromeres, suggest that the distance between the flanking inverted repeat units is not critical. *In vivo* insertion of ~ 9 kb of pBR322-derived plasmid DNA into the cc domain of a native chromosome has no detectable effect on chromosome stability, indicating that substantially extending the space between the inverted repeats, or the insertion of heterologous sequences into the central core, does not significantly alter or inhibit centromere activity (Baum *et al.* 1994). The orientation of the central core DNA relative to the repeated units is also unlikely to be important since the left and right critical regions flanking the central core are mirror images, and also as evident from *cen3* of the different *S. pombe* strains (Fig. 3.3C).

K″/K-type repeats are essential for *cen* function

Removal of both copies of the K″ repeat from *cen1* completely abolishes mitotic and meiotic *cen* activities, whereas removal of one copy has no significant effect, suggesting that at least one copy of the K″ element is required for centromere function (Hahnenberger *et al.* 1991). On the other hand, minichromosomes containing multiple copies of K″ exert an additive effect, in that each additional copy of K′, regardless of orientation and position, results in an increase in mitotic stability of the constructs and a corresponding decrease in 0 + :4– meiotic products, although increasing K″ copy number alone (in the absence of K′ + L repeats, see below) does not alter the high degree of precocious sister chromatid separation in meiosis I (Baum *et al.* 1994). Sequence determination indicates that a 2.1 kb region within the K″ repeat is essential for centromere function, and that this essential region is present in the larger K repeats of *cen2* and *cen3* (Baum *et al.* 1994). The ability of the K″ repeat to act in an orientation-independent way in minichromosome assays is not surprising since in the native genomic state the orientation of the K″ repeat in *cen1* is reversed with respect to those of its counterpart (K repeats) in *cen2* and *cen3* (Fig. 3.2). Furthermore, the additive effects of these repeats, and the failure of heterologous DNA inserted at various locations in the *S. pombe* centromere regions, including within the inverted repeat and the central core, to impair centromere function (Fishel *et al.* 1988; Clarke and Baum 1990), suggest that the K′/K-type sequences can function at different positions relative to the central core or the other repeats.

K′ + L repeats are required for meiotic sister chromatid attachment

Hahnenberger *et al.* (1991) studied the DNA elements that may be responsible for the sister chromatid attachment property using a construct containing the entire *cen1* structure except for the left-side K′ and L repeats. This minichromosome exhibits mitotic stability and segregates predominantly 2 + :2– through meiosis but undergoes precocious sister chromatid separation in the first meiotic division in more than 50% of the tetrads analysed. Restoring to the minichromosome the missing K′ + L sequences reinstates full meiotic activity. Thus, at least a portion of the K′ + L repeat is required for full meiotic centromere function, probably through its direct role in ensuring proper attachment of sister chromatids in meiosis I. The critical K′ + L region for this meiotic function must be present on both sides of the central core region, since the retention of the right-side structure alone is insufficient.

The role of K′ + L on mitosis is likely to be very minimal, as deleting one or both copies of K′ and L in the above and other minichromosomes (Hahnenberger *et al.* 1991) has little or no effect on mitotic centromere function. The L repeat is not functional as a centromere on its own, nor does it substitute for the function of the K″/K-type sequences or restore centromere function to constructs which carry central core DNA and only a small amount of flanking inverted repeat (Clarke *et al.* 1993).

J-type, B-type, and central core-associated repeats

Preliminary evidence suggests that these sequences may not be functionally important. For example, J′ and B′ of *cen1*, and the central core-associated repeats of *cen2* are not necessary for centromere function in a circular minichromosome since constructs lacking these repeats are mitotically stable and segregate in meiosis, albeit aberrantly at times (Baum *et al.* 1994). These repeats appear to be dispensable and cannot substitute for the critical central core sequences and K″/K-type repeats. However, further detailed studies are necessary to elucidate the role, if any, these repeat sequences may have in centromere function on native chromosomes.

tRNA

The three centromeres of *S. pombe* are characterized by small clusters of tRNA genes located within the B′, B, and possibly M repeats, and within the central core-associated repeats and the central core itself of *cen3* (Kuhn *et al.* 1991; Takahashi *et al.* 1991, 1992). The density of the tRNA genes in these centromeres is at least an order of magnitude higher than would be predicted for a random distribution within the genome of *S. pombe* (Kuhn *et al.* 1991). It is not known whether these genes are transcribed, and there is no evidence of transcription of any other centromeric repeats or central core sequences in *S. pombe* (Fishel *et al.* 1988; Polizzi and Clarke 1991). It is possible that the tRNA genes may in some ways be involved in centromere function, such as forming a nucleosome-disrupting transition zone between the central core (which lacks normal nucleosomes) and sequences within the inverted repeats (which are packaged into standard nucleosomal arrays; see below). Alternatively, the tRNA genes could be utilizing some existing property of the centromere, such as the reduced level of meiotic recombination (Clarke *et al.* 1986; Nakaseko *et al.* 1986; Chikashige *et al.* 1989) to facilitate maintenance of multiple copies of important tRNAs (Kuhn *et al.* 1991).

ARS sequences and replication of *cen* DNA

Putative *a*utonomously *r*eplicating *s*equence or *ARS* activity is abundantly found throughout the central core structure. This activity is identified by inserting a series of DNA fragments derived from an entire 20 kb region of *cen1* that includes the cc1 region plus one complete side of the *cen1* inverted repeats, into a vector that contains no *ARS*, and demonstrating propagation of the resulting constructs in *S. pombe* (Takahashi *et al.* 1992). Although the exact locations of *ARS* within the different cloned fragments are not known, *S. pombe* *ARS* consensus sequences (Maundrell *et al.* 1988) have been identified in each of the cc1, K″, and B′ domains, and there appear to be at least 3, and may be as many as 9, *ARS* within the 20 kb portion of *cen1* (Takahashi *et al.* 1992). This density is unusually high since *ARS* fragments are present at an average of approximately one in every 20 kb of *S. pombe* genomic DNA. A high density of DNA segments containing *ARS* has similarly been demonstrated in *cen2* (Smith *et al.* 1995). Since each of the *ARS* sequences defined within these centromeres can potentially act as an origin of DNA replication, it would appear that more potential initiation sites are required in centromeric DNA than in bulk DNA, perhaps because the centromeric DNA, as a result of altered conformation and possible condensation, may be less amenable to replication initiation. Alternatively, the prevalence of *ARS* elements in the centromere might indicate that at least some DNA-binding proteins involved in replication are also involved in centromere function.

Protein components

Unlike the situation in *S. cerevisiae*, very little is known about the protein components of the *S. pombe* centromere. Electrophoretic mobility shift assay (Garner and Revzin 1981) has been used to identify protein binding sites within the K″ and adjacent J′ repeats of *S. pombe cen1*. Preincubation of DNA subfragments of the K″ + J′ region with *S. pombe* crude nuclear extract results in retardation of the electrophoretic mobility of a number of fragments and allows identification of five putative nuclear protein-binding sites (Clarke *et al.* 1993; Baum *et al.* 1994). Two of these sites are present within the functionally critical 2.1 kb K″ region, while three are found in the J′ sequence. The K″ + J′ region has also been analysed for the presence of nuclease-hypersensitive sites, which are often indicative of nearby protein-binding domains (Gross and Garrard 1988). Nine hypersensitive sites have been located, with five of these sites (including one very strong site) in the 2.1 kb K″ region, and the remaining sites tightly flanking the two sides of this region (Clarke *et al.* 1993). These results indicate that proteins of yet undetermined nature interact with DNA within and immediately surrounding the 2.1 kb K″ sequence.

Swi6 protein

Swi6 (for **switching** gene product 6) is a 50 kDa protein that was recently localized to the repetitive, heterochromatin-like centromeric and telomeric regions of *S. pombe* (Ekwall *et al.* 1995). It contains a region of 48 amino acids that is homologous to a chromo (or **chr**omatin **o**rganization **mo**difier) domain sequence motif found in various chromatin-associated proteins: HP-1 and Polycomb in *Drosophila melanogaster*; M31, M32 and M33 in mouse; and HSM1 in humans (Lorentz *et al.* 1994). The protein is required to maintain transcriptional repression at the centromeres (Lorentz *et al.* 1994; Allshire *et al.* 1995). Mutations in the *swi6* gene result in a highly increased rate (30- to 100-fold) of chromosome loss during mitosis (Allshire *et al.* 1995; Ekwall *et al.* 1995). Analysis of late anaphase cells indicates that, while centromeres and chromosomes have undergone completion of segregation in wild-type strains at this stage, strains bearing a null *swi6* allele show a high incidence of lagging centromeres (Allshire *et al.* 1995). Defective centromere function in these *swi6* mutants is presumably due to the absence of this protein at the centromeres and the resulting weakening of kinetochores that become defective in microtubule interaction and poleward movement. Swi6 is likely to be an important component of *S. pombe* centromeres that has a role in mediating the formation of a functional kinetochore and heterochromatin-like repressive structure (see PEV below).

Centromere chromatin

Examination of the chromatin structure of *S. pombe* native centromeres using nuclease digestion demonstrates that the central core and core-proximal associated repeats form an unusual chromatin structure essentially devoid of typical nucleosomal packaging, whereas the remaining centromeric repeats, including the functionally critical K-type sequences, are packaged into nucleosomes (with average nucleosomal repeat lengths of 155 bp and linker lengths of about 10 bp) that are typical of bulk chromatin (Chikashige *et al.* 1989; Polizzi and Clarke 1991; Takahashi *et al.* 1992; Clarke *et al.* 1993; Marschall and Clarke, 1995). Several lines of evidence indicate that this unique chromatin structure absolutely correlates with centromere function. (i) Analysis of a host of functional minichromosomes containing portions of native centromeric DNA from all three *S. pombe* chromosomes reveals all to contain a central core chromatin structure lacking a regular periodic nucleosomal array. (ii) In nonfunctional artificial minichromosomes containing an entire central core region but lacking most of the inverted repeats, the central core sequences are packaged into nucleosomes typical of bulk chromatin, suggesting that the central core DNA itself does not preclude formation of an ordered nucleosomal structure. (iii) When minichromosomes carrying a complete *S. pombe cen* are introduced into *S. cerevisiae*, a host in which the *S. pombe* centromere is nonfunctional (Hahnenberger *et al.* 1989), no unique central core chromatin structure is formed (due presumably to the lack of those proteins in the budding yeast that specify the distinct central core chromatin found in *S. pombe*).

A DNA element that resides within the functionally important 2.1 kb fragment of the K″/K repeat is necessary and sufficient to confer the unique chromatin structure to central core sequences (Marschall and Clarke 1995). Examination of the positions of this DNA in native *S. pombe* centromeres and on various minichromosomes verifies that the element exerts its effect from a distance, and acts in an orientation-independent manner with respect to the central core. In addition, the DNA element appears to interact specifically with central core sequences, as indicated by the presence of a periodic nucleosomal array for the intervening DNA sequences between the DNA element and central core, and by the lack of any apparent effect of the DNA on sequences other than those of the central core. These properties have led to the proposition of the term 'centromere enhancer' for this DNA element. Sequences within other centromeric repeats have been shown to be unable to substitute for this enhancer element in altering central core chromatin structure. It has been suggested that interaction between the centromere enhancer and the central core is achieved by folding of centromeric DNA into a functional higher order structure, facilitated by proteins bound within the enhancer and the central core sequences (Marschall and Clarke 1995).

Centromere chromatin at different stages of the cell cycle

The chromatin structure of the *S. pombe* central core region during the cell cycle has been investigated using temperature-sensitive *cell division cycle* mutants to arrest cells at G_1 (Nurse *et al.* 1976), G_2 (Fantes 1979), and M (Nurse *et al.* 1976). At all these stages, the unique chromatin structure is shown to be present with no marked changes, suggesting that the *S. pombe* centromere regions maintain specific chromatin configurations through the cell cycle (Polizzi and Clarke 1991). Similar studies in *S. cerevisiae* have shown that the distinct centromeric chromatin structure of this organism also remains constant during the cell cycle (Chapter 2).

Position effect variegation (PEV) at the *S. pombe* centromere

PEV (defined in Chapter 4) has been demonstrated in *S. pombe* centromeres (Allshire *et al.* 1994). When normally active genes are experimentally placed within the central core region, the transcription levels of the genes become variably, but reversibly, repressed. As in *Drosophila* (Gowan and Gay, 1933; Spofford 1976), the repressed state of the inserted genes is temperature dependent, in that higher temperatures disturb the repression process, leading to increased expression, compared with lower temperatures. The different levels of repression–expression, once established, can be stably maintained for at least 20 generations. Such intermediate states of expression are not normally found in strains of *Drosophila* exhibiting variegation, suggesting an intrinsic difference in the mechanism of variegation imposed by heterochromatin in *Drosophila*, which may only allow on and off states to be established.

PEV occurs only when genes are placed in central core sequences that are associated with other centromeric elements in an active centromere. The central core domain alone cannot impose a position effect on the expression of genes placed within them (Allshire *et al.* 1994). This feature is reminiscent of the situation in which the formation of centromere-specific chromatin in the central core regions and the establishment of centromere function both necessitate the participation of flanking inverted repeats, and suggests a link between the unique chromatin structure of the centromere and variegated expression of genes placed within these structures. Consistent with such a link is the demonstration that inserting a gene within the *cen* region results in the gene taking on an appearance similar to the surrounding irregular chromatin.

Allshire *et al.* (1994) proposed a model suggesting that the repression of genes placed within the central cores is caused by diffusion of the unusual centromeric chromatin into the gene and interfering with transcription. Establishment of different levels of repression–expression is explained by the unusual chromatin extending into and being maintained at different distances within the genes; invasion of a small part of the 5′ transcription initiation end of the genes would cause a small decrease in gene activity, while further encroachment of this unusual structure would have a more severe effect on expression. Variegation is explained by fluctuations in the amount of the genes occupied by this chromatin during colony growth. Lower temperatures may encourage assembly of this unusual chromatin around the inserted genes, while higher temperatures could disrupt the packaging of extensive regions of this chromatin into the genes. Although the nature of this chromatin is at present unknown, the proteins that bind to the central cores to make them incompatible with the formation of regular nucleosomal structures must also be refractory to the formation or passage of active RNA polymerase II transcription complexes.

Epigenetic effect on centromere function

An epigenetic effect has been identified in *S. pombe* that can regulate *in vivo* the centromere function of various *in vitro* constructed plasmids carrying abbreviated centromeric DNA structures (including ones that either lack one full arm of the inverted repeats or have severely diminished inverted repeat lengths) (Steiner and Clarke 1994). For these plasmids, the epigenetic effect leads to the production of two phenotypic classes of transformants—one class showing an unstable plasmid, while another class showing a relatively stable plasmid. In the stable class of transformants, the epigenetic system has resulted in the conversion of a nonfunctional centromere to a functional one without changes in the content, structural arrangement, or chemical modification state of the plasmid DNA. The conversion from a centromere-inactive to an active state occurs at a relatively high frequency during mitotic cell division.

This centromere-targeted epigenetic system supports a model for centromere function involving specific *de novo* folding of centromeric components into a higher order chromatin structure. Conceivably, every time the *S. pombe* DNA replicates, the kinetochore structure is perturbed and needs to be re-established correctly for subsequent chromosome segregation during mitotic cell division. For the native chromosomes and plasmids that contain complete centromeric DNA sequences, this process should occur readily. However, the *S. pombe* plasmids that are subject to epigenetic alteration of the centromeric state carry incomplete centromere structures. It would appear that while these plasmids clearly have the potential for acting as minichromosomes with functional centromeres, their minimized DNA structures jeopardize the acquisition of the higher order chromatin structure or folding that is required for centromere function and thus lead to the generation, in some cases, of the unstable, nonfunctional state.

Spatial distribution of centromeres through the cell cycle

During mitosis

S. pombe centromeres display a specific spatial arrangement in the nucleus that is dependent on the cell cycle (Uzawa and Yanagida 1992; Funabiki *et al.* 1993) (Fig. 3.4). Throughout interphase, the three centromeres are tightly associated and located adjacent to the spindle pole body, at the periphery of the nucleus. This interphase centromere clustering requires the normal function of a highly conserved nuclear protein encoded by the *crm1*[+] gene (Adachi and Yanagida 1989; Toda *et al.* 1991, 1992; Funabiki *et al.* 1993). During mitosis, the centromeres interact with spindle microtubules, move away from the spindle pole bodies, and dissociate from one another. Since the spindle is present only in mitosis (which comprises about one-fifth of the cell cycle), centromere–microtubule interaction occurs only during this period. [This is in contrast to the situation in *S. cerevisiae*, where spindle microtubules are present and permanently attached to the chromosomes throughout cell division (Byers 1981; Bloom *et al.* 1989).] Sister chromatid separation in S. *pombe* occurs in an analogous way to higher eukaryotes. In higher eukaryotes, the distance between the spindle poles and the centromeres decreases in anaphase A, whereas the distance between the spindle poles increases in anaphase B (Alberts *et al.* 1994). In *S. pombe*, steps

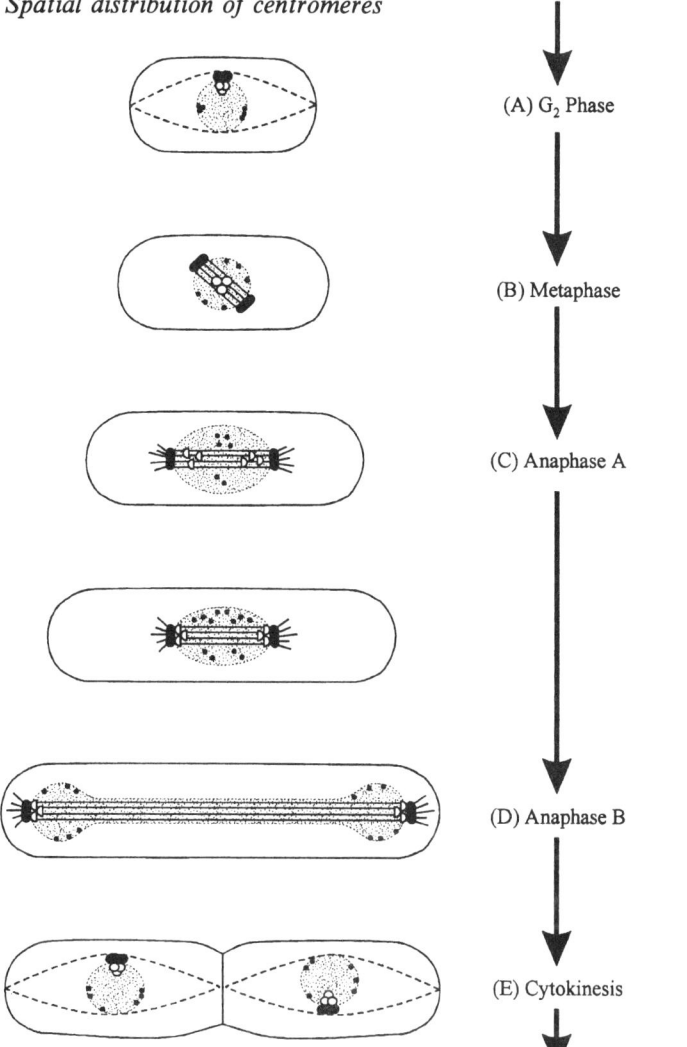

Fig. 3.4 Spatial arrangement of *S. pombe* centromeres (and telomeres) through the mitotic cell cycle (Funabiki *et al.* 1993). (A) Throughout interphase up to and including G_2 phase, the three centromeres (open circles) tightly cluster together adjacent to the spindle pole body (SPB; black cylindrical rod) at the periphery of the nucleus (stippled area). The telomeres (black circles) form one or two clusters near the nuclear periphery. Broken lines denote cytoplasmic microtubule arrays. (B) During metaphase, the centromeres dissociate from SPB and from one another. The centromeres interact with spindle microtubules (parallel lines) but remain undivided until the spindle develops to a critical length. The telomeres are dissociated and remain so until cytokinesis. (C) In anaphase A, the sister centromeres (half open circles) separate, move towards the SPBs, and re-cluster at the SPBs. Flared lines associated with SPB represent astral microtubules seen in anaphase. (D) In anaphase B, the distance between the SPBs increases. (E) The cell undergoes cytokinesis, during which the cleavage of the cytoplasm is completed by the deepening of a partition that is drawn inward, around the middle of the cell, perpendicular to the spindle axis and between the two daughter nuclei.

analogous to both anaphase A (Uzawa and Yanagida 1992; Funabiki *et al.* 1993) and anaphase B (McCully and Robinow 1971; Hiraoka *et al.* 1984; Tanaka and Kanbe 1986) have been defined. It has further been shown that during these processes, while topoisomerase II (reviewed by Yanagida and Sternglanz 1990; see Chapter 6) is necessary for the separation of sister chromatids, the enzyme is not essential for sister centromere separation (Uemura and Yanagida 1984, 1986; Uemura *et al.* 1987; Funabiki *et al.* 1993).

During meiosis

Centromere (and telomere) dynamics during meiosis have also been examined in *S. pombe* cells (Chikashige *et al.* 1994). A surprising finding is that, during the period preceding meiosis, and unlike mitotic chromosome movement in which the centromeres lead, the telomeres remain clustered at the spindle pole bodies and lead chromosome movement (Fig. 3.5). Once meiotic chromosome segregation starts, however, centromeres resume the leading position in chromosome movement, as they do in mitosis. These results indicate that drastic nuclear reorganization switching the apparatus that leads chromosome movement takes place during meiosis. Because telomeres lead in chromosome movement only during the karyogamy and horse-tail stages, telomeres may function specifically in premeiotic chromosomal events such as the association and recombination of homologous chromosomes. It may be speculated that the linearly aligned configuration of chromosomes pulled by telomeres may facilitate the pairing of homologous chromosomes by shuffling and resolving interlocked chromosomes.

Fig. 3.5 Movement of centromeres (and telomeres) during pre- and early meiotic stages, showing the behaviour of one of the three *S. pombe* chromosomes (Chikashige *et al.* 1994). (A) At the time of induction of meiosis, centromeres (open circles) form a single cluster near the spindle pole body (SPB; black cylindrical rod), while telomeres (black circles) are located at several sites away from the SPB. (B and C) During karyogamy when parental nuclei approach each other, centromeres become separated from the SPB, while telomeres form a single cluster adjacent to the SPB and lead in chromosome movement. (D and E) Association between homologous chromosomes begins at the telomeres led by the SPB. After several rounds of oscillation, parental sets of chromosomes are suddenly pulled by the SPB moving in one direction and turning into a horse-tail shape. (F) By this motion, parental sets of chromosomes now have the same orientation of the polarized telomere–centromere configuration in a horse-tail nucleus. Association of homologous chromosomes continues, leading to the pairing of centromeres. (G and H) Prior to or during meiosis I, centromeres resume their proximity to the SPB, while telomeres take up positions away from the SPB. Arrows indicate directions of SPB movement.

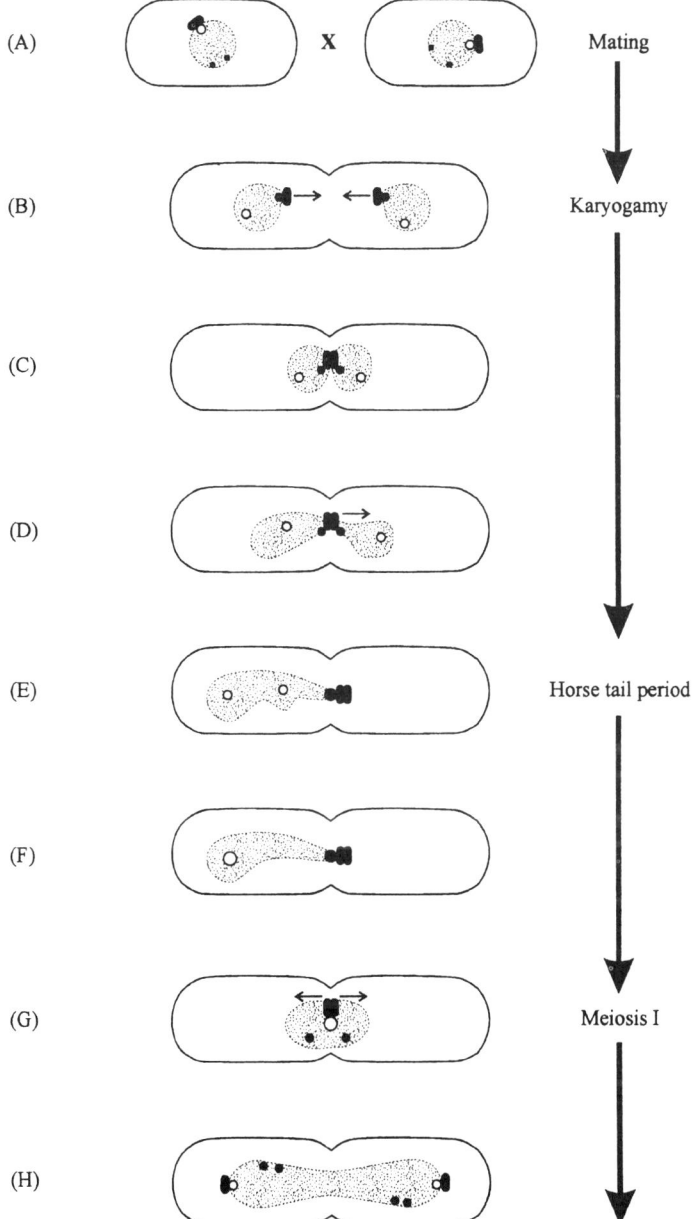

Cell cycle checkpoints and other factors affecting mitosis

As with the budding yeast, fission yeast is an excellent model organism for the analysis of the many complex factors that are invloved in the control of the eukaryotic cell division cycle. A number of checkpoints have been identified in *S. pombe* that arrest or delay the cell cycle in response to incomplete DNA synthesis or DNA damage (reviewed by Murray 1995; D'Urso and Nurse 1995). For example, a checkpoint exists that can prevent G_1 cells that have not passed Start, and thus have not begun DNA replication, from entering mitosis. This checkpoint requires the Rum1 protein, which may be responsible for signalling the presence in the nucleus of a licensing factor that binds to DNA and allows it to replicate (Moreno and Nurse 1994). In another example, a separate pathway appears to prevent cells that have passed Start, but not yet initiated DNA replication, from entering mitosis. One of the components of this pathway is the Cdc18 protein since cells that lack this protein can enter mitosis without initiating DNA replication (Kelly *et al.* 1993). As well as checkpoints that monitor unreplicated and damaged DNA, a number of genes have been cloned that are required for spindle assembly, assembly of a septum, and cell division. The products of some of these genes have multiple functions and define additional checkpoints important for coordinating the events of mitosis and cytokinesis. Although specific checkpoints for *S. pombe* kinetochore structure have not been described, experience from *S. cerevisiae* and the mammalian systems (Chapters 2 and 6) suggests that these are likely to be present.

In addition to cell cycle checkpoint controls, mutations have been identified in *S. pombe* for various steps of the cell division cycle that result in defective mitosis. A good example is the *cut* (*c*ell *u*ntimely *t*orn) mutants (Hirano *et al.* 1986). These are temperature-sensitive mutants which, at the restrictive temperature, exhibit defective nuclear division but cytokinesis proceeds to bisect the nucleus at the middle of the cell. Approximately 20 *cut*[+] genes have been identified, suggesting variable causes for deregulating the cell cycle (Hirano *et al.* 1986; Uemura and Yanagida 1986; Samejima *et al.* 1993). Some of the *cut*[+] gene products have been defined and are shown to be related to DNA replication, DNA topology, mitotic chromosome condensation and segregation, protein phosphorylation and anaphase progression, spindle formation, or spindle pole body duplication (Hirano *et al.* 1986;. Uemura and Yanagida 1986; Hagan and Yanagida 1990; Uzawa *et al.* 1990; Saka and Yanagida 1993; Samejima and Yanagida 1994). In other mutants such as *nda2*, *nda3*, *dis1*, *dis2*, *dis3*, *nuc2*, *sds22*, *pim1*, and *mts2*, cells display defective mitosis but not the *cut* phenotype (Toda *et al.* 1983; Hiraoka *et al.* 1984; Hirano *et al.* 1988; Ohkura *et al.* 1988, 1989, Matsumoto and Beach 1991; Gordon *et al.* 1993; Stone *et al.* 1993). For example, in the *nuc2* mutant (Hirano *et al.* 1988), the septum is made but cytokinesis is blocked. Clearly, further detailed study of these mutants will continue to enhance our understanding of the mechanisms that regulate the processes of eukaryotic cell division.

Conclusion

In addition to the advantage offered by *S. pombe* as a simple and tractable experimental system, a compelling reason for its use in centromere studies is in order to allow comparison with the better studied centromere of *S. cerevisiae* and with those of the higher eukaryotes. It is clear from the information presented in this and the previous chapter that, whilst the study of centromere proteins is severely lagging in *S. pombe* compared with *S. cerevisiae*, the delineation of the structural and functional components of centromeric DNA has progressed sufficiently well in both organisms to allow a meaningful comparison. In this comparison, the single most outstanding feature is the apparent lack of similarity between the DNA of the two centromeres. Such dissimilarity extends from a gross difference in the size and overall organization of the DNA elements, to a complete absence of homology at the nucleotide sequence level. As will become clear when the centromeric DNA of a host of other organisms are examined in Chapter 5, absence of conservation of this DNA is not unique amongst the yeasts, but is common throughout the phylogeny.

Since the centromeres of *S. pombe* and *S. cerevisiae* appear so different from each other, the question may be asked as to which will serve better as a model for the understanding of the higher eukaryotic centromere. The relatively simple DNA structure of the *S. cerevisiae* centromere suggests that this centromere may represent a more primitive organelle that may not accurately reflect the properties of the higher eukaryotic centromere. In contrast, the centromere of *S. pombe*, an organism thought to have diverged from *S. cerevisiae* many hundreds of millions of years ago, shares some important features with the centromeres of higher eukaryotes, the most notable of which are its much larger size and the presence of a substantial amount of repeated DNA sequences. Significantly, at least one (the K″/K-type DNA) of the repeated sequences in *S. pombe* has been shown to be essential for centromere chromatin organization and centromere function. This implies that perhaps at least some components of the repeated DNA found in the higher eukaryotic centromere may also have similar direct functional roles. Another common property between the *S. pombe* and higher eukaryotic centromeres is the binding of multiple microtubules per kinetochore, which contrasts with the simpler one kinetochore – one microtubule situation seen in *S. cerevisiae*, and again makes the *S. pombe* centromere a better system for studying the more complex microtubule-binding properties of the higher eukaryotic centromere. These features, together with the fact that the mitotic cell cycle and the requirement for chromosomal condensation in the fission yeast are also more typical of the higher eukaryotes than the budding yeast, clearly make *S. pombe* an excellent model system for the study of the higher eukaryotic centromere. However, notwithstanding the importance of these comparisons, the very simplicity of the *S. cerevisiae* centromere is itself uniquely invaluable and advantageous for experimentation. Furthermore, the recent identification of centromeric proteins that are conserved between *S. cerevisiae* and mammalian cells further asserts *S. cerevisiae* as a useful model for the higher eukaryotes. Thus, in view of the unique advantages offered by each of the two yeast systems, it would appear that any choice is likely to be based on academic preference and that both systems will continue to be deployed to full advantage in advancing our understanding of the centromere.

References

Adachi, Y. and Yanagida, M. (1989). Higher order chromosome structure is affected by cold-sensitive mutations in a *Schizosaccharomyces pombe* gene crm1$^+$ which encodes a 115-kD protein preferentially localized in the nucleus and at its periphery. J. Cell Biol. *108*, 1195–1207.

Alberts, B., Bray, D., Lewis, J., Raff, M., Roberts, K., and Watson, J. (1994). Molecular Biology of the Cell,. 3rd Edn. Garland Publishing, New York.

Allshire, R., Javerzat, J., Redhead, N., and Cranston, G. (1994). Position effect variegation at fission yeast centromeres. Cell *76*, 157–169.

Allshire, R., Nimmo, E., Ekwall, K., Javerzat, J., and Cranston, G. (1995). Mutations derepressing silent centromeric domains in fission yeast disrupt chromosome segregation. Genes Dev. *9*, 218–233.

Bahler, J., Schuchert, P., Grimm, C., and Kohli, J. (1991). Synchronized meiosis and recombination in fission yeast: observations with *pat1-114* diploid cells. Curr. Genet. *19*, 445–451.

Bahler, J., Wyler, T., and Loidl, J. (1993). Unusual nuclear stuctures in meiotic prophase of fission yeast: a cytological analysis. J. Cell Biol. *121*, 241–256.

Baum, M., Ngan, V., and Clarke, L. (1994). The centromeric K-type repeat and the central core are together sufficient to establish a functional *Schizosaccharomyces pombe* centromere. Mol. Biol. Cell *5*, 747–61.

Bloom, K., Hill, A., Kenna, M., and Saunders, M. (1989). The structure of a primitive kinetochore. Trends Biochem. Sci. *14*, 223–227.

Burke, D., Carle, G., and Olson, M. (1987). Cloning of large segments of exogenous DNA into yeast by means of artificial chromosome vectors. Science *236*, 806–812.

Byers, B. (1981). Cytology of the yeast life cycle. In The Molecular Biology of the Yeast *Saccharomyces*. I. Life Cycle and Inheritance, eds. Strathern, J.N., Jones, E.W. and Broach, J.R. Cold Spring Harbor Laboratory Press, Cold Spring Harbor, 59–96.

Chikashige, Y., Kinoshita, Y., Nakaseko, T., Matsumoto, T., Murakami, S., Niwa, O., *et al.* (1989). Composite motifs and repeat symmetry in *S. pombe* centromeres: direct analysis by integration of *Not*I restriction sites. Cell *57*, 739–751.

Chikashige, Y., Ding, D., Funabiki, H., Haraguchi, T., Mashiko, S., and Yanagida, M. (1994). Telomere-led premeiotic chromosome movement in fission yeast. Science *264*, 270–273.

Clarke, L., Amstutz, H., Fishel, B., and Carbon, J. (1986). Analysis of centromeric DNA in the fission yeast *Schizosaccharomyces pombe*. Proc. Natl Acad. Sci. USA *83*, 8253–8257.

Clarke, L. and Baum, M. (1990). Functional analysis of a centromere from fission yeast: a role for centromere-specific repeated DNA sequences. Mol. Cell. Biol. *10*, 1863–1872.

Clarke, L., Baum, M., Marschall, L., Ngan, V., and Steiner, N. (1993). Structure and Function of *Schizosaccharomyces pombe* Centromeres. Cold Spring Harbor Laboratory Press, Cold Spring Harbor, Vol. LVIII, 687–695.

D'Urso, G. and Nurse, P. (1995). Checkpoints in the cell cycle of fission yeast. Curr. Opin. Genet. Dev. *5*, 12–16.

Ding, P., McDonald, K. L., and MacIntosh, J. R. (1993). Three-dimensional reconstruction and analysis of mitotic spindles from the yeast *Schizosaccharomyces pombe*. J. Cell Biol. *120*, 141–150.

Egel, R. (1973). Commitment to meiosis in fission yeast. Mol. Gen. Genet. *121*, 277–284.

Egel, R., and Egel-Mitani, M. (1974). Premeiotic DNA synthesis in fission yeast. Exp. Cell. Res. *88*, 127–134.

Ekwall, K., Javerzat, J.-P., Lorentz, A., Schmidt, H., Cranston, G., and Allshire, R. (1995). The chromodomain protein: Swi6: a key component at fission yeast centromeres. Science *269*, 1429–1431.

Fan, J., Chikashige, Y., Smith, C., Niwa, O., Yanagida, M., and Cantor, C. (1988). Construction of a *Not* I restriction map of the fission yeast *Schizosaccharomyces pombe* genome. Nucleic Acids Res. *17*, 2801–2810.

Fantes, P. (1979). Epistatic gene interactions in the control of division in fission yeast. Nature (Lond.) *279*, 428–430.

Fishel, B., Amstutz, H., Baum, M., Carbon, J., and Clarke, L. (1988). Structural organization and functional analysis of centromeric DNA in the fission yeast *Schizosaccharomyces pombe*. Mol. Cell. Biol. *8*, 754–763.

Funabiki, H., Hagan, I., Uzawa, S., and Yanagida, M. (1993). Cell cycle-dependent specific positioning and clustering of centromeres and telomeres in fission yeast. J. Cell Biol. *121*, 961–976.

Garner, M. and Revzin, A. (1981). A gel electrophoresis method for quantifying the binding of proteins to specific DNA regions: application to components of the *E.coli* lactose operon regulatory system. Nucleic Acids Res. *9*, 3047–3060.

Gordon, C., McGurk, G., Dillon, P., Rosen, C., and Hastie, N. (1993). Defective mitosis due to a mutation in the gene for a fission yeast 26S protease subunit. Nature (Lond.) *366*, 355–357.

Gowen, J. and Gay, E. (1933). Effect of temperature on ever-sporting eye color in *Drosphila melanogaster*. Science *77*, 312.

Gross, D. and Garrard, W. (1988). Nuclease hypersensitive sites in chromatin. Annu. Rev. Biochem. *57*, 159.

Hagan, I. and Yanagida, M. (1990). Novel potential mitotic motor protein encoded by the fission yeast $cut7^+$ gene. Nature (Lond.) *347*, 563–566.

Hahnenberger, K. M., Baum, M. P., Polizzi, C. M., Carbon, J., and Clarke, L. (1989). Construction of functional artificial minichromosomes in the fission yeast *Schizosaccharomyces pombe*. Proc. Natl Acad. Sci. USA *86*, 577–581.

Hahnenberger, K., Carbon, J., and Clarke, L. (1991). Identification of DNA regions required for mitotic and meiotic functions within the centromere of *Schizosaccharomyces pombe* chromosome I. Mol. Cell. Biol. *11*, 2206–2215.

Hirano, T., Funahashi, S., Uemura, T., and Yanagida, M. (1986). Isolation and characterization of *Schizosaccharomyces pombe cut* mutants that block nuclear division but not cytokinesis. EMBO J. *5*, 2973–2979.

Hirano, T., Hiraoka, Y., and Yanagida, M. (1988). A temperature sensitive mutation of the *S. pombe* gene $nuc2^+$ that encodes a nuclear scaffold-like protein blocks spindle elongation in mitotic anaphase. J. Cell Biol. *106*, 1171–1183.

Hiraoka, Y., Toda, T., and Yanagida, M. (1984). The *NDA3* gene of fission yeast encodes B-tubulin; a cold sensitive *nda3* mutation reversibly blocks spindle formation and chromosome movement in mitosis. Cell *39*, 349–358.

Kaufer, N., Simanis, V., and Nurse, P. (1985). Fission yeast *Schizosaccharomyces pombe* correctly excises a mammalian RNA transcript intervening sequence. Nature *318*, 78–80.

Kelly, T., Martin, G., Forsburg, S., Stephen, R., Russo, A., and Nurse, P. (1993). The fission yeast cdc18[+] gene product couples S phase to START and mitosis. Cell *74*, 371–382.

Kuhn, R., Clarke, L., and Carbon, J. (1991). Clustered tRNA genes in *Schizosaccharomyces pombe* centromeric DNA sequence repeats. Proc. Natl Acad. Sci. USA *88*, 1306–1310.

Lee, M. G. and Nurse, P. (1987). Complementation used to clone a human homologue of the fission yeast cell cycle control gene cdc2. Nature *327*, 31–35.

Leupold, U. (1950). Die vererbung von homothallie und heterothallie bei *Schizosaccharomyces pombe*. C.R. Trav. Lab. Carlsberg Ser. Physiol. *24*, 381.

Lorentz, A., Ostermann, K., Fleck, O., and Schmidt, H. (1994). Switching gene *swi6*, involved in repression of silent mating-type loci in fission yeast, encodes a homologue of chromatin-associated proteins from *Drosophila* and mammals. Gene *143*, 139–143.

Marschall, L. G. and Clarke, L. (1995). A novel *cis*-acting centromeric DNA element affects *S. pombe* centromeric chromatin structure at a distance. Cell Biol. *128*, 445–454.

Matsumoto, T. and Beach, D. (1991). Premature initiation of mitosis in yeast lacking *RCC1* or an interacting GTPase. Cell *66*, 347–360.

Matsumoto, T., Murakami, S., Niwa, O., and Yanagida, M. (1990). Construction and characterization of centric circular and acentric linear chromosomes in fission yeast. Curr. Genet. *18*, 323–330.

Maundrell, K., Hutchison, A., and Shall, S. (1988). Sequence analysis of ARS elements in fission yeast. EMBO J. *7*, 2203–2209.

McCully, E. and Robinow, C. (1971). Mitosis in the fission yeast *Schizosaccharomyces pombe:* a comparative study with light and electron microscopy. J. Cell Sci. *9*, 475–507.

Moreno, S. and Nurse, P. (1994). Progression through the G1 phase of the cell cycle is regulated by the rum1[+] gene. Nature *367*, 236–242.

Murakami, S., Matsumoto, T., Niwa, O., and Yanagida, M. (1991). Structure of the fission yeast centromere *cen3* direct analysis of the reiterated inverted region. Chromosoma *101*, 214–221.

Murray, A. (1995). The genetics of cell cycle checkpoints. Curr. Opin. Genet. Dev. *5*, 5–11.

Nakaseko, Y., Adachi, Y., Funahashi, S., Niwa, O., and Yanagida, M. (1986). Chromosome walking shows a highly homologous repetitive sequence present in all the centromere regions of fission yeast. EMBO J. *5*, 1011–1021.

Nakaseko, Y., Kinoshita, N., and Yanagida, M. (1987). A novel sequence common to the centromere regions of *Schizosaccharomyces pombe* chromosomes. Nucleic Acids Res. *15*, 4705.

Nurse, P., Thuriaux, P., and Nasmyth, K. (1976). Genetic control of the cell division in the fission yeast *Schizosaccharomyces pombe*. Mol. Gen. Genet. *146*, 167–178.

Ohkura, H., Adachi, Y., Kinoshita, N., Niwa, O., Toda, T., and Yanagida, M. (1988). Cold-sensitive and caffeine supersensitive mutants of the *Schizosaccaromyces pombe dis* gene implicated in sister chromatid separation during mitosis. EMBO J. *7*, 1465–1473.

Ohkura, H., Kinoshita, N., Miyatani, S., Toda, T., and Yanagida, M. (1989). The fission yeast *dis2*⁺ gene required for chromosome disjoining encodes one of two putative type 1 protein phosphatases. Cell *57*, 997–1007.

Peterson, J. and Ris, H. (1976). Electron-microscopic study of the spindle and chromosome movement in the yeast *Saccharomyces cerevisiae*. J. Cell Sci. *22*, 219.

Polizzi, C. and Clarke, L. (1991). The chromatin structure of centromeres from fission yeast: differentiation of the central core that correlates with function. J. Cell Biol. *112*, 191–201.

Saka, Y. and Yanagida, M. (1993). Fission yeast *cut5*⁺ gene required for the onset of S-phase and the restraint of M-phase is identical to the radiation-damage repair gene *rad4*⁺. Cell *74*, 383–393.

Samejima, I. and Yanagida, M. (1994). Identification of *cut8*⁺ and *cek1*⁺, a novel protein kinase gene, which complement a fission yeast mutation that blocks anaphase. Mol. Cell. Biol. *14*, 6361–6371.

Samejima, I., Matsumoto, T., Nakaseko, Y., Beach, D., and Yanagida, M. (1993). Identification of seven new *cut* genes involved in *Schizosaccharomyces pombe* mitosis. J. Cell Sci. *105*, 135–143.

Smith, G. P. (1976). Evolution of repeated DNA sequences by unequal crossover. Science *191*, 528–535.

Smith, J., Caddle, M., Bulboaca, G., Wohlgemuth, J., Baum, M., Clarke, L.*et al.* (1995). Replication of centromere II of *Schizosaccharomyces pombe*. Mol. Cell. Biol. *15*, 5165–5172.

Spofford, J. (1976). Position effect variegation in *Drosophila*. In The Genetics and Biology of *Drosophila,* eds. Ashburner M. and Novitski, E. Academic Press, New York, Vol. 1C, 955–1018.

Steiner, N., Hahnenberger, K., and Clarke, L. (1993). Centromeres of the fission yeast *Schizosaccharomyces pombe* are highly variable genetic loci. Mol. Cell. Biol. *13*, 4578–4587.

Steiner, N. C. and Clarke, L. (1994). A novel epigenetic effect can alter centromere function in fission yeast. Cell *79*, 865–874.

Stone, E., Yamano, H., Kinoshita, N., and Yanagida, M. (1993). Mitotic regulation of protein phosphates by the fission yeast sds22 protein. Curr. Biol. *3*, 13–26.

Takahashi, K., Murakami, S., Chikashige, Y., Niwa, O., and Yanagida, M. (1991). A large number of tRNA genes are symmetrically located in fission yeast centromeres. J. Mol. Biol. *218*, 13.

Takahashi, K., Murakami, S., Chikashige, Y., Funabiki, Y., and Yanagida, M. (1992). A low copy number central sequence with strict symmetry and unusual chromatin structure in fission yeast centromere. Mol. Biol. Cell *3*, 819–835.

Tanaka, K. and Kanbe, T. (1986). Mitosis in the fission yeast *Schizosaccharomyces pombe* as revealed by freeze-substitution electron microscopy. J. Cell Sci. *80*, 253–268.

Toda, T., Umesono, K., Hirata, A., and Yanagida, M. (1983). Cold-sensitive nuclear division arrest mutants of the fission yeast *Schizosaccaromyces pombe*. J. Mol. Biol. *168*, 251–270.

Toda, T., Shimanuki, M., and Yanagida, M. (1991). Fission yeast genes that confer resistance to staurosporine encode an AP-1-like transcription factor and a protein kinase related to the mammalian ERK1/MAP2 and budding yeast FUS3 and KSS1 kinases. Genes Dev. *5*, 60–73.

Toda, T., Shimanuki, M., Saka, Y., Yamano, Y., Adachi, Y., Shirakawa, M. *et al.* (1992). Fission yeast pap1-dependent transcription is negatively regulated by an essential nuclear protein, crm1. Mol. Cell. Biol. *12*, 5474–5484.

Uemura, T. and Yanagida, M. (1984). Isolation of type I and II DNA topoisomerase mutants from fission yeast: single and double mutants show different phenotypes in cell growth and chromatin organization. EMBO J. *3*, 1737–1744.

Uemura, T. and Yanagida, M. (1986). Mitotic spindle pulls but fails to separate chromosomes in type II DNA topoisomerase mutants: uncoordinated mitosis. EMBO J. *5*, 1003–1010.

Uemura, T., Ohkura, H., Adachi, Y., Morino, K., Shiozaki, K., and Yanagida, M. (1987). DNA topoisomerase II is required for condensation and separation of mitotic chromosomes in *S. pombe*. Cell *50*, 917–925.

Uzawa, S. and Yanagida, M. (1992). Visualization of centromeric and nucleolar DNA in fission yeast by fluorescence *in situ* hybridization. J. Cell Sci. *101*, 267–275.

Uzawa, S., Samejima, I., Hirano, T., Tanaka, K., and Yanagida, M. (1990). The fission yeast *cut1*[+] gene regulates spindle pole body duplication and has homology to the budding yeast *ESP1* gene. Cell *62*, 913–925.

Yanagida, M. and Sternglanz, R. (1990). Genetics of DNA topoisomerases. In DNA Topology and its Biological Effects, eds. Wang, J.C. and Cozzarelli, N. Cold Spring Harbor Laboratory Press, Cold Spring Harbor, 299–320.

4

Structural organization and general properties of the higher eukaryotic centromere

Standard cytogenetic staining of the chromosomes of higher eukaryotes reveals the centromere as a constricted and relatively homogeneous region of densely packed heterochromatin, but gives little clue to the underlying complexity of this structure. With improved techniques of cell fixation for electron microscopic analysis and the use of specific antibodies that allow subregional staining of the centromere, it is now clear that the centromere of higher eukaryotes is structurally composed of a number of distinct domains. Within these domains are located the basic centromeric DNA molecule and an increasing number of known protein components that participate in the activities of the centromere. In this chapter, the structural organization as well as a number of general properties of the higher eukaryotic centromere are described. Where appropriate, the names of relevant DNA and protein components found in the different centromeric domains are mentioned, but detailed discussion of these components will be presented in Chapters 5 and 6.

Pairing domain

At the ultrastructural level, the constricted region of the higher eukaryotic chromosome is made up of a centromere–kinetochore complex that can be subdivided into at least three definable domains. These domains have been named the pairing domain, central domain, and kinetochore domain (Fig. 4.1). The pairing domain is thought to consist of both DNA and protein structures that promote linkage between sister chromatids. The DNA structure that has been postulated to link centromeres is made up of sister chromatid DNA strands that become intertwined or catenated during DNA replication (Sundin and Varshavsky 1980, 1981; Murray and Szostak 1985). A number of proteins that are found at the point of contact between sister centromeres at metaphase have been suggested as candidates for promoting centromere cohesion. These proteins include $INCENP_I$ and $INCENP_{II}$ (or inner centromere proteins **I** and **II**) and CLiP (or chromatid linking protein). In addition, three cohesion proteins (Mei-S332, Ord, and Cor1) have been described on the centromeres of meiotic chromosomes.

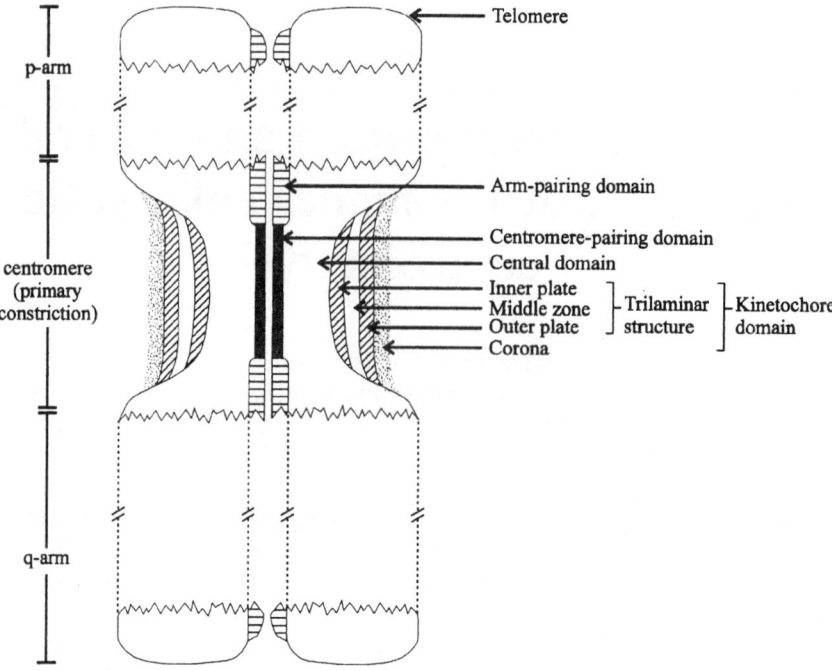

Fig. 4.1 Structural domains of the human centromere–kinetochore complex. These include a centromere-pairing domain (as distinct from the sister chromatid arm-pairing domains), a central domain consisting of densely packed heterochromatin, and a disc-shaped kinetochore domain. The kinetochore domain is made up of an outermost fibrous corona and a trilaminar structure composed of an electron-dense inner and outer plate and an electron-lucent middle zone.

Central domain

This domain contains a densely packed chromatin mass on which the kinetochore is anchored. The chromatin mass, which forms the cytologically defined 'constitutive heterochromatin', is composed primarily of various families of repetitive DNA and their associated proteins. In primates, the most abundant family of these centromeric repeats is the α-satellite DNA. High-resolution *in situ* hybridization has shown that α-satellite DNA is distributed throughout the central domain (reviewed by Pluta *et al.* 1990). Immunoelectron microscopy using anti-CENP-B antibodies demonstrates that CENP-B, a protein that binds specifically to a subset of α-satellite DNA, is present in the heterochromatin of the central domain and is distinctly located outside the kinetochore domain. In addition to CENP-B, two other proteins (HMG-I and pJα) have been implicated in binding α-satellite DNA and may also co-localize within the central domain. α-Satellite chromatin has been proposed to have a number of structural roles, including supporting the formation of kinetochore structure, regulation of chromatid pairing, and establishment of a centromere-specific chromatin domain that serves as a barrier to transcription or recombination.

Kinetochore domain

The kinetochore of higher eukaryotes is a trilaminar structure consisting of a poorly defined inner dense plate closely apposed to the centromeric heterochromatin (and occassionally indistinguishable from it), a middle electron-lucent zone, and a well-defined outer dense plate where the majority of microtubules appear to end (Comings and Okada 1971; Witt *et al.* 1980; Reider 1982) (Fig. 4.1). In the absence of kinetochore microtubules, a fibrous corona radiates from the surface of the outer plate facing away from the chromosomes. The inner plate has been shown to contain DNA material and a constitutive centromere binding protein, CENP-C. This protein could play a role in DNA–kinetochore interfacing as well as determination of the structural integrity of the kinetochore, such as its size. For a while, the middle zone was believed to be a space largely devoid of materials, but recently a tension-sensitive protein, 3F3/2, has been localized to this zone. This protein is believed to be a cell cycle checkpoint signal for metaphase to anaphase transition. At least two proteins, CENP-E (a motor protein) and CENP-F, have been found in the outer plate of the kinetochore structure. Both these proteins could interact with the microtubule and have a direct role in chromosome movement. Finally, the fibrous corona is known to contain another molecular motor, cytoplasmic dynein, that is likely to be intimately involved in chromosome movement. In addition to these proteins, a host of other proteins, such as $p34^{cdc2}$, NuMA, and MPM2 to name just a few, have also been tentatively localized to the kinetochore region.

Direct study of the 3D structure of the outer kinetochore plate in rat kangaroo PtK cells has indicated a highly complex, 35–40 nm thick structure composed of 10–20 nm-thick fibres that are highly variable in length, orientation, and spacing (McEwen *et al.* 1993). When viewed from the side, the outer plate is seen to be divided into three distinct structural domains in many regions (see 'Compound kinetochore' below) and is connected to the inner plate by parallel rows of 10–20 nm thick fibres that run almost perpendicularly to the chromosomal long axis. Several models have suggested that the outer plate consists of chromatin loops (Rattner 1987; Zinkowski *et al.* 1991; McEwen *et al.* 1993), although a concerted effort has failed to detect DNA in this structure (Cooke *et al.* 1993). It is thus possible that the outer kinetochore plate is strictly a proteinaceous structure.

The trilaminar kinetochore structure has been observed in species as diverse as mammals (Comings and Okada 1971; Witt *et al.* 1980; Reider 1982), insects (Wolf *et al.* 1991, 1994) and algae (Godward 1985), and therefore appears to be conserved throughout evolution. An interesting exception has, however, been reported in the unicellular biflagellate alga

Polytoma papillatum. Instead of a trilaminar organization, the kinetochore of this organism is made up of five layers, consisting of three dense plates interpersed by two transparent zones (Wolf 1995). The polemost dense layer serves as the attachment site for kinetochore microtubules and the innermost dense layer is intimately associated with the chromatin. This five-layered kinetochore is quite unique, especially when the usual trilaminar organization is found in closely related alga species (Godward 1985). It may be speculated that this unique kinetochore has arisen by duplication of an element already present in the widespread trilaminar structure.

Compound centromere

It has long been recognized that the centromere of higher eukaryotes could be broken up into smaller units, each of which would continue to provide centromere function (McClintock 1938; Sears 1952). More recent work described below involving a variety of organisms has provided further support for such an observation. These studies have led to the concept that the higher eukaryotic centromere is essentially a compound structure composed of redundant subunits that are capable of functioning either collectively or separately as individual entities (Zinkowski *et al.* 1991; Brinkley *et al.* 1992).

A lucid illustration of compound kinetochores has come from the study of Indian muntjac cells. Indian muntjac is an asiatic deer that is unique among mammals because it has the smallest karyotype [$2n = 6$ (female) or 7 (male)] with the largest chromosomes, that are thought to have evolved from smaller Chinese muntjac chromosomes ($2n = 46$) through multiple centric and tandem fusions (Brinkley *et al.* 1984). The presence of compound kinetochores on Indian muntjac chromosomes has been demonstrated using two different techniques, one involving caffeine-induced fragmentation of unreplicated kinetochores, and the other involving physically stretching kinetochores up to 20 times their normal length by hypotonic and/or shear forces generated in a cytocentrifuge (Zinkowski *et al.* 1991). The fragmentation experiment results in the generation of 80–100 kinetochore subfragments from the large kinetochores of the Indian muntjac (or a similar number from the 46 primordial Chinese muntjac kinetochores), with all of the subfragments able to interact with spindle microtubules and progress through the entire repertoire of mitotic movements. The stretching experiment, on the other hand, results in a kinetochore structure displaying a linear array of anti-centromere antibody-binding subunits arranged in a repetitive pattern along a centromeric DNA fibre. In addition to antibody binding, each repetitive subunit also interacts with exogenous tubulins (which are building blocks for spindle microtubules) and contain cytoplasmic dynein (a microtubule motor localized in the corona zone; Chapter 6).

Using the above techniques, similar kinetochore structures with repetitive subunits have been detected in Chinese hamster ovary and human HeLa cells

(Zinkowski *et al.* 1991). Other studies, again relying on chromosome fragmentation, have provided further evidence for the multiplicity of functional subunits on human centromeres. In these studies, when human chromosomes are truncated at the centromeric heterochromatin as a result of *de novo* centric fission (Chapter 7) or chromosomal rearrangement involving a break within the centromere (Wevrick *et al.* 1990), or through *in vitro* centromere dissection using telomeric DNA (Brown *et al.* 1994) (Chapter 8), the two resulting truncated products both contain active centromeres that are capable of proper segregation during cell division.

Centromere in interphase

Whereas it is beyond doubt that the centromere is a highly differentiated and active organelle through the different stages of mitosis and meiosis, its structure and function during interphase is less well understood. Three lines of evidence suggest that this organelle, rather than being quiescent and amorphous, may actually have an organized structure as well as some functional roles during interphase (reviewed by Pluta *et al.* 1995):

(i) *Centromeres remain condensed during interphase*. The persistence of condensed centromeres during interphase is revealed by the staining of discrete foci with anti-centromere autoimmune antibodies (Moroi *et al.* 1981). The demonstration that these foci both correspond to specialized heterochromatins (Cooke *et al.* 1990) and co-localize with centromeric α-satellite DNA suggests a tight association of at least some proteins with the centromere through interphase (Masumoto *et al.* 1989). Three of these proteins are likely to be CENP-A, -B, and -C, since these are known to bind to the centromere constitutively throughout the cell cycle. The association of these, and possibly other, proteins with the centromere during interphase appears to be required for the subsequent assembly of a functional kinetochore, since microinjection of centromere antibodies during this phase disrupts kinetochore formation and inhibits mitotic progression (Bernat *et al.* 1990; Simerly *et al.* 1990; Tomkiel *et al.* 1994). Thus, the interphase centromere interacts specifically with different proteins to maintain a distinct chromosomal structure whose integrity is crucial to its ensuing role in chromosome segregation during mitosis.

(ii) *Centromeres occupy distinct nuclear positions during interphase*. As will be described below (see Rabl configuration), within the interphase nuclei of mitotic cells of *Drosophila* and many other species (but not mammals), the centromeres are clustered or polarized at one end of the nucleus, while the telomeres are positioned adjacent to the nuclear envelope at the other. In mammalian interphase cells, instead of clustering, the centromeres tend to congregate near the nuclear periphery and around nucleoli, although they can be found throughout the nuclear volume. These interphase centromere positions are not static, nor does their arrangement conform to any specific pattern. Rather, within a cell type, centromere distribution in relation to the two nuclear landmarks reproducibly changes during cell cycle progression (Ferguson and Ward 1992) and in response to the functional (Park and De Boni 1992) and transcriptional states of the cell (Leger *et al.* 1994; Janevski *et al.* 1995). Thus, it is possible that the centromere, perhaps through the influence of its repetitive DNA, may organize the interphase nucleus in ways that influence gene expression (Manuelidis and Borden 1988).

(iii) *Centromeres associate with the nucleoli during interphase.* The association of interphase centromeres with nucleoli in human cells is true not only for the NOR (or nucleolus organizer region)-bearing acrocentric chromosomes, but also for the interphase centromeres of non-NOR-bearing chromosomes (Manuelidis and Borden 1988; Leger *et al.* 1994). Furthermore, centromere proteins have been detected in isolated human nucleoli (Ochs and Press 1992), and biochemical interaction between centromere proteins and proteins that are localized to the nucleoli has been demonstrated (Pluta *et al.* 1995). These observations suggest that centromeres may have as yet undiscovered functions connected with the activity of the nucleoli during interphase.

Holocentric chromosomes

Caenorhabditis elegans

The chromosomes of this nematode are unusual in that they are holocentric or holokinetic, i.e. the centromeres are diffuse along the chromosomes, attaching to spindle microtubules at many points along their length rather than at a single differentiated centromere point (Herman *et al.* 1979; Albertson and Thompson 1982; McKim and Rose 1990). Evidence for this dispersed attachment is provided by light and electron microscopic examination of the chromosomes (Albertson and Thompson 1982) and by analysis of the genetic stability of chromosome fragments (Herman *et al.* 1979; Albertson and Thompson 1982). This holocentric property suggests that the *C. elegans* genome contains many sequences (Chapter 5) distributed along the chromosomes that can act analogously to the localized centromeres of other types of organisms.

Parascaris univalens (chromosomes with three different centromere states)

The centromere of this horse parasitic nematode is of interest because of some rather unusual properties. Firstly, like *C. elegans*, the centromere of *P. univalens* is holocentric in nature. Secondly, under physiological conditions, the kinetochore can be regulated to produce three distinct centromere states in different cell types; briefly stated, this organelle takes a different location, size, and structure in germline mitotic, meiotic, and presomatic early embryonic cells. The mechanism for the manifestation of these different centromere states is closely related to the unusual chromosome structures of this organism. In germline cells, the genome of *P. univalens* is organized into two large chromosomes each containing a central euchromatic region surrounded by two large blocks of heterochromatic AT-rich satellite DNA which constitute about 80% of genomic DNA (Fig. 4.2A) (Moritz and Roth 1976; Roth 1979; Goday and Pimpinelli 1984). In presomatic blastomeres starting in the second or third embryonic division, the cells undergo a process known as chromatin diminution or heterochromatin elimination (Boveri 1910). In these cells, the chromosomes are fragmented into as many as 60 chromosomes, and all of the heterochromatic DNA is eliminated. Neither the molecular mechanisms nor the cytoplasmic factors (Tobler 1986) involved in chromatin diminution have so far been identified. Thus, in *P. univalens* the two germline chromosomes retain their integrity and maintain their heterochromatin, but somatic chromosomes are numerous, much smaller, and totally euchromatic.

The three different states of the *Parascaris* centromere are shown in Fig. 4.2B. The first two states are seen in the mitotic and meiotic germline cells. In the mitotic germline cells (state 1), the two large chromosomes are holocentric. The continuous kinetochore and the holocentric behaviour imply kinetic activity of both euchromatic and heterochromatic regions. In contrast, in the meiotic germline cells (state 2), centromeric activity is restricted to the terminal heterochromatic areas where although no kinetochore structures are seen at the microtubule–chromatin binding point, 'direct insertion' of spindle microtubules into the heterochromatin has been observed (Goday and Pimpinelli 1989). The third state of the *Parascaris* centromere occurs during the chromatin diminution process in early embryogenesis. Immunochemical and electron microscopic studies demonstrate that mitotic spindle microtubules associate only with euchromatic chromosomal regions in all prediminute and diminute chromatin, while heterochromatic regions do not bind spindle microtubules (Goday *et al.* 1992). Kinetochore plates are absent in the heterochromatic regions while euchromatic regions possess continuous kinetochore plates similar to those described in germline mitotic chromosomes. After chromosome fragmentation and heterochromatin elimination, the newly formed chromosomes segregate independently and exhibit continuous kinetochore plates.

In all three centromere states, there is no evidence for any modification of the DNA content of the chromosomes. This suggests that centromeric DNA sequences scattered along the chromosomes are probably differently available to build up a kinetochore along the chromosomes, and that there is a specific chromosomal organization in the different cell types that would explain not only the regional restriction of the kinetochores in mitotic germline cells and early embryonic cells, but also the lack of a visible kinetochore structure in meiotic cells.

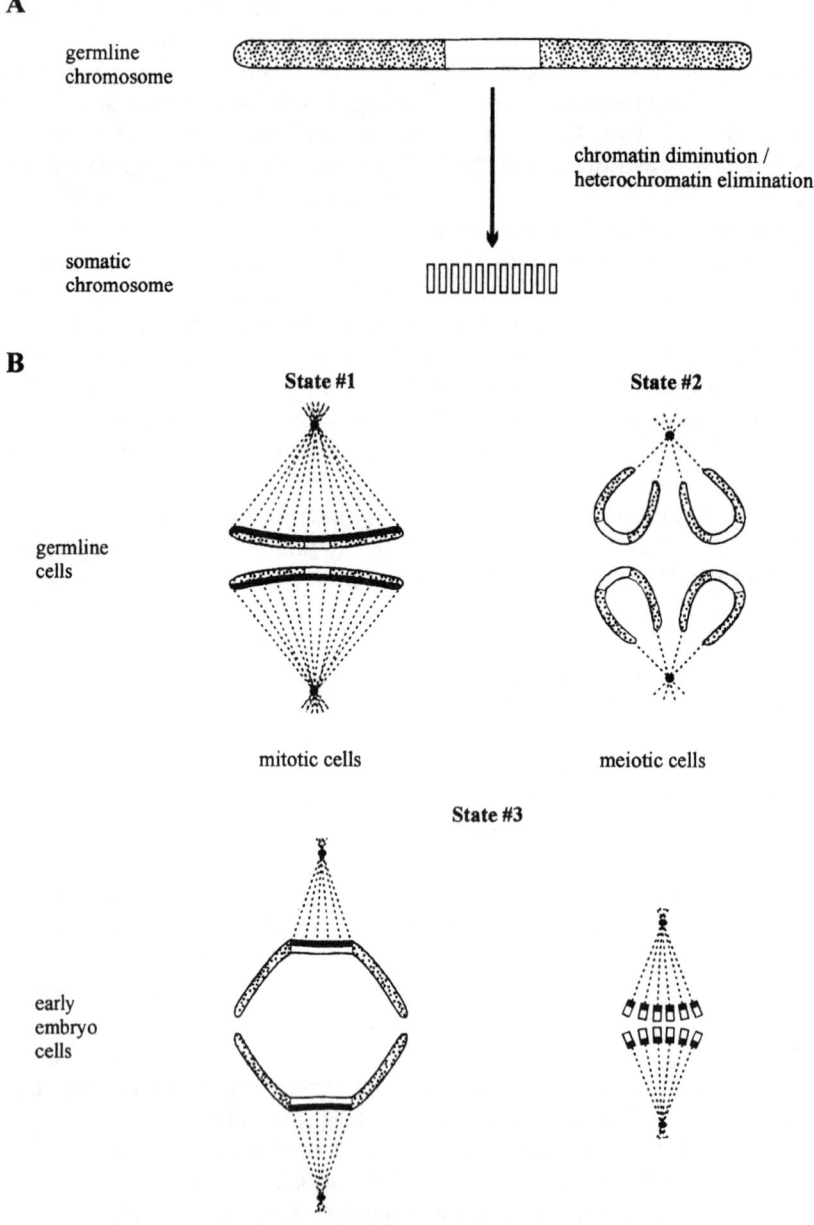

Fig. 4.2 (A) *P. univalens* chromosomes before and after chromatin diminution and heterochromatin elimination. Clear and stippled areas represent euchromatin and heterochromatin, respectively. (B) The three centromere states of *P. univalens* chromosomes. Clear, stippled, and black areas on chromosomes represent euchromatin, heterochromatin, and kinetochore, respectively. Functional kinetochores are shown attached to microtubule spindle (dotted lines) and spindle poles (black circles).

Spatial arrangement of centromeres

Rabl configuration

The rather undistinguished appearance of the mitotic interphase nucleus of many species viewed under the microscope might give the illusion that the DNA contained within exists as a 'ball of string' that is distributed randomly throughout the volume of the nucleus. However, there is considerable evidence to suggest that the interphase nucleus is far from lacking in large-scale order (reviewed by Hilliker and Appels 1989; Spector 1993). For example, in the interphase nuclei of mitotic cells of *Drosophila* salivary gland, the centromeres of the polytene chromosomes are seen clustered at one end of the nucleus, with the telomeres adjacent to the nuclear envelope at the other (Agard and Sedat 1983; Mathog *et al.* 1984; Hochstrasser *et al.* 1986; Hiraoka *et al.* 1990), forming a so-called Rabl orientation (Rabl 1885) (Fig. 4.3A). This polarized orientation of chromosomes, apparently established at the end of the previous telophase and largely maintained through the intervening interphase (Hiraoka *et al.* 1989), reflects the force applied at the centromere by interaction with the mitotic spindle during anaphase chromosome movement. Rabl configurations have been reported in many eukaryotic species (reviewed by Comings 1980) although it is by no means universal. For example, it is generally not found in mammalian cells (Manuelidis and Borden 1988; Billia and De Boni 1991).

Chromosome rosette

A chromosome rosette is a spatial arrangement of mitotic chromosomes that is formed when all the chromosomes aggregate briefly during prometaphase into a wheel-shaped ring. The territorial distribution of individual chromosomes within such a rosette has recently been investigated in human cells using chromosome-specific centromeric α-satellite and whole-chromosome painting probes in fluorescence *in situ* hybridization (Nagele *et al.* 1995). The results demonstrate that chromosomal homologues are consistently positioned on opposite sides of a rosette, suggesting that chromosomes are separated into two haploid sets, each derived from one parent (Fig. 4.3B). Furthermore, the chromosome orders within the two haploid sets are antiparallel. This arrangement of chromosomes appears to be both species specific and independent of cell type, and may influence chromosome topology throughout the cell cycle.

The mechanism underlying the formation of such an exact spatial order of chromosomes within the prometaphase rosette is presently unknown. One possibility is that chromosomes are attached permanently to one another in a precise order at the level of their centromeric domains. Although centromeric interactions have not yet been observed, the tendency for centromeres to line up into a 'string-of-pearls' arrangement during G_2 (Haaf and Schmid 1989)

(A) Rabl Configuration

Side view   View from centromere end

(B) Prometaphase rosette

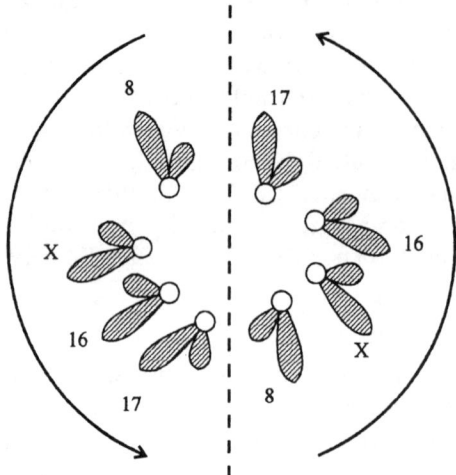

Fig. 4.3 (A) Polarized (Rabl) orientation of chromosomes in the interphase cell nucleus, showing all centromeres (black circles) facing one nuclear pole and all telomeres (black triangles) pointing toward the opposite pole. (B) Precise spatial positioning of chromosomes during prometaphase, showing antiparallel ordering (arrows) of human chromosomes 8, X, 16, and 17, with reference to any arbitrary line (dashed line) that bisects the chromosome complement into two haploid sets (Nagele *et al.* 1995).

and to form a ring at the hub of the prometaphase rosette provides indirect support for their existence (Nagele *et al.* 1995). Permanent associations among chromosomes through centromeric interconnections may offer advantages such as the synchronization of chromosome movements during congression toward the forming metaphase plate and the maintenance of a precise topological order of the chromosomal structures both through the segregational stages and throughout the cell cycle. A consistent chromosome arrangement in human cells may also have implications on the mechanisms of chromosomal translocation and aneuploidy among certain chromosomes (Nagele *et al.* 1995).

Temporal order of centromere separation

The order in which the centromeres of different human chromosomes undergo mitotic division is reported to be nonrandom in that a 'normal sequence' of centromere separation exists (Vig and Wodniczki 1974; Vig 1981a, 1981b, 1983). Thus, chromosome 18 is the first to separate and chromosomes 2, 3, 4, 5, 17, and X also divide very early, while chromosomes 1, 16, Y, and the acrocentrics are the last to separate. The existence of such an order suggests that any drastic 'out-of-phase' division of the centromeres may lead to chromosome missegregation. For example, it has been shown that late out-of-phase centromere division of chromosome 18 may lead to trisomy 18 (Méhes 1978; Bajnóczky and Méhes 1988). As will be discussed in Chapter 7, early out-of-phase or premature centromere division (PCD) is also known to cause aneuploidies of chromosomes 21 and X.

Evolution of different centromere sizes

When the genome size of an organism is compared with the number of kinetochore microtubules, the amount of genomic DNA per microtubule is found to be fairly constant (50-fold range) throughout phylogeny (reviewed by Bloom 1993) (Table 4.1). In contrast, the amount of centromeric DNA (CEN value) spans 3–4 orders of magnitude [e.g. 1.25×10^2 bp in *S. cerevisiae* (Chapter 2) and $2-4 \times 10^6$ bp in humans (Chapter 5)] over a 300-fold difference in genome size (1.4×10^7 bp in *S. cerevisiae* and 3.9×10^9 bp in humans). Thus, even considering the kinetochore in the budding yeast as a primitive structure, Bloom (1993) speculates that the dramatic increase in CEN value in evolution must have a significant advantage that cannot be accounted for simply by the increased number of microtubule-binding sites. The author suggests that as cell volume increases, augmenting the size of the centromere might enhance the efficiency of microtubule capture, thus providing a selective advantage for organisms with larger CEN values.

Meiotic recombination suppression

It has long been recognized that the centromere exerts a direct, negative effect on meiotic recombination both within itself and in flanking chromosomal sequences, a phenomenon known as the 'centromere effect' or 'spindle fiber effect' (Beadle 1932; Mather 1938). This centromere effect has been documented in a variety of organisms, including *Drosophila* (Beadle 1932; Mather 1938; Roberts 1965; Lefevre 1971; Yamamoto and Miklos 1977; Szauter 1984),

Table 4.1 Genome size versus number of kinetochore microtubules (Bloom 1993)

Type of organism	Species	Genomic DNA content[1] (bp)	Chromosome number[1]	Microtubules per chromosome	Genomic DNA per microtubule (bp)
Protozoan	C. reinhardtii	1.09×10^8	19	1	5.7×10^6
Budding yeast	S. cerevisiae	1.4×10^7	16	1	0.87×10^6
Budding yeast	K. lactis	1.4×10^7	6	1	2.3×10^6
Fission yeast	S. pombe	1.4×10^7	3	2–4	1.5×10^6
Fruit fly	D. melanogaster	1.65×10^8	4	6–21	4.1×10^6
Grasshopper	L. migratoria	6.5×10^9	11	18–23	2.8×10^7
Human	Homo sapiens	3.9×10^9	23	20–30	6.7×10^6
Plant	H. katharinae	5.3×10^{10}	9	120	4.9×10^7

[1] Haploid values.

mouse (Kipling *et al.* 1994), *Neurospora* (Centola and Carbon 1994), tomato plant (Khush and Rick 1967, 1968; Rick 1969, 1972; Tanksley *et al.* 1992), and humans (Wevrick and Willard 1989; Mahtani and Willard 1990; Marcais *et al.* 1991). In these organisms, the level of recombination suppression can be as high as 10- to 40-fold those of the rest of the genome (Roberts 1965; Tanksley *et al.* 1992; Centola and Carbon 1994). Centromere effect has also been described in the fission yeast (Nakaseko 1986) and budding yeast (Clarke and Carbon 1980; Malone *et al.* 1980; Meddle *et al.* 1984; Panzeri and Philippsen 1982; Stinchcomb *et al.* 1982; Mortimer and Schild 1985; Lambie and Roeder 1986, 1988), but conflicting evidence indicating that such an effect is lacking at the budding yeast centromere has been reported (Symington and Petes 1988). Relatively little is known about whether mitotic recombination is similarly affected by the centromere effect, although the result of one study in the budding yeast indicates that mitotic recombination may not be suppressed at the centromere (Liebman *et al.* 1988).

The mechanisms responsible for the suppression of meiotic recombination at the centromere are not well understood. Such a suppression is believed to be the result of the effect of pericentric heterochromatin which, due presumably to its more condensed state in meiosis at the time of crossing over, would experience reduced levels of recombination compared with euchromatin (Roberts 1965; Khush and Rick 1967, 1968; Rick 1969, 1972). However, evidence indicating that influences other than heterochromatinization may be involved has come from study in the budding yeast, where a cloned centromere, although lacking any visible form of heterochromatin, has been shown to decrease recombination when it is artificially integrated into new sites in the genome (Lambie and Roeder 1986).

Position effect variegation (PEV)

PEV is the unstable expression that results when a gene is placed at particular positions in the genome (reviewed by Henikoff 1990; Cook and Karpen 1994; Karpen 1994; Weiler and Wakimoto 1995; Elgin 1996). This phenomenon appears to be very widespread, since it has been observed in many different organisms, including insects, mammals and plants (reviewed by Lima-de-Faria 1983). It was first discovered in *Drosophila* when a chromosomal inversion placing a wild-type *white* gene (required for red eye pigmentation) adjacent to centromeric heterochromatin resulted in flies with a mixture of red and white patches of cells within the compound eye, due to either expression or repression of the gene within individual cells (Muller 1930). Subsequently, it was noted that most cases of PEV in *Drosophila* are associated with centromeric heterochromatin (Schultz 1936), and that the shorter regions of heterochromatin at telomeres also induce variegation (Hazelrigg *et al.* 1984; Levis *et al.* 1985; reviewed by Kipling 1995; Shore 1995). In addition to higher eukaryotes, PEV has similarly been demonstrated in the heterochromatin-like domains of *S. pombe* centromeres (Allshire *et al.* 1994).

Although the PEV phenomenon has been known and studied for well over 60 years, its molecular mechanism remains unexplained. One hypothesis is that transcriptionally inactive heterochromatin spreads into adjacent DNA, so that a gene placed nearby is inactivated (Demerec 1940; Hartmann-Goldstein 1967; Locke *et al.* 1988; Eissenberg 1989; Tartof *et al.* 1989; Tartof and Bremer 1990). The mechanism underlying this spread of hetero-chromatin is not fully understood. At a higher-order organizational level, heterochromatin-induced gene inactivation has been shown to correlate with a change in chromatin structure both in *Drosophila* (Wallrath and Elgin 1995) and in *S. pombe* (Allshire *et al.* 1994) (Chapter 3). At the molecular level, the activity of modifiers or *trans*-acting proteins has been implicated as regulators of PEV. One of these modifiers is HP-1, a protein that localizes predominantly to the *Drosophila* chromocenter and appears to have a role in the formation of heterochromatin (James and Elgin 1986; James *et al.* 1989; Eissenberg *et al.* 1990). (A chromocenter is a central aggregation of hetero-chromatic chromosomal elements found in the *Drosophila* larval salivary gland cell nucleus from which euchromatic chromosome arms extend.) The HP-1 modifier behaves like a suppressor, since increased copy number of the *HP-1* gene results in enhancement of *white* gene repression and variegation (Eissenberg *et al.* 1992). Other suppressors with a similar effect on PEV have been described in *Drosophila* (Spofford 1967; Reuter *et al.* 1982, 1990; Locke *et al.* 1988) and in *S. pombe* (Ekwall *et al.* 1995; see Swi6 protein in Chapter 3). A mammalian homologue of HP-1 that binds to pericentromeric heterochromatin has also been identified (Nicol and Jeppesen 1994). The emerging model is one in which fluctuations in the level of heterochromatin proteins influence the amount of heterochromatin formed, which results in more or less spreading from an initiation point and ultimately increased or decreased repression of a gene (Locke *et al.* 1988; Eissenberg 1989; Tartof *et al.* 1989; Henikoff 1990; Spradling and Karpen 1990).

Caution should, however, be exercised against any simple view of hetero-chromatin merely as a molecularly compact structure that is both transcrip-tionally inert and capable of only steadfast repression of genes placed in its proximity. The reason has come from the discovery of numerous essential functions that reside in heterochromatic regions. These functions include transcriptional activity for the ribosomal RNA genes (e.g. as found on the heterochromatic short arms of the human acrocentric chromosomes), and for genes in *Drosophila* that are required for viability (e.g. *lethal mutable* genes) and fertility (e.g. Y-linked male fertility factors) (Gatti and Pimpinelli 1992). Thus, it is likely that a complex relationship exists between hetero-chromatin and gene expression (Weiler and Wakimoto 1995).

Conclusion

The higher eukaryotic centromere is a highly differentiated structure of the chromosome. In order to perform the various essential functions, a number of unique properties have evolved on this structure. In this chapter, we have outlined the spatial organization of the various known structural components of the higher eukaryotic centromere. We have also discussed a number of interesting phenomena that are associated with this centromere. The study of these phenomena is an important part of our attempt to gain a full understanding of the biology of the higher eukaryotic centromere. With the advent of new technology and our rapidly expanding knowledge of the DNA and protein constituents of the centromere, progress towards such an understanding should come expeditiously in the near future.

References

Agard, D. and Sedat, J. W. (1983). Three-dimensional architecture of a polytene nucleus. Nature *302*, 676–681.

Albertson, D. and Thomson, J. (1982). The kinetochores of *Caenorhabditis elegans*. Chromosoma *86*, 409–428.

Allshire, R., Javerzat, J., Redhead, N., and Cranston, G. (1994). Position effect variegation at fission yeast centromeres. Cell *76*, 157–169.

Bajnoczky, K. and Mehes, K. (1988). Parental centromere separation sequence and aneuploidy in the offspring. Hum. Genet. *78*, 286–288.

Beadle, G. (1932). A possible influence of the spindle fibre on crossing-over in *Drosophila*. Proc. Natl Acad. Sci. USA *18*, 160–165.

Bernat, R. L., Borisy, G. G., Rothfield, N. F., and Earnshaw, W. C. (1990). Injection of anticentromere antibodies in interphase disrupts events required for chromosome movement in mitosis. J. Cell Biol. *111*, 1519–1533.

Billia, F. and DeBoni, U. (1991). Localization of centromeric satellite and telomeric DNA sequences in dorsal root ganglion neurons, *in vitro*. J. Cell Sci. *100*, 219–226.

Bloom, K. (1993). The centromere frontier: kinetochore components, microtubule-based motility, and the CEN-value paradox. Cell *73*, 621–624.

Boveri, T. (1910). Die potenzen der *Ascaris* blastomeren bei abgeanderter furchung. Zugleich ein beitrag zur frage qualitativ-ungleicher chromosomen-teilung. In Festschrift für R. Hertwig. Fischer, Jena, Vol. III, 131–214.

Brinkley, B. R., Valdivia, M. M., Tousson, A., and Brenner, S. L. (1984). Compound kinetochores of the Indian muntjac: evolution by linear fusion of unit kinetochores. Chromosoma (Berl.) *91*, 1–11.

Brinkley, B. R., Ouspenski, I., and Zinkowski, R. P. (1992). Structure and molecular organization of the centromere–kinetochore complex. Trends Cell Biol. *2*, 14–21.

Brown, K. E., Barnett, M. A., Burgtorf, C., Shaw, P., Buckle, V. J., and Brown, W. R. A. (1994). Dissecting the centromere of the human Y chromosome with cloned telomeric DNA. Hum. Mol. Genet. *3*, 1227–1237.

Centola, M. and Carbon, J. (1994). Cloning and characterization of centromeric DNA from *Neurospora crassa*. Mol. Cell. Biol. *14*, 1510–1519.

Clarke, L. and Carbon, J. (1980). Isolation of the centromere-linked *CDC10* gene by complementation in yeast. Proc. Natl Acad. Sci. USA *77*, 2173–2177.

Comings, D. E. and Okada, T. A. (1971). Fine structure of kinetochore in Indian Muntjac. Exp. Cell Res. *67*, 97–110.

Comings, D. (1980). Arrangement of chromatin in the nucleus. Hum. Genet. *53*, 131–143.

Cook, K. R. and Karpen, G. H. (1994). A rosy future for heterochromatin. Proc. Natl Acad. Sci. USA *91*, 5219–5221.

Cooke, C. A., Bernat, R. L., and Earnshaw, W. C. (1990). CENP-B: a major human centromere protein located beneath the kinetochore. J. Cell Biol. *110*, 1475–1488.

Cooke, C. A., Bazett-Jones, D. P., Earnshaw, W. C., and Rattner, J. B. (1993). Mapping DNA within the mammalian kinetochore. J. Cell Biol. *120*, 1083–1091.

Demerec, M. (1940). Genetic behaviour of euchromatic segments inserted into heterochromatin. Genetics *25*, 618–627.

Eissenberg, J. (1989). Position effect variegation in *Drosophila*: towards a genetics of chromatin assembly. BioEssays *11*, 14–17.

Eissenberg, J., James, T., Foster-Hartnett, D., Hartnett, T., and Ngan, V. (1990). Mutation in a heterochromatin specific chromosomal protein is associated with suppression of position effect variegation in *Drosophila melanogaster*. Proc. Natl Acad. Sci. USA *87*, 9923–9927.

Eissenberg, J., Morris, G., Reuter, G., and Hartnett, T. (1992). The heterochromatin-associated protein HP-1 is an essential protein in *Drosophila* with dosage-dependent effects on position effect variegation. Genetics *131*, 345–352.

Ekwall, K., Javerzat, J.-P., Lorentz, A., Schmidt, H., Cranston, G., and Allshire, R. (1995). The chromodomain protein: Swi6: a key component at fission yeast centromeres. Science *269*, 1429–1431.

Elgin, S. C. R. (1996). Heterochromatin and gene regulation in *Drosophila*. Curr. Opin. Genet. Dev. *6*, 193–202.

Ferguson, M. and Ward, D. C. (1992). Cell cycle dependent chromosomal movement in pre-mitotic human T-lymphocyte nuclei. Chromosoma *101*, 557–565.

Gatti, M. and Pimpinelli, S. (1992). Functional elements in *Drosophila melanogaster* heterochromatin. Annu. Rev. Genet. *26*, 239–275.

Goday, C. and Pimpinelli, S. (1984). Chromosome organization and heterochromatin elimination in *Parascaris*. Science *224*, 411–413.

Goday, C. and Pimpinelli, S. (1989). Centromere organization in meiotic chromosomes of *Parascaris univalens*. Chromosoma *98*, 160–166.

Goday, C., Gonzalez, G. J., Esteban, M. R., Giovinazzo, G., and Pimpinelli, S. (1992). Kinetochores and chromatin diminution in early embryos of *Parascaris univalens*. J. Cell Biol. *118*, 23–32.

Godward, M. (1985). The kinetochore. Int. Rev. Cytol. *94*, 77–105.

Haaf, T. and Schmid, M. (1989). Centromeric association and non-random distribution of centromeres in human tumour cells. Hum. Genet. *81*, 137–143.

Hartmann-Goldstein, I. J. (1967). On the relationship between heterochromatinization and variegation in *Drosophila* with special reference to temperature sensitive periods. Genet. Res. *10*, 143–159.

Hazelrigg, T., Levis, R., and Rubin, G. (1984). Transformation of *white* locus DNA in *Drosophila*: dosage compensation, *zeste* interaction, and position effects. Cell *36*, 469–481.

Henikoff, S. (1990). Position effect variegation after 60 years. Trends Genet. *6*, 422–426.

Herman, R., Madl, J., and Kari, C. (1979). Duplications in *Caenorhabditis elegans*. Genetics *92*, 419–435.

Hilliker, A. and Appels, R. (1989). The arrangement of interphase chromosomes: structural and functional aspects. Exp. Cell. Res. *185*, 297–318.

Hiraoka, Y., Minden, J., Swedlow, J., Sedat, J., and Agard, D. (1989). Focal points for chromosome condensation and decondensation revealed by three-dimensional *in vivo* time-lapse microscopy. Nature *342*, 293–296.

Hiraoka, Y., Agard, D., and Sedat, J. (1990). Temporal and spatial coordination of chromosome movement, spindle formation, and nuclear envelope breakdown during prometaphase in *Drosophila melanogaster* embryos. J. Cell Biol. *111*, 2815–2828.

Hochstrasser, M., Mathog, D., Gruenbaum, Y., Saumweber, H., and Sedat, J. (1986). Spatial organisation of chromosomes in the salivary gland nuclei of *Drosophila melanogaster*. J. Cell Biol. *102*, 112–123.

James, T. and Elgin, S. (1986). Identification of a nonhistone chromosomal protein associated with heterochromatin in *Drosophila melanogaster* and its gene. Mol. Cell. Biol. *6*, 3862–3872.

James, T., Eissenberg, J., Craig, C., Dietrich, V., Hobson, A., and Elgin, S. (1989). Distribution patterns of HP-1, a heterochromatin-associated nonhistone chromosomal protein in *Drosophila*. Eur. J. Cell Biol. *50*, 170–180.

Janevski, J., Park, P. C., and Boni, U. D. (1995). Organization of centromeric domains in hepatocyte nuclei: rearrangement associated with de *novo* activation of the vitellogen gene family in *Xenopus laevis*. Exp. Cell Res. *217*, 227–239.

Karpen, G. (1994). Position-effect variegation and the new biology of heterochromatin. Curr. Opin. Genet. Dev. *4*, 281–291.

Khush, G. and Rick, C. (1967). Studies on the linkage map of chromosome 4 of the tomato and on the transmission of induced deficiencies. Genetica *38*, 74–94.

Khush, G. and Rick, C. (1968). Cytogenetic analysis of the tomato genome by means of induced deficiencies. Chromosoma *23*, 452–484.

Kipling, D. (1995). The Telomere. Oxford University Press, Oxford.

Kipling, D., Wilson, H., Mitchell, A., Taylor, B., and Cooke, H. (1994). Mouse centromere mapping using oligonucleotide probes that detect variants of the minor satellite. Chromosoma *103*, 46–55.

Lambie, E. and Roeder, G. (1986). Repression of meiotic crossing over by a centromere (*CEN3*) in *Saccharomyces cerevisiae*. Genetics *114*, 769–789.

Lambie, E. and Roeder, G. (1988). A yeast centromere acts in *cis* to inhibit meiotic gene conversion of adjacent sequences. Cell *52*, 863–873.

Lefevre, G. (1971). Salivary chromosome bands and the frequency of crossing over in *Drosophila melanogaster*. Genetics *67*, 497–513.

Leger, I., Guillaud, M., Krief, B., and Brugal, G. (1994). Interactive computer-assisted analysis of chromosome 1 colocalization with nucleoli. Cytometry *16*, 313.

Levis, R., Hazelrigg, T., and Rubin, G. (1985). Effects of genomic position on the expression of transduced copies of the *white* gene of *Drosophila*. Science *229*, 558–561.

Liebman, S., Symington, L., and Petes, T. (1988). Mitotic recombination within the centromere of a yeast chromosome. Science *241*, 1074–1077.

Lima-de-Faria, A. (1983). Molecular Evolution and Organization of the Chromosome. Elsevier, Amsterdam.

Locke, J., Kotarski, M., and Tartof, K. (1988). Dosage-dependent modifiers of position effect variegation in *Drosophila* and a mass action model that explains their effect. Genetics *120*, 181–198.

Mahtani, M. M. and Willard, H. F. (1990). Pulsed-field gel analysis of alpha satellite DNA at the human X chromosome centromere: high-frequency polymorphisms and array size estimate. Genomics *7*, 607–613.

Malone, R., Golin, J., and Esposito, M. (1980). Mitotic versus meiotic recombination in *Saccharomyces cerevisiae*. Curr. Genet. *1*, 241–248.

Manuelidis, L. and Borden, J. (1988). Reproducible compartmentalization of individual chromosome domains in human CNS cells revealed by *in situ* hybridization and three-dimensional reconstruction. Chromosoma *96*, 397–410.

Marcais, B., Bellis, M., Gerard, A., Pages, M., Boublik, Y., and Roizes, G. (1991). Structural organization and polymorphism of the alpha satellite DNA sequences of chromosomes 13 and 21 as revealed by pulse field gel electrophoresis. Hum. Genet. *86*, 311–316.

Masumoto, H., Sugimoto, K., and Okazaki, T. (1989). Alphoid satellite DNA is tightly associated with centromere antigens in human chromosomes throughout the cell cycle. Exp. Cell Res. *181*, 181–196.

Mather, K. (1938). Crossing over and heterochromatin in the X chromosome of *Drosophila melanogaster*. Genetics *24*, 413–435.

Mathog, D., Hochstrasser, M., Gruenbaum, Y., Saumweber, H., and Sedat, J. (1984). Characteristic folding pattern of polytene chromosomes in *Drosophila* salivary gland nuclei. Nature *308*, 414–421.

McClintock, B. (1938). The production of homozygous deficient tissues with mutant characteristics by means of the aberrant mitotic behavior of ring-shaped chromosomes. Genetics *23*, 315–376.

McEwen, B. F., Arena, J. T., Frank, J., and Rieder, C. L. (1993). Structure of the colcemid-treated PtK1 kinetochore outer plate as determined by high voltage electron microscopic tomography. J. Cell Biol. *120*, 301–312.

McKim, K. and Rose, A. (1990). Chromosome I duplications in *Caenorhabditis elegans*. Genetics *124*, 115–132.

Meddle, C., Kumar, P., Ham, J., Hughes, D., and Johnston, I. (1984). Cloning of the *CDC7* gene of *Saccharomyces cerevisiae* in association with centromeric DNA. Gene *34*, 179–186.

Méhes, K. (1978). Non-random centromere division: a mechanism of non-disjunction causing aneuploidy? Hum. Hered. *28*, 255–260.

Moritz, K. and Roth, G. (1976). Complexity of germ-line and somatic DNA in *Ascaris*. Nature *259*, 55–57.

Moroi, Y., Hartman, A. L., Nakane, P. K., and Tan, E. M. (1981). Distribution of kinetochore (centromere) antigen in mammalian cell nuclei. J. Cell Biol. *90*, 254–259.

Mortimer, R. and Schild, D. (1985). Genetic map of *Saccharomyces cerevisiae*. Microbiol. Rev. *49*, 181–212.

Muller, H. (1930). Types of visible variations induced by x-rays in *Drosophila*. J. Genet. *22*, 299–334.

Murray, A. W. and Szostak, J. W. (1985). Chromosome segregation in mitosis and meiosis. Annu. Rev. Cell Biol. *1*, 289–315.

Nagele, R., Freeman, T., McMorrow, L., and Lee, H. (1995). Precise spatial positioning of chromosomes during prometaphase: evidence for chromosomal order. Science *270*, 1831–1835.

Nakaseko, Y., Adachi, Y., Sunahashi, S.-I., Niwa, O., and Yanagida, M. (1986). Chromosome walking shows a highly homologous repetitive sequence present in all the centromere regions of fission yeast. EMBO J. *5*, 1011–1021.

Nicol, L. and Jeppesen, P. (1994). Human autoimmune sera recognize a conserved 26 kD protein associated with mammalian heterochromatin that is homologous to heterochromatin protein 1 of *Drosophila*. Chromosome Res. *2*, 245–253.

Ochs, R. L. and Press, R. I. (1992). Centromere autoantigens are associated with the nucleolus. Exp. Cell Res. *200*, 339.

Panzeri, L. and Philippsen, P. (1982). Centromeric DNA from chromosome VI in *Saccharomyces cerevisiae* strains. EMBO J. *1*, 1605–1611.

Park, P. C. and DeBoni, U. (1992). Spatial rearrangement and enhanced clustering of interphase nuclei of dorsal root ganglion neurons *in vitro*: association with nucleolar fusion. Exp. Cell Res. *203*, 222–229.

Pluta, A. F., Cooke, C. A., and Earnshaw, W. C. (1990). Structure of the human centromere at metaphase. Trends Biochem. Sci. *15*, 181–185.

Pluta, A., Mackay, A., Ainsztein, A., Goldberg, I., and Earnshaw, W. (1995). The centromere: hub of chromosomal activities. Science *270*, 1591–1594.

Rabl, C. (1885). Uber Zelltheilung. Morphol. Jahrbuch *10*, 214–330.

Rattner, J. B. (1987). The organization of the mammalian kinetochore: scanning electron microscope study. Chromosoma (Berl.) *95*, 175–181.

Reuter, G., Werner, W., and Hoffman, H. (1982). Mutants affecting position-effect heterochromatinization in *Drosophila melanogaster*. Chromosoma *85*, 539–551.

Reuter, G., Giarre, M., Farah, J., and Spierer, A. (1990). Dependence of position effect variegation in Drosophila on dose of a gene encoding an unusual zinc-finger protein. Nature *344*, 219–223.

Rick, C. (1969). Controlled introgression of chromosomes of *Solanum pennellii* into *Lycopersicon esculentum*: segregation and recombination. Genetics *62*, 753–768.

Rick, C. (1972). Further studies on segregation and recombination in backcross derivatives of a tomato species hybrid. Biol. Zentralbl *91*, 209–220.

Rieder, C. L. (1982). The formation, structure and composition of the mammalian kinetochore and kinetochore fiber. Int. Rev. Cytol. *79*, 1–58.

Roberts, P. (1965). Difference in the behavior of eu- and hetero-chromatin: crossing over. Nature *205*, 725–726.

Roth, E. (1979). Satellite DNA properties of the germ-line limited DNA and the organization of somatic genomes in the nematodes *Ascaris suum* and *Parascaris equorum*. Chromosoma *74*, 355–371.

Schultz, J. (1936). Variegation in *Drosophila* and the inert chromosome regions. Proc. Natl Acad. Sci. USA *22*, 27–33.

Sears, E. R. (1952). Misdivision of univalents in common wheat. Chromosoma (Berl.) *4*, 535–550.

Shore, D. (1995). Telomere position effects and transcriptional silencing in the yeast *Saccharomyes cerevisiae*. In: Telomeres. Cold Spring Harbor Laboratory Press, Cold Spring Harbor. eds. Blackburn, E.H. and Greider, C.W.

Simerly, C., Balczon, R., Brinkley, B. R., and Schatten, G. (1990). Microinjected kinetochore antibodies interfere with chromosome movement in meiotic and mitotic mouse embryos. J. Cell Biol. *111*, 1491–1504.

Spector, D. (1993). Macromolecular domains within the cell nucleus. Annu. Rev. Cell Biol. *9*, 265–315.

Spofford, J. (1967). Single-locus modification of position-effect variegation in *Drosophila melanogaster*. I. White variegation. Genetics *57*, 751–766.

Spradling, A. and Karpen, G. (1990). 60 years of mystery. Genetics *126*, 779–784.

Stinchcomb, D., Mann, C., and Davis, R. (1982). Centromeric DNA from *Saccharomyces cerevisiae*. J. Mol. Biol. *158*, 157–179.

Sundin, O. and Varshavsky, A. (1980). Terminal stages of SV40 DNA replication proceed via multiply intertwined catenated dimers. Cell *21*, 103–114.

Sundin, O. and Varshavsky, A. (1981). Arrest of segregation leads to accumulation of highly intertwined catenated dimers: dissection of the final stages of SV40 DNA replication. Cell *25*, 659–669.

Symington, L. and Petes, T. (1988). Meiotic recombination within the centromere of a yeast chromosome. Cell *52*, 237–240.

Szauter, P. (1984). An analysis of regional constraints on exchange in *Drosophila melanogaster* using recombination defective meiotic mutants. Genetics *106*, 45–71.

Tanksley, S., Ganal, M., Prince, J., Vicente, M. d., Bonierbale, M., Broun, P. *et al.* (1992). High density molecular linkage maps of the tomato and potato genomes. Genetics *132*, 1141–1160.

Tartof, K. and Bremer, M. (1990). Mechanisms for the construction and development control of heterochromatin formation and imprinted chromosome domains. Development (Suppl.), 35–45.

Tartof, K., Bishop, C., Jones, M., Hobbs, C., and Locke, J. (1989). Towards an understanding of position effect variegation. Dev. Genet. *10*, 162–176.

Tobler, H. (1986). The differentiation of germ and somatic cell lines in Nematodes. In Germ Line–Soma Differentiation, ed. Henning, W. Springer-Verlag, Berlin, 1–58.

Tomkiel, J., Cooke, C. A., Saitoh, H., Bernat, R. L., and Earnshaw, W. C. (1994). CENP-C is required for maintaining proper kinetochore size and for a timely transition to anaphase. J. Cell Biol. *125*, 531–545.

Vig, B. (1981a). Sequence of centromere separation: analysis of mitotic chromosomes in man. Hum. Genet. *57*, 247–253.

Vig, B. (1981b). Centromere separation: existence of sequences. Experientia *37*, 566–567.

Vig, B. (1983). Sequence of centromere separation: occurrence, possible significance and control. Cancer Genet. Cytogenet. *8*, 249–274.

Vig, B. and Wodnitzki, J. (1974). Separation of sister centromeres in some chromosomes from cultured human lymphocytes. J. Hered. *65*, 149–152.

Wallrath, L. and Elgin, S. (1995). Position effect variegation in *Drosophila* is associated with an altered chromatin structure. Genes Dev. *9*, 1263–1277.

Weiler, K. and Wakimoto, B. (1995). Heterochromatin and gene expression in *Drosophila*. Annu. Rev. Genet. *29*, 577–605.

Wevrick, R. and Willard, H. F. (1989). Long-range organization of tandem arrays of alpha-satellite DNA at the centromeres of human chromosomes: high-frequency array-length polymorphism and meiotic stability. Proc. Natl Acad. Sci. USA *86*, 9394–9398.

Wevrick, R., Earnshaw, W., Howard-Peebles, P., and Willard, H. (1990). Partial deletion of alpha satellite DNA associated with reduced amounts of the centromere protein CENP-B in a mitotically stable human chromosome rearrangement. Mol. Cell. Biol. *10*, 6374–6380.

Witt, P. L., Ris, H., and Borisy, G. G. (1980). Origin of kinetochore microtubules in Chinese hamster cells. Chromosoma *81*, 483–505.

Wolf, K. W. (1995). Centromere structure and chromosome number in mitosis of the colourless phytoflagellate *Polytoma papillatum* (Chlorophyceae, Volvocales, Chlamydomonadaceae). Genome *38*, 1249–1254.

Wolf, K. W., Mertl, H. G., and Traut, W. (1991). Structure, mitotic and meiotic behaviour, and stability of centromere-like elements devoid of chromosome arms in the fly *Megaselia scalaris* (Phoridae). Chromosoma *101*, 99–108.

Wolf, K. W., Mitchell, A., Nicol, L., and Jeppesen, P. (1994). Analysis of centromere structure in the fly *Megaselia scalaris* (Phoridae, Diptera) using CREST sera, anti-histone antibodies, and a repetitive DNA probe. Biol. Cell *80*, 11–23.

Yamamoto, M. and Miklos, G. (1977). Genetic studies on heterochromatin in *Drosophila melanogaster* and their implications for the function of satellite DNA. Chromosoma *66*, 71–98.

Zinkowski, R. P., Meyne, J., and Brinkley, B. R. (1991). The centromere–kinetochore complex: a repeat subunit model. J. Cell Biol. *113*, 1091–1110.

5

Centromere DNA of higher eukaryotes

Paul Kalitsis and K. H. Andy Choo

Perhaps the most striking known structural characteristic of the centromeres of the higher eukaryotes is the universal presence of a great abundance of tandemly repeated DNA. The amount of this DNA can range from hundreds of kilobases to tens of megabases on each centromere. This property suggests *a priori* that the centromere of the higher eukaryotes may be a molecularly complex structure that is difficult to study. However, despite this suggestion, many workers have been attracted to its study due to its perceived biological as well as potential practical importance (Chapter 8). On the other hand, the centromeric repeats, by their very abundant nature, can be relatively easily isolated for study. Many such repeats have now been identified and cloned from different organisms. Through the structural analysis of these repeats, it has become apparent that there is a widespread lack of sequence conservation of centromeric DNA throughout the phylogeny. Furthermore, although a substantial amount of structural information is now available for these centromeric repeats, the delineation of the functional role of these highly divergent sequences yet presents a challenge of a somewhat different magnitude. In this chapter, the current state of our knowledge on the molecular properties of the centromeres of a selected number of higher eukaryotes is described.

Heterochromatic constitution of higher eukaryotic centromeres

Cytogenetically visible heterochromatin is generally present in the centromeric regions of the chromosomes of higher eukaryotes. Although the term heterochromatin was originally defined cytologically as regions of mitotic chromosomes that remain condensed in interphase of the cell cycle, it is now more loosely applied to include regions of chromosomes that show characteristic properties such as gene repression and silencing, and has been extended to include chromosomal regions of lower eukaryotes (reviewed by Lohe and Hilliker 1995; Elgin 1996). Constitutive heterochromatin is invariably associated with relatively short DNA sequences that are highly repeated in long tandem arrays, commonly referred to as satellite DNA. In addition to centromeric or juxtacentromeric locations, heterochromatin can be found in other chromosomal positions, sometimes covering large portions of the chromosomes (such as in the human Yq and the acrocentric p arm regions), but sometimes occupying relatively small domains (such as in the telomeres, except in plants). Within a heterochromatic region, a specific type of repetitive DNA may predominate, two or more sizeable blocks of heterochromatins may be juxtaposed together, or there may be interspersion of different types of minor repeats. Furthermore, the location of a particular family of repeat may not have to be confined to only one type of chromosomal location. For example, telomeric repeats have been found in the centromere, and satellite 3 DNA has been localized to the pericentric as well as the long and short arms of different human chromosomes.

Humans

α-Satellite DNA

α-Satellite is the most abundant family of repeated DNA found at the centromeres of all human chromosomes, and chromosomes of primates in general (Manuelidis 1978a, 1978b; Mitchell *et al.* 1985; Willard 1985). Subchromosomally, α-satellite has been localized to the central domain of the centromere (Chapter 4). Since its discovery 25 years ago in African green monkey (Maio 1971) and subsequently in humans (Manuelidis 1976), it has become the most extensively studied of the highly repetitive DNA families and, as such, provides a conceptual model for the large number of different repeat families that characterize the genomes of higher eukaryotes.

Basic structure and chromosome-specific subfamilies

α-Satellite is made up of fundamental repeat units (or monomers) of approximately 171 bp (see Fig. 5.1 for a consensus sequence) that is tandemly arranged in a head-to-tail fashion (Rosenberg *et al.* 1978; Wu and Manuelidis 1980). The long arrays of adjacent monomers can stretch, largely if not completely uninterrupted by other sequences, for millions of base pairs and constitute 1–5% of individual chromosomes. Superimposed on this seemingly monotonous pattern is a hierarchy of intra- and interchromosomal organization, involving complex relationship of both sequence and structure (reviewed by Willard and Waye 1987; Willard 1991). It is through this complex relationship that α-satellite is organized into distinct subfamilies in a largely chromosome-specific manner, a feature that has become the hallmark of this family of repeats within the human genome (Table 5.1).

The DNA of an α-satellite subfamily tends to be relatively homogeneous in sequence and structure and is clearly distinguishable (> 15–30% sequence divergence) from related sequences of all other subfamilies. These subfamilies display a higher-order organization based on multiple copies of the 171 bp monomer (Fig. 5.2A). The monomers within a higher-order structure

Fig. 5.1 A consensus sequence for the 171 bp human α-satellite DNA monomer based on 293 cloned individual monomers originating from all 24 different human chromosomes (Choo et al. 1991). (−) indicates a gap introduced in the monomer to allow optimal sequence alignment.

Table 5.1 Chromosomal distribution and higher-order structures of human α-satellite DNA subfamilies[1]

Chromosome no.	GDB no.[2]	α-Satellite subfamily[3]	Found on chromosome no.	Size of higher-order structure (no. of monomers)	Enzyme[4]	Reference
1	D1Z5	pSD1-1, 1-2	1	1.9 kb (11-mer)	HdIII, Xb, Sp	Waye et al. 1987b
	D1Z7	pαRI-12, pC1.8	1, 5, 19	0.34 kb (2-mer)	Ec, Hf, Ha	Willard 1985; Baldini et al. 1989
2	D2Z1	pBS4D, p2-7, p2-8, p2-11	2	0.68 kb (4-mer)	Hf, Xb, HdIII, Ha, St	Rocchi et al. 1990; Haaf and Willard 1992
3	D3Z1	p3-9, pHSO5, pα3-5	3	2.9 kb (17-mer)	Ac, Dr, HdIII, Ps, Pv, Sa	Yurov et al. 1987; Waye and Willard 1989b
	n.a.	VIIB4	3	2.75 kb (16-mer)	HdIII, Tq	Delattre et al. 1988
4	D4Z1	D4Z1, pYAM11-39	4	3.2 kb (19-mer)	Ms	Alexandrov et al. 1989a
	n.a.	p4n1/4	4	3.4 kb (20-mer)	Ss	D'Aiuto et al. 1993
	n.a.	pG-XbaII/340, pZ4.1	4, 9	1.2 kb (7-mer)	Ps, Rs, Tq	Hulsebos et al. 1988; D'Aiuto et al. 1993
5	D5Z2	pαRI-12, pC1.8	1, 5, 19	0.34 kb (2-mer)	Ec, Hf, Ha	Willard 1985; Baldini et al. 1989
	n.a.	pG-A16	5, 19	2.25 kb (13-mer)	Ms, Ps, Rs, Ss, Ec	Hulsebos et al. 1988
6	D6Z1	p308	6	2.86 kb (17-mer)	Ba, Tq	Jabs et al. 1984; Jabs and Perisco 1987
7	D7Z1	pα7-d1, 7-t1	7	0.34 kb (2-mer), 0.68 kb (4-mer)	Ec	Waye et al. 1987a
	D7Z2	pMGB7	7	2.7 kb (16-mer)	HdIII, Ha, Tq	Waye et al. 1987a

8	D8Z2	pBS8-164, λ285-1225	8	1.3–5.3 kb (7-, 11-, 15-, 19-, 23-, & 31-mer)	HdIII	Donlon et al. 1987; Ge et al. 1992
9	D9Z4	pMR9A	9	2.7 kb (16-mer)	Ps, Rs, Ms	Rocchi et al. 1991
	n.a.	pG-XbaII/340, pZ4.1	4, 9	1.2 kb (7-mer)	Ps, Rs, Tq	Hulsebos et al. 1988; D'Aiuto et al. 1993
10	D10Z1	pα10RP8,RR6	10	1.35 kb (8-mer)	Rs, Ac, Ps, Sa, St, Ha	Devilee et al. 1988; Wu and Kidd 1990; Carson and Simpson 1990
11	D11Z1	pLC11-A, 11-B, pHS53	11	0.85 kb (5-mer)	Xb	Waye et al. 1987c; Yurov et al. 1987
12	D12Z3	pBR12, pSP12-1, pα12H8	12	1.4 kb (8-mer)	HdIII, Pv, Tq, Xb	Baldini et al. 1990; Looijenga et al. 1990; Greig et al. 1991
13	D13Z1	αRI, L1.26	13, 21	0.68 kb (4-mer)	Ec, Xb	Jorgensen et al. 1987; Devilee et al. 1986a, 1986b
	D13Z2	pTRA-1	13, 14, 21	n.a.	n.a.	Choo et al. 1988; Vissel and Choo 1991
	D13Z3	pTRA-4	13, 14, 21	n.a.	n.a.	Choo et al. 1988; Vissel and Choo 1991
	D13Z4	pTRA-2	13, 14, 21	3.95 kb (23-mer)	HdIII, Rs, Ms	Choo et al. 1988; Vissel and Choo 1991
	D13Z5	pTR9-H2	13, 14, 21	n.a.	n.a.	Vissel et al. 1992
	D13Z6	pTRA-7	13, 14, 21	n.a.	n.a.	Choo et al. 1988; Vissel and Choo 1991
14	D14Z4	pTRA-1	13, 14, 21	n.a.	n.a.	Choo et al. 1988; Vissel and Choo 1991
	D14Z5	pTRA-4	13, 14, 21	n.a.	n.a.	Choo et al. 1988; Vissel and Choo 1991

Table 5.1 *contd*

Chromosome no.	GDB no.[2]	α-Satellite subfamily[3]	Found on chromosome no.	Size of higher-order structure (no. of monomers)	Enzyme[4]	Reference
14 *contd*	D14Z6	pTRA-2	13, 14, 21	3.95 kb (23-mer)	HdIII, Rs, Ms	Choo et al. 1988; Vissel and Choo 1991
	D14Z7	pTR9-H2	13, 14, 21	n.a.	n.a.	Vissel et al. 1992
	D14Z8	pTRA-7	13, 14, 21	n.a.	n.a.	Choo et al. 1988; Vissel and Choo 1991
	D14Z9	αXT	14, 22	1.36 kb (8-mer)	Xb, Tq	Jorgensen et al. 1988
	n.a.	pTRA-54	14	n.a.	n.a.	Vissel and Choo 1991
	n.a.	p82H	14	n.a.	n.a.	Mitchell et al. 1985; Waye et al. 1988
15	D15Z3	pTRA-20	15	2.5 kb (14-mer)	Rs, Dr	Choo et al. 1990b
	D15Z4	pTRA-25	15	4.5 kb (26-mer)	Ac, Bg, Bs, HdII, Ns	Choo et al. 1990b
16	D16Z2	pSE16	16	1.7 kb (10-mer)	Sa, EcV	Greig et al. 1989
17	D17Z1	p17H8, TR17	17	2.05–2.74 kb (12- to 16-mer)	Ec, Pv	Waye and Willard 1986; Choo et al. 1987
18	D18Z1	L1.84, pYAM9-60	18	0.68 kb (4-mer), 1.36 kb (8-mer)	Ec	Devilee et al. 1986a, 1986b; Alexandrov et al. 1989b, 1991
	D18Z2	pYAM4-22	18	1.7 kb (10-mer)	HdIII	Alexandrov et al. 1989b, 1991
19	D19Z3	pαRI-12, pC1.8	1, 5, 19	0.34 kb (2-mer)	Ec, Hf, Ha	Willard 1985; Baldini et al. 1989
	n.a.	pG-A16	5, 19	2.25 kb (13-mer)	Ms, Ps, Rs, Ss, Ec	Hulsebos et al. 1988

20	D20Z1	p3-4	20	n.a.	n.a.	Waye and Willard 1987
	D20Z2	pZ20	20	1.0 kb (6-mer)	Hf	Baldini et al. 1992
21	D21Z1	αRI, L1.26	13, 21	0.68 kb (4-mer)	Ec, Xb	Jorgensen et al. 1987; Devilee et al. 1986a, 1986b
	D21Z3	pTRA-1	13, 14, 21	n.a.	n.a.	Choo et al. 1988; Vissel and Choo 1991
	D21Z4	pTRA-4	13, 14, 21	n.a.	n.a.	Choo et al. 1988; Vissel and Choo 1991
	D21Z5	pTRA-2	13, 14, 21	3.95 kb (23-mer)	HdIII, Rs, Ms	Choo et al. 1988; Vissel and Choo 1991
	D21Z6	pTR9-H2	13, 14, 21	n.a.	n.a.	Vissel et al. 1992
	D21Z7	pTRA-7	13, 14, 21	n.a.	n.a.	Choo et al. 1988; Vissel and Choo 1991
22	D22Z2	p22/1:2.1, 2.8, 0.73	22	2.1 kb (12-mer), 2.8 kb (16-mer)	Ec	McDermid et al. 1986
X	n.a.	αXT	14, 22	1.36 kb (8-mer)	Xb, Tq	Jorgensen et al. 1988
	DXZ1	pXBR-1, pBamX7	X	2.0 kb (12-mer)	Ps, Ba, Ss	Yang et al. 1982; Willard et al. 1983; Waye and Willard 1985
Y	DYZ3	CY73, CY84, CY97	Y	5.5 kb (32-mer), 5.7 kb (33-mer), 6.0 kb (35-mer)	Ec, Ps, HdIII	Wolfe et al. 1985; Tyler-Smith and Brown 1987

n.a. = not available.

[1] Updated from an earlier compilation (Choo et al. 1991).

[2] Genome database designations.

[3] Original names of cloned α-satellite DNA sequences.

[4] Enzyme defining the higher-order structure or periodicity of a subfamily. Ac = *AccI*, Ba = *BamHI*, Bg = *BglII*, Bs = *BstNI*, Dr = *DraI*, Ec = *EcoRI*, EcV = *EcoRV*, Ha = *HaeIII*, HdII = *HindII*, HdIII = *HindIII*, Hf = *HinfI*, Ms = *MspI*, Ns = *NsiI*, Ps = *PstI*, Pv = *PvuII*, Rs = *RsaI*, Sa = *Sau3A*, Sp = *SphI*, Ss = *SstI*, St = *StuI*, Tq = *TaqI*, Xb = *XbaI*.

generally show substantial DNA sequence divergence, in the order of 20–40%. Thus, in the schematic model shown in Fig. 5.2B, monomers 1 and 2, or 2 and 3, or 1 and 3, etc. will differ greatly in primary sequence and will not necessarily be any closer in sequence homology than each is to monomers from different subfamilies. However, after some characteristic and integral number of monomers in the linear array ('n' monomers), the sequence of the next monomer ('n + 1', 'n + 2', etc.) will be virtually identical (< 2% sequence divergence) to monomer 1, 2, etc. The higher-order repeat unit, which is operationally defined by the long-range periodicities revealed by restriction endonucleases (Fig. 5.2C, and Table 5.1), is characteristic for each subfamily and corresponds in theory to an actual multimeric unit of amplification and sequence homogenization during recent evolution of the array (Smith 1976).

The distribution of the relatively large number of α-satellite subfamilies on the different human chromosomes is quite complex. At least three basic patterns have been recognized (Table 5.1). In the first and simplest pattern, each subfamily is specific for a single chromosome. In the second pattern, a subfamily is shared by two or more chromosomes. In the third pattern, multiple subfamilies co-exist on a single chromosome. On any one chromosome, one, two or all three of these patterns can be present. For example, the centromere of chromosome 1 contains two different subfamilies (D1Z5 and D1Z7) (pattern 3), but one of these (D1Z5) is specific for chromosome 1 only (pattern 1) while the other (D1Z7) is shared by three different chromosomes (chromosomes 1, 5, and 19) (pattern 2). The precise mechanisms for the origin of these complex distribution patterns are not known but they probably involve recent intra- and interchromosomal exchange and homogenization events (reviewed by Willard and Waye 1987; Choo 1990; Willard 1991).

An important structural feature of the α-satellite DNA that has received considerable attention is a 17 bp [CT(T/A)(C/T)G(T/G)TGGAAA(C/A)GG(G/A)A] sequence that has been identified within a significant population of α-satellite monomers. This sequence is known as the CENP-B box motif because of its demonstrated ability to bind the constitutive centromere protein, CENP-B (Masumoto *et al.* 1989; Muro *et al.* 1992). Further discussion of the properties of this motif and those of its associated protein CENP-B will be presented in Chapter 6.

Long-range organization

To date, the long-range pericentric maps for human chromosomes 7 (Wevrick and Willard 1991; Wevrick *et al.* 1992), 10 (Jackson *et al.* 1993, 1996), 13 (Trowell *et al.* 1993), 14 (Trowell *et al.* 1993), 21 (Trowell *et al.* 1993; Ikeno *et al.* 1994), and Y (Tyler-Smith 1987; Cooper *et al.* 1993a, 1993b) have been determined (Fig. 5.3). Each of these maps displays a predominant α-satellite DNA domain that is composed of a large and homogeneous array of repeating units consisting of members of a single

105

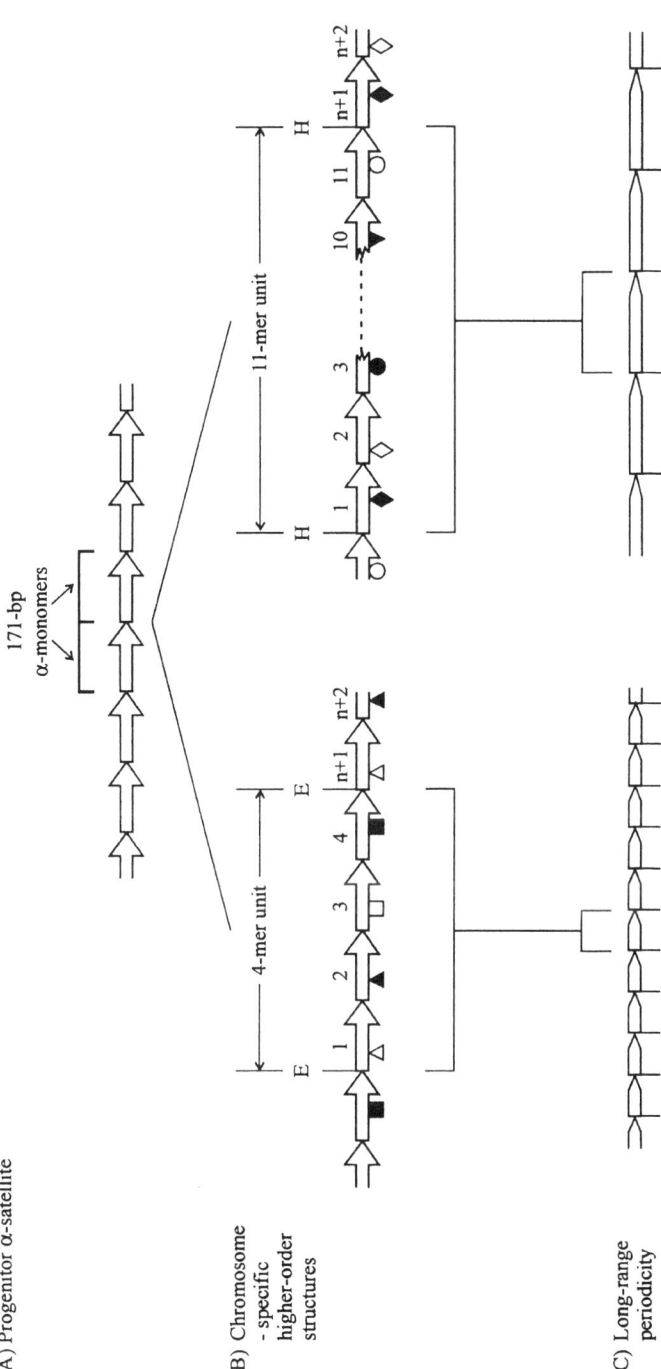

Fig. 5.2 Evolution and structural organization of α-satellite DNA. (A) A progenitor α-satellite DNA array showing head-to-tail tandem arrangement of the 171 bp monomeric units. (B) Evolution of two different subfamilies, one with a 4-mer (left diagram) and the other a 11-mer (right diagram) higher-order structure. Geometric symbols denote mutational changes that have accompanied the evolution of these two α-satellite subfamilies. 'n + 1' and 'n + 2' denote the first and second monomers of an adjacent higher-order repeat unit. (C) Long-range periodicity of α-satellite DNA arrays for the two subfamilies shown in B. The restriction enzymes used to define the periodicities of the 4-mer and 11-mer subfamilies are *Eco*RI (E) and *Hind*III (H), respectively.

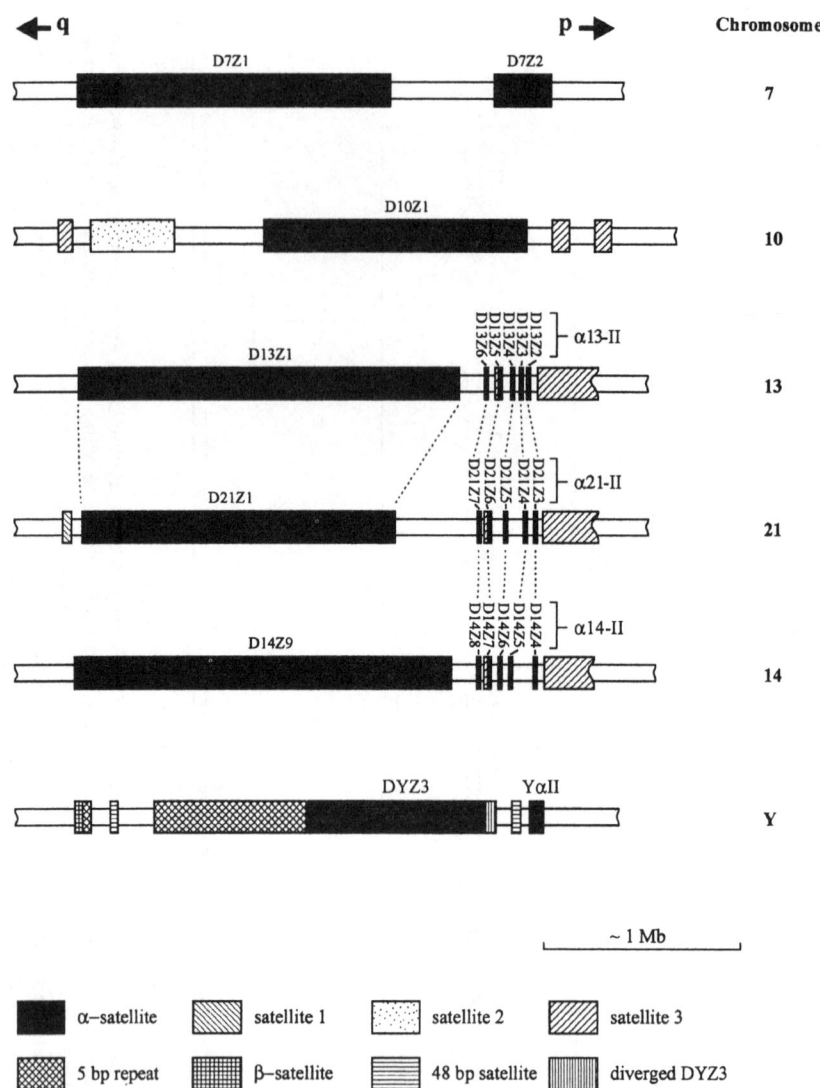

Fig. 5.3 Long-range pericentric maps of human chromosomes 7, 10, 13, 21, 14, and Y. Refer to Tables 5.1 and 5.2 for further information on the properties of the different satellite DNA sequences. All chromosomes are shown with their p arms pointing right and q arms pointing left. α13-II, α21-II and α14-II denote a second domain on chromosomes 13, 21, and 14, respectively, that contains a group of heterogeneous α-satellite subfamilies found on the p arm-side of the major α-satellite DNA block. Dotted lines connect chromosomal domains containing common α-satellite DNA subfamilies, and illustrate an identical overall pattern of organization between chromosomes 13 and 21, as well as a partially shared pattern between chromosomes 13, 14, and 21. The '5 bp repeat' family is probably classical satellite 2 or 3.

α-satellite subfamily. This α-satellite DNA array is highly variable in size both amongst the nonhomologous chromosomes and homologous chromosomes, and can range from < 200 kb to > 3 Mb. In addition to this major block of α-satellite DNA, each chromosome usually contains one or more islands of α-satellite sequences that are made up of different α-satellite DNA subfamilies or diverged sequences of the main α-satellite DNA block.

In addition to α-satellite DNA, the pericentric regions appear to have a propensity to associate with other types of simple repeat DNA (Fig. 5.3, discussed below). A number of studies have also directly identified cloned sequences that contain a junction between α-satellite and non-α-satellite repeats (Grimaldi and Singer 1982, 1983; McCutchan *et al.* 1982; Potter and Jones 1983; Thayer and Singer 1983; Jackson *et al.* 1992; Vissel *et al.* 1992), suggesting that at least in some situations α-satellite DNA arrays can be interrupted by or joined directly onto non-α-satellite repeats.

Role of α-satellite DNA

The results of a number of studies have provided evidence in favour of α-satellite DNA having a role in centromere function. (i) Introduction of a 0.34 kb fragment of human α-satellite DNA into hamster cells followed by amplification of the incoming DNA leads to an increase in the number of dicentric and ring chromosomes per cell (Heartlein *et al.* 1988). (ii) Transfection of a cloned 17 kb uninterrupted α-satellite array (Haaf *et al.* 1992) or a YAC containing 120 kb of α-satellite DNA (Larin *et al.* 1994) into cultured human, simian, or hamster cells confers centromere functions at the sites of integration, including the formation of a primary constriction, disruption of normal chromosome segregation, and binding to CENP-B. (iii) Analysis of re-arranged Y chromosomes (Tyler-Smith *et al.* 1993), and dissection of the human Y centromere with cloned telomeric DNA (Brown *et al.* 1994; see Chapter 8), demonstrate that sequences necessary for Y centromere function can be localized to an interval that includes 150–200 kb of α-satellite DNA. (iv) The generation of a fully stable human X-derived minichromosome (< 10 Mb) by fragmentation within the centromere indicates that preservation of 2.5 Mb of α-satellite DNA accompanies the retention of centromere function (Farr *et al.* 1995).

However, a number of critical observations have raised uncertainty on the role of α-satellite DNA in centromere function. (i) In all the above transfection and/or fragmentation studies, it is unclear if other non-α-satellite DNA sequences are embedded within the centromeric sites and operating independently of, or in concert with, the α-satellite DNA. (ii) In human dicentric chromosomes, α-satellite DNA is present on both the active and inactive centromeres (Earnshaw *et al.* 1989), suggesting that the presence of α-satellite *per se* is insufficient to prescribe centromere function. (iii) A stable and morphologically normal human chromosome 21 that is devoid of any detectable α-satellite has been described, indicating that α-satellite, at least a great abundance of it, is functionally obsolete for the reported chromosome (Verma and Luke 1992). (iv) In a number of constitutional marker chromosomes, neocentromeres with no detectable α-satellite DNA have formed within hitherto non-centromeric locations of the human genome (Callen *et al.* 1992; Crolla *et al.* 1992; Rauch *et al.* 1992; Magnani *et al.* 1993; Voullaire *et al.* 1993; Blennow *et al.* 1994; Ohashi *et al.* 1994; Bukvic *et al.* 1996) (Chapter 7), again suggesting that the presence of α-satellite DNA is not mandatory to centromere formation. Thus, in view of the above conflicting evidence, the role of α-satellite DNA in centromere function remains unclear at present (see also Chapter 6 for a related discussion on the role of CENP-B).

Other pericentric repetitive DNA

As well as α-satellite DNA, a host of other families of repetitive DNA have been localized to the pericentric regions of human chromosomes (reviewed by Choo 1990; Tyler-Smith and Willard 1993). These repetitive sequences may be of the type that are organized into variable length tandem arrays from smaller repeating units, or they may be present as localized interspersed repeats (Table 5.2). Since the mapping of these sequences on chromosomes has mostly been performed using *in situ* hybridization on highly condensed metaphase chromosomes, their precise subregional localization is generally poor or not available. Where long-range maps have been constructed employing pulsed field gel electrophoretic analysis (Fig. 5.3), these repetitive DNA are shown to reside predominantly outside the boundaries of the major α-satellite DNA domains, although some relatively minor components of these repeats can also exist within the α-satellite arrays.

Satellites 1–3

The classical satellites, satellites 1–3, are among the first tandomly repeated DNA characterized in the human genome. These satellite sequences are made up of short repeats that are A + T rich and are distinguishable from the rest of the genomic DNA because they form 'satellite' bands on caesium chloride density gradients (Corneo *et al.* 1968). The monomeric repeating unit for satellite 1 DNA is 42 bp in length (Prosser *et al.* 1986). Although earlier studies using gradient-banded DNA materials have demonstrated the presence of this satellite sequence on most human chromosomes (Gosden *et al.* 1975), subsequent *in situ* hybridization localization using cloned satellite 1 or oligonucleotide probes prepared from conserved regions of satellite 1 has shown this DNA to be present on chromosomes 3 and 4 and all the acrocentric chromosomes, with little or no detectable signals on the remaining chromosomes (Kalitsis *et al.* 1993; Meyne *et al.* 1994; Tagarro *et al.* 1994). On chromosomes 3 and 4, this DNA has been mapped to the proximal q arm side of the centromeric α-satellite DNA block (Tagarro *et al.* 1994),

Table 5.2 Non-α-satellite sequences that have been identified at the pericentric regions of human chromosomes

Name	Size (bp) of repeat unit	Found on chromosome no.	References
Tandemly repeated satellite DNA			
Satellite 1	42	3, 4, 13, 14, 15, 21, 22	Kalitsis *et al.* 1993; Meyne *et al.* 1994; Tagarro *et al.* 1994
Satellites 2 and 3	5	Probably all chromosomes	Grady *et al.* 1992; Gosden *et al.* 1975
β-Satellite (or Sau3A repeat)	68	1, 9, 13, 14, 15, 21, 22, Y	Meneveri *et al.* 1985; Agresti *et al.* 1987; Waye and Willard 1989a; Cooper *et al.* 1992
γ-Satellite	220	8, X	Lin *et al.* 1993; Lee *et al.* 1995
48 bp repeat	48	21, 22, Y, possibly others	Metzdorf *et al.* 1988; Müllenbach *et al.* 1992; Cooper *et al.* 1992
Localized interspersed repeat DNA			
sn5		13, 14, 15, 21, 22	Johnson *et al.* 1992
724 and 7D		13, 14, 15, 21, 22	Kurnit *et al.* 1984, 1986
*Hae*III repeat		Adjacent to β-satellite	Agresti *et al.* 1989
ATRS		1, 2, 7, 9, 13, 14, 15, 16, 17, 21, 22	Wevrick *et al.* 1992
H7		1, 13, 14, 15, 21, 22	Devine *et al.* 1985; Graham *et al.* 1985
YII1.1, YII2.1, 64b, KFC11, KFC37, KFC43, KFC52		Y and others	Cooper *et al.* 1992; 1993a

whereas on the acrocentric chromosomes it is located at the pericentric as well as distal p13 (cytological satellite) regions (Kalitsis *et al.* 1993). Like α-satellite DNA, satellite 1 sequences are capable of forming chromosome-specific subfamilies that are definable by characteristic higher-order structures (Kalitsis *et al.* 1993).

The tandemly reiterating arrays of satellites 2 and 3 are based on variants of the 5 bp (GGAAT) monomeric units (Frommer *et al.* 1982; Prosser *et al.* 1986). These satellite sequences are present in highly variable amounts at the pericentric regions of most (possibly all) human chromosomes and appear to be highly conserved throughout evolution. These properties, together with their unusual hydrogen bonding and possibly protein-binding characteristics, have led to the suggestion that they may be functionally important components of the centromere (Grady *et al.* 1992). In addition to pericentric locations, these satellites are prominently found in several other chromosomal locations, including the q12 heterochromatic regions of chromosomes 1, 9, and Y, and the p arm regions of all the acrocentric chromosomes (Gosden *et al.* 1981; Dale *et al.* 1989; Earle *et al.* 1989; Rocchi *et al.* 1991). To date, a number of chromosome-specific satellite 3 subfamilies have been reported for several human chromosomes (Cooke and Hindley 1979; Deninger *et al.* 1981; Higgins *et al.* 1985; Moyzis *et al.* 1987; Choo *et al.* 1990a, 1992; Rocchi *et al.* 1991; Vissel *et al.* 1992). Interestingly, the chromosomal distribution profiles for these satellite 3 subfamilies appear to conform to the same three basic patterns described above for the α-satellite subfamilies.

β-*Satellite DNA*

β-Satellite is a G + C-rich repetitive DNA with monomeric units of ∼ 68 bp (Waye and Willard 1989a). This satellite family is also known as Sau3A repeats since each monomer characteristically contains a restriction site for the enzyme *Sau*3A (Meneveri *et al.* 1985). The monomers are tandemly arranged into arrays of at least several hundred kilobases that collectively constitute approximately 0.2% of the human genome. This family of repeats has been localized to the pericentric regions of chromosomes 1, 13, 14, 15, 21, 22, and Y. In addition, it is found on the heterochromatic region of the long arm of chromosome 9, as well as on the distal cytological satellites of the five acrocentric chromosomes (Agresti *et al.* 1987; Waye and Willard 1989a; Cooper *et al.* 1992; Greig and Willard 1992). Like the previously described satellite DNA families, the basic β-satellite monomer can be fixed into distinct higher-order repeat units that show characteristic chromosomal distribution patterns (Waye and Willard 1989a; Greig and Willard 1992).

γ-*Satellite*

γ-Satellite is the newest family of repeats that has been added to the centromeric domain. Specific subfamilies of this repeat have so far been described for chromosomes 8 (Lin *et al.* 1993) and X (Lee *et al.* 1995). The

DNA is made up of tandem monomers of $\sim$ 220 bp that comprise $\sim$ 0.015% of each of the two chromosomes. Fluorescence *in situ* hybridization has sublocalized this family of repeats to the lateral sides of the primary constriction, suggesting that the DNA could be more closely located to the kinetochore domain than the α-satellite DNA.

48 bp repeat

This family of repeats consists of tandemly repeating units of 48 bp in size. Subfamilies of these repeats specific for human chromosomes 21 (Müllenbach *et al.* 1992) and 22 (Metzdorf *et al.* 1988) have been identified. On these chromosomes, the repeats are organized into array sizes of 700–900 kb that are positioned within several hundred kilobases of the main α-satellite DNA domain. In addition, members of this repeat family are found in the pericentric regions of the Y chromosome (Fig. 5.3) (Cooper *et al.* 1992, 1993a). No cross-hybridization with other mammalian genomes except for those of the apes has been observed, suggesting that the genesis of this family of repeats has occurred after the phylogenic separation of higher primates from other monkeys about 30 million years ago. Specific protein-binding properties based on gel electrophoretic mobility shift assays have been reported for this DNA (Müllenbach *et al.* 1992).

Localized interspersed repeats

In addition to the tandemly repeated satellite DNA, a number of dispersed repeat sequences have been found at or near centromeres (Table 5.2). An 'unclonable' sequence known as sn5 has been isolated using polymerase chain reaction from a chromosome 20-derived minichromosome and shown to be most abundant on the acrocentric chromosomes (Johnson *et al.* 1992). The 724 and 7D families are found on the acrocentric chromosomes, as well as at some non centromeric positions on other chromosomes (Kurnit *et al.* 1984, 1986). A diverse set of sequences has been identified adjacent to the β-satellite, including a family of repeats called the HaeIII family (Agresti *et al.* 1989). The ATRS (or *A + T-* rich *s*equence) originated as a sequence adjacent to α-satellite DNA on chromosome 7, but is also detected in different amounts at the centromeres of 10 other chromosomes including all the acrocentric chromosomes (Wevrick *et al.* 1992). A sequence (H7), which hybridizes to the pericentric region of chromosome 1 and the proximal short arm regions of all the acrocentric chromosomes has been described (Devine *et al.* 1985; Graham *et al.* 1985). Analysis of the Y chromosome has revealed seven distinct repeat families (YII1.1, YII2.1, 64b, KFC11, KFC37, KFC43, and KFC52) that have one or two copies around the centromere, and multiple additional copies of unknown locations on other chromosomes (Cooper *et al.* 1992, 1993a). In addition to these sequences, it is probable that other short interspersed nucleotide elements (SINEs) and long interspersed nucleotide elements (LINEs) are present in the centromeric region and await discovery.

Mouse

A typical mouse *Mus musculus* chromosome is generally described as being acrocentric, i.e. having its centromere located at one end of the chromosome and thus lacking a cytologically detectable short arm (Evans 1981). An exception is the Y chromosome, which has a minute short arm, carrying amongst other genes, the testis-determining gene *Tdy* (Roberts *et al.* 1988; Gubbay *et al.* 1990). In this mouse species, two classes of repetitive DNA, termed major and minor satellites, are known to be associated within or close to the centromeres.

Major satellite DNA

Mouse major satellite DNA is made up of roughly a million tandem repeats of a 234 bp monomer per haploid genome (Horz and Altenburger 1981; Manuelidis 1981). This DNA is found at the pericentric, heterochromatic and C-band-positive regions of all chromosomes except Y (Pardue and Gall 1970). The Y chromosome rarely shows C-banding when standard technique is used, but sometimes a very small C-band can be detected with the restriction enzyme banding method, indicating the presence of repetitive DNA sequences of other sorts (Kaelbling *et al.* 1984). Long-range DNA mapping indicates that the bulk of major satellite monomers are organized into largely uninterrupted arrays that vary from a minimum of 240 kb to greater than 2000 kb in length (Vissel and Choo 1989). The major satellite of *M. musculus* is found in most, but not all (e.g. not in *M. cervicolor* and *M. cookii*), species and subspecies of the genus *Mus* (Garagna *et al.* 1993; Kipling *et al.* 1995). Failure to detect the satellite DNA in some of these species and subspecies may be due to a total absence, low copy presence, or substantial divergence from the probe DNA used.

Sequencing of 30 randomly derived clones of mouse major satellite DNA demonstrates a mean monomer sequence deviation of 3.9% (range = 0.9–9.1%; Vissel and Choo 1989) from the consensus sequence (Holmquist and Dancis 1979). This result indicates that, unlike human centromeric α-satellite DNA where individual 171 bp monomers may vary by 20–40% (Willard and Waye 1987), mouse major satellite DNA from both inter- and intrachromosomal origins appear well conserved. This high degree of conservation suggests frequent recombinational exchanges between nonhomologous chromosomes—an occurrence possibly facilitated by the acrocentric nature of the mouse chromosomes (Vissel and Choo 1989).

Minor satellite DNA

Mouse minor satellite DNA, detectable in quantities 10–20 times less than the major satellite DNA, has a 120-bp basic repeating unit that organizes into tandem arrays of approximately 300 kb, or roughly 2500 copies, per chromosome (Piertras *et al.* 1983; Wong and Rattner 1988; Joseph *et al.* 1989; Broccoli *et al.* 1990; Moens and Pearlman 1990; Wong *et al.* 1990; Kipling *et al.* 1991, 1994; Narayanswami *et al.* 1992; Garagna *et al.* 1993). Analysis of the distribution of minor satellite in 11 species and subspecies of the genus *Mus* indicates the presence of this DNA in all except *M. caroli* (see below), *M. cervicolor,* and *M. cookii* (Garagna *et al.* 1993). In all *Mus* members which carry minor satellite DNA, *in situ* hybridization has localized the DNA exclusively to the centromeres of all chromosomes except Y (Pietras *et al.* 1983; Wong and Rattner 1988; Joseph *et al.* 1989; Broccoli *et al.* 1990; Moens and Pearlman 1990; Wong *et al.* 1990). As with the human α-satellite DNA, variants of minor satellite DNA that map preferentially to the centromere of one or a small group of chromosomes have been identified (Broccoli *et al.* 1991; Kipling *et al.* 1991, 1994).

CENP-B box

In *M. musculus,* minor satellite is present at the primary constriction and co-localizes with centromere-associated proteins detectable by sera derived from patients with CREST autoimmune syndrome (Chapter 6) (Rattner 1991). Furthermore, despite a lack of overall homology to human α-satellite DNA, some monomers of mouse minor satellite DNA contain the consensus 17 bp CENP-B box motif that is found in the human α-satellite DNA (Masamoto *et al.* 1989). This 17 bp motif has been shown *in vitro* to be the recognition site for the mammalian centromere-binding protein CENP-B (Chapter 6). These observations suggest that CENP-B box-containing minor satellite sequences may provide binding sites for CENP-B proteins and that these sequences may have a role in the assembly and/or other functions of the *M. musculus* centromere.

Asian mouse M. caroli *centromeres lack minor satellite but can bind CENP-B*

The Asian mouse *Mus caroli,* which separated from *M. musculus* about 5–7 million years ago, has little if any detectable minor satellite in its genome (Piertras *et al.* 1983; Wong *et al.* 1990; Garagna *et al.* 1993) (although it does carry a small amount of a sequence family which cross-hybridizes to the major satellite) (Kipling *et al.* 1995). Cells from *M. caroli* also show no detectable CENP-B proteins but interestingly, contain copies of the CENP-B gene. Two abundant satellite DNA sequences, designated 60 and 79 bp satellites, have been identified in these mice (Kipling *et al.* 1995). These satellites, which are not found in *M. musculus,* are apparently organized into noninterspersed tandem repeat arrays of over 1 Mb. All *M. caroli* autosomes carry both satellites, whereas the Y chromosome contains only the 60 bp satellite, and the X chromosome contains neither. Interestingly, despite the absence of the canonical 17 bp CENP-B box motif in the *M. caroli* genome, use of anti-CENP-B antisera has demonstrated CENP-B binding by *M.*

caroli centromeres in *M. caroli* × *M. musculus* interspecific cell lines. Sequence analysis of the 79 bp *M. caroli* satellite reveals a motif that contains all nine bases that have been shown to be necessary for *in vitro* binding of CENP-B. This motif binds CENP-B from HeLa cell nuclear extract and has been proposed as the cause of the association of CENP-B with the *M. caroli* centromere in the interspecific hybrid cells. This observation of the preservation of CENP-B binding potential in a species that no longer produces CENP-B protein is intriguing and difficult to explain.

Subchromosomal location of major and minor satellite DNA

5-Azacytidine, which is thought to inhibit mammalian methylases, has been used to decondense the heterochromatin around the centromeric domains of mouse chromosomes to study the subregional locations of major and minor satellites. *In situ* hybridization of such preparations indicates that these satellite sequences are organized into two separate domains, with the minor satellite domain situated towards the telomeric end of the short arm, and the major satellite domain localizing to the pericentric heterochromatin and separating the minor satellite from the euchromatic long arm (Joseph *et al.* 1989) (Fig. 5.4). Pulsed field gel electrophoretic analysis further demonstrates direct physical linkage of minor satellite arrays with telomere repeat sequences and places many minor satellite arrays on terminal restriction fragments smaller than 1 Mb (Kipling *et al.* 1991). Cloned DNA sequences have also been found that contain both the major and minor satellites, indicating that on at least some chromosomes the arrays are directly adjacent (Wong and Rattner 1989).

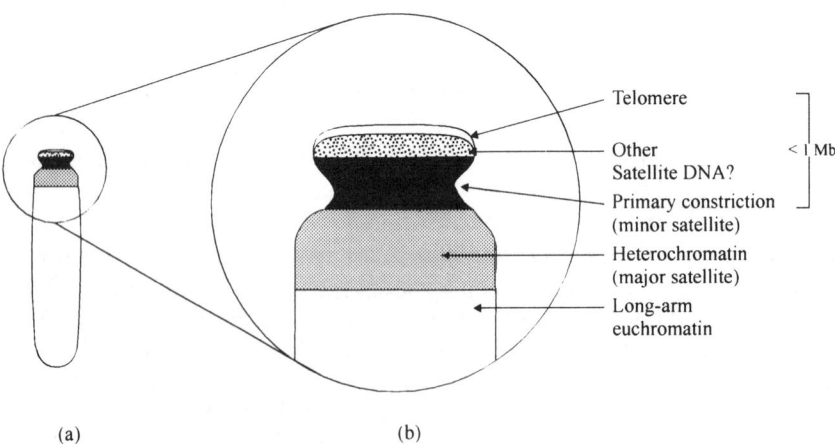

(a)   (b)

Fig. 5.4 Schematic representation of (a) a typical acrocentric mouse chromosome, and (b) close-up of the centromeric region.

Role of major and minor satellite DNA

The localization of major satellite in the long arm heterochromatin distal to the primary constriction rules out a direct involvement of this DNA in centromere function. In contrast, a number of observations have suggested a possible centromere role for minor satellite. (i) It is mapped to the primary constriction. (ii) It has a CENP-B box that is capable of binding the mammalian centromere protein CENP-B. (iii) In a marker chromosome in which the entire heterochromatic C-band has been translocated but left with an active centromere, minor satellite is present at this active centromere (Broccoli *et al.* 1990). (iv) In house-mouse metacentric chromosomes generated by Robertsonian translocation, minor satellite DNA is retained in the resulting functional centromere (Garagna *et al.* 1995). (v) Whilst regions containing major satellite decondense upon treatment with bisbenzimidazole, the chromatin carrying minor satellite resists decondensing, suggesting that the latter behaves similarly to the protein-compacted structure of the centromere (Neuer-Nitsche *et al.* 1988; Vig *et al.* 1994).

However, a number of observations have questioned the suggestion that minor satellite is important for centromere function: (i) Minor satellite DNA is shown not to be associated with the centromeres of some chromosomes in a mouse cell line (C11D) even though the DNA is present on these chromosomes (Vig *et al.* 1994). (ii) Some species of mice (e.g. *M. caroli*) lack detectable minor satellite and/or CENP-B protein and obviously do not require these elements for the function of their centromere. (iii) In mouse species where minor satellite is present on all other chromosomes, the DNA is consistently absent on the Y chromosome. (iv) In some marker and translocation mouse chromosomes, minor satellite is found at interstitial C-band regions that are not associated with centromeric activity (Broccoli *et al.* 1990). (v) In multicentric mouse chromosomes, only the site of the active centromere reacts with anti-kinetochore antibodies (Vig and Zinkowski 1986) even though minor satellite DNA is present at the inactive centromeres (Vig and Richards 1992).

Thus, as with the α-satellite DNA, the available conflicting evidence has made it difficult to draw firm conclusions regarding the functional significance of the mouse minor satellite DNA. Clearly, more work will be required to resolve whether this DNA has a true role in mouse centromere function.

The amplified long genomic sequence

In addition to the major and minor satellite DNA, at least one other sequence, ALGS (*a*mplified *l*ong *g*enomic *s*equence), has been identified near the mouse centromere (Koide *et al.* 1992). The haploid genome of standard strains of laboratory mice contains approximately 70 copies of this sequence. The length of each repeating unit is greater than 60 kb, and the sequences of the repeating units are highly conserved. This repeat family has been mapped to the pericentric regions of all mouse chromosomes except 1, 17 and Y. It is found in all subspecies of *M. musculus* and is abundant in *M. spicilegus*, but is absent in *M. spretus*, *Rattus*, and other closely related genera. The exact chromosomal location of ALGS DNA in relation to the major and minor satellite DNA has not been described.

Rat

Satellite I

The first family of repetitive DNA localized to the pericentric region of rat chromosomes has been designated as satellite I but the DNA is unrelated to the human classical satellite 1 sequence. This DNA constitutes 1–3% of the rat genome, and has a tandemly repeating element of 370 bp which is made up of four units of alternative 92 and 93 bp *Eco*RI sequences that show 62–73% homology (Pech *et al.* 1979; Lapeyre *et al.* 1980; Sealy *et al.* 1981). A variant form of this satellite has also been described in different rat species (Witney and Furano 1983). This variant, designated rat satellite I', contains 185 bp tandemly repeated sequences, and represents a truncated version, containing only two of the four 92/93 bp units, of satellite I. An 86% homology is observed between satellite I' and the corresponding 185 bp region of the longer satellite I. The satellite I family of repeats is found primarily at centromeres and telomeres of metaphase chromosomes, though some interstitial sites are also observed (Sealy *et al.* 1981).

RNA sequences complementary to the rat satellite I DNA have been detected in cultured rat hepatoma cells (Sealy *et al.* 1981). This is somewhat unexpected since highly reiterated DNA sequences within heterochromatic regions are thought not to be transcribed. It is quite likely that, instead of the satellite I DNA within centromeric heterochromatin directly serving as templates for transcription, the observed RNA may have resulted from transcriptional read-through errors of satellite I sequences that are inter-spersed in the genome and juxtaposed downstream of some transcriptionally active genes.

200 bp repeat

The second family of rat tandem repeats is a DNA segment that contains no homology to known repetitive DNA sequences but was isolated by reassociation of sheared total rat genomic DNA (Essers *et al.* 1995). This family of repeats has a monomeric unit of approximately 200 bp and shows a strong presence in the centromeric regions of chromosomes 3 and 12, on the q arm of Y, as well as (in greatly reduced amounts) in the centromeric regions of chromosomes 11, 19, and X. A subfamily member of this repeat DNA that is specific for the Y chromosome has also been identified (Essers *et al.* 1995).

α-Satellite DNA

Digestion of total rat DNA with the restriction enzyme *Hind*III yields a repeat sequence of 179 bp (Gupta 1983). This sequence appears to be alphoid in nature, since it shows significant homology with the human α-satellite DNA (41%) and African green monkey α-satellite DNA (37%). Although no chromosomal localization information has been described, this is assumed to be centromeric based on knowledge of the human α-satellite DNA. The identification of this DNA in rat should allow interesting structural and functional comparison with the α-satellite sequences found in the primates, including the investigation of its CENP-B binding status.

South American rodents of the genus *Ctenomys*

The South American octodontid rodents of the genus *Ctenomys* ('tuco-tucos') are of interest to evolutionary genetic studies since they represent the most specious of all subterranean mammals (Reig *et al.* 1990). *Ctenomys* species have one of the highest rates of karyotypic differentiation known in mammals, with diploid numbers ranging from 10 to 70 (Reig and Kiblisky 1969; Anderson *et al.* 1987; Cook *et al.* 1990; Ortells *et al.* 1990; Gallardo 1991). Chromosomal rearrangements are believed to have triggered the explosive speciation process in *Ctenomys* (Reig 1989).

A high-abundance repeat DNA, designated RPCS (for *r*epetitive *Pvu*II *Ctenomys* *s*equence), has been described in *Ctenomys* (Rossi *et al.* 1990; Massarini *et al.* 1991; Reig *et al.* 1992; Rossi *et al.* 1993a, 1993b). This DNA has a monomeric size of 348 bp and shows a tandem arrangement typical of satellite sequences. Examination of the chromosomal distribution of this DNA in eight species by *in situ* hybridization shows different numbers of chromosomes with positive signals on pericentric and/or complete short arm regions (Rossi *et al.* 1995). In some species, a positive signal is scarce (or not detectable) and when present is usually located in the pericentric areas. In those species where the repeats are highly amplified, their chromosomal localization tends to encompass the entire length of the short arm. It is suggested that this DNA has evolved from a strictly pericentric position to comprising the whole of the short arms of some chromosomes (Rossi *et al.* 1995). An unusual feature of the RPSC DNA sequence is that its monomer possesses identity with the U_3 region of the long terminal repeat (LTR) of Rous sarcoma virus, as well as the consensus sequences involved in transcriptional regulation that are characteristic of LTRs (Rossi *et al.* 1993a). In addition, these consensus sequences present in the RPCS monomers bind transcription factors *in vitro* (Pesce *et al.* 1994). It is proposed that the

clustering of RPCS in heterochromatin, according to the cytological properties of this genomic fraction, might prevent the millions of target sites present
in the overall satellite from binding to the transcription factors *in vivo* (Rossi
et al. 1995).

Chinese hamster

The only DNA sequence known to hybridize to the centromere region of the
chromosomes of Chinese hamster cells is the 'telomere-specific' TTAGGG
repeat (Meyne *et al.* 1990). It is thought that these telomeric repeats are relics
of Robertsonian fusion of ancient acrocentric chromosomes, and that they
have undergone subsequent amplification. The role of telomeric repeats in
the pericentric region, if any, is obscure. Interestingly, there are no large,
amplified blocks of telomeric repeats at the pericentric region of the larger
metacentric chromosomes 1 and 2, but a 6 kb long, single-copy DNA that is
species specific and appears to have features suggestive of a role in
centromeric chromatin packaging has been mapped to the pericentric region
of chromosome 1 (Ouspenski and Brinkley 1993). A nontelomere-like
satellite DNA (HC2sat) that is specific to the pericentric region of chromosome 2 has also been reported (Fatyol *et al.* 1994). This DNA has a repeating
unit of 2.8 kb, forms a long array of 7–14 Mb, and constitutes a major
component of the pericentric region of chromosome 2.

Bovine

At least eight distinguishable satellite DNA sequences have been identified in
the bovine genome using density gradient centrifugation (Schildkraut *et al.*
1962; Polli *et al.* 1966; Kurnit *et al.* 1973; Cortadas *et al.* 1977; Macaya *et al.*
1978). These satellite sequences together constitute about 23% of the bovine
genome (Macaya *et al.* 1978). Comparison of the primary structure of five of
these satellites reveals a common ancestral origin consisting of a dodecanucleotide sequence of GATCGGGCTAC(T/C) (Taparowsky and Gerbi
1982). Based on this dodecanucleotide sequence, each of the satellites is
characterized by a bigger repeat unit of up to 2.5 kb (reviewed by Beridze
1986).

The C-banding technique reveals constitutive heterochromatin in the
centromeres of all the bovine autosomes, but not X and Y chromosomes.
The location of four of the satellite sequences (I, II, III, and IV) on these
chromosomes has been determined (Kurnit *et al.* 1973). Satellite I is
encountered in the centromeres of all autosomes. Two-thirds of all satellite
II DNA are observable in centromeres of the autosomes, with the remaining
one-third present in telomeric or interstitial regions. Satellites III and IV are
localized in the centromeres of most, but not all, autosomes. There are few, if
any, satellite DNA on the sex chromosomes.

Deer

Several workers have described centromeric satellite DNA sequences from different species of the deer family *Cervidae* (Bogenberger *et al.* 1985; Yu *et al.* 1986; Lin *et al.* 1991; Lee *et al.* 1994). For example, Lee *et al.* (1994) reported a highly repetitive DNA from the Canadian woodland caribou (*Rangifer tarandus caribau*). This DNA has a 991 bp tandemly repeated monomer that comprises 5.7% of the genome. Comparison of this sequence to centromeric satellite DNA from several other deer species reveals considerable DNA sequence conservation, e.g. the first 800 bp of this DNA sharing about 65–75% of sequence similarity with the European roe deer and Chinese and Indian muntjac DNA.

Pig

The domestic pig (*Sus scrofa domestica*) has a bimodal karyotype consisting of 12 meta-/submetacentric and 6 acrocentric pairs of autosomes. All autosomes and the X chromosome have centromeric heterochromatin. The Y chromosome, in addition, has heterochromatin on both arms. There are two major families of satellite DNA in this organism, designated Ac2 and Mc1 (Jantsch *et al.* 1990). *In situ* hybridization reveals that each satellite is confined to a morphologically distinct chromosomal subgroup in the *S. s. domestica* karyotype. Members of the Ac2 family are homogeneous and are built up from tandem arrays of a 14 bp monomeric consensus sequence of GATT(G/C)AATGCAG(C/T)G. The DNA is found on all the near-terminal centromeric heterochromatin of the acrocentric chromosome. Members of the Mc1 family are more heterogeneous and hybridize in a chromosome-specific or chromosome group-specific fashion to the centromeric heterochromatin of all the biarmed-chromosomes. There is no overlap in the chromosomal location of these two families of satellites. These results indicate that there has been independent evolution of satellite DNA in the two subgenomes, probably for many millions of years. The greater homology of the Ac2 satellite DNA in the acrocentric chromosomes might reflect a more recent amplification event in this subgenome and/or a higher rate of sequence homogenization between the acrocentric chromosomes.

In another study, Miller *et al.* (1993) reported the isolation of a 340 bp tandem repeat family from porcine DNA that hybridizes to the centromeres of all pig chromosomes except Y. When hybridization stringency is increased, probe signals persist on the majority of metacentric chromosomes. This DNA shows approximately 70% sequence homology to the metacentric-specific clones derived from the Mc1 family.

Fish

A tandemly repeated DNA family has been identified in the widely marine-cultured fish species *Sparus aurata* (Garrido-Ramos *et al.* 1994). This DNA is composed of monomeric units of 186 bp in length and accounts for 2% of the fish genome. The repeating units are AT rich (67%) and are characterized by short stretches of consecutive AT base pairs, as well as short direct and inverted repeats. *In situ* hybridization under both low- and high-stringency conditions map this sequence to the centromeres of all the chromosomes in this fish. The presence of this repeat family in the genome of other *Sparidae* species, some of which are relatively distant from *S. aurata*, suggests that this DNA could be an important component of the centromere (Garrido-Ramos *et al.* 1995).

Drosophila

Heterochromatin constitutes about 30% of the *Drosophila* genome. Most of this heterochromatin is found in the pericentric regions of the chromosomes of most *Drosophila* species (Beridze 1986). The heterochromatin in *D. melanogaster* is comprised predominantly of tandem repeats that are arranged in long, homogeneous stretches. These highly repeated sequences band as four major satellites in gradient centrifugation, although many other tandemly repeated sequences that are minor in amount have also been discovered by molecular cloning. Most of the repeats are simple and consist of 5, 7, or 10 bp units, but several are more complex, such as the 359 bp repeat family (Lohe and Roberts 1988).

Abad *et al.* (1992) reported the isolation of a repetitive sequence that is located predominantly in the centromeric region of a specific *D. melanogaster* chromosome. This DNA contains tandemly repeated, GC-rich 11-mer and 12-mer units designated as dodeca satellite DNA. The sequence is found to cross-hybridize to the DNA of five *Drosophila* species (*D. simulans, D. buzzatii, D. hydei, D. immigrans,* and *D. virilis*) that have been separated for > 60 million years. Cross-hybridization is also seen with one plant species (*Arabidopsis thaliana*) and three mammalian species (*Cricetulus griseus, Mus musculus,* and *Homo sapiens*). However, despite the apparent conservation of this DNA in a wide range of organisms, its detection in only a subset of the *D. melanogaster* chromosomes argues against an indispensable role in centromere function, but the possibility of the presence of smaller amounts or diverged members of this DNA on other chromosomes cannot be excluded.

A different GC-rich, telomere-like $(GGGTCAT)_n$ satellite has been described in *D. hydei* (Burgtorf and Bunemann 1993). This DNA is located mainly in the centromeric heterochromatin of all large acrocentric autosomes. It comprises 4% of genomic DNA and is organized as megabase-size clusters. A telomere-related function is rather unlikely because the chromosomal ends of *D. hydei* seem to be free of this DNA motif and several other species of *Drosophila* do not contain any cross-hybridizing sequences at all. The role of this DNA in centromere function, if any, is not clear. In addition to the above sequences, the chromocentre (centromere) region of *Drosophila* polytene chromosomes has been reported to cross-hybridize with the primary motif of human classical satellite 3 (Grady *et al.* 1992; see above).

Localization of centromere function in a minichromosome

Using a minichromosome, *Dp1187*, Murphy and Karpen (1995) recently described important information on the functional DNA components of the *Drosophila* centromere. This minichromosome is approximately 1 Mb in size, contains heterochromatin and euchromatin, and is transmitted with high fidelity through meiosis and mitosis. Fig. 5.5 shows the DNA substructure of the centromeric heterochromatin of this minichromosome (Le *et al.* 1995). There are at least three regions of 'complex DNA' (so called because the DNA contains clustered restriction sites, thus representing middle-repetitive and/or single copy sequences), separated by blocks of simple satellite DNA (which has relatively few restriction sites) such as the 359 bp and 5 bp (AATAT) repeats. The three 'islands' of complex DNA have been named *Tahiti*, *Moorea*, and *Bora Bora*, and at least a small part of one island, *Tahiti*, has been sequenced and shown to contain retroposons (O'Hare *et al.* 1991) juxtaposed with satellite DNA. Analysis of the transmission behaviour of deleted derivatives of this minichromosome has localized sequences necessary for chromosome inheritance within the heterochromatic structure (Murphy and Karpen 1995). The essential core of the centromere, which is

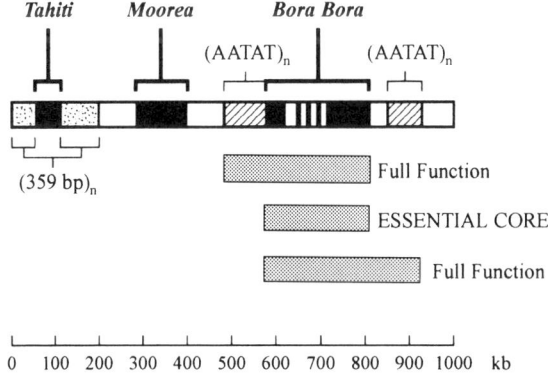

Fig. 5.5 Organization of centromeric heterochromatin in *Drosophila* minichromosome *Dp1187*. Black boxes are the islands of complex DNA: *Tahiti*, *Moorea*, and *Bora Bora*. The inter island simple satellites are indicated by open, stippled, and hatched boxes. The grey bars below denote a region that constitutes the essential centromere core and two regions that are capable of conferring full centromere function.

necessary to achieve good levels of chromosomal stability, is contained within a 220 kb region composed predominantly of the complex DNA island *Bora Bora*. Completely normal inheritance also requires $\sim$ 200 kb of flanking heterochromatic DNA on either one of the two sides of *Bora Bora*. This flanking DNA contains the functionally redundant AT-rich, AATAT satellite sequences. The flanking domains are not, by themselves, sufficient for centromere activity, since the AATAT sequence is also detected at noncentromeric sites of *Drosophila* chromosomes (Lohe *et al.* 1993). The required amount of the flanking DNA for normal transmission is further shown to differ between the *Drosphila* sexes. It is proposed that the essential core is the site of kinetochore formation and that flanking DNA provides two functions: sister chromatid cohesion and indirect assistance in kineto-chore formation or function (Murphy and Karpen 1995).

Comparison with centromeres in other organisms

The overall organization of the centromere of the *Dp1187* minichromosome is similar to those described for fission *yeast S. pombe* (Chapter 3). In both organisms, normal segregation requires a single copy central core plus adjacent repeated sequences. The central core is proposed to be the site of spindle attachment. Nondisjunction and loss increases as more *S. pombe* flanking repeats are deleted, similar to the behaviour of *Dp1187* derivatives. However, the *S. pombe* essential core is significantly smaller (4–7 kb), and the flanking repeats are middle repetitive, not simple satellite, DNA. Compared with the budding yeast *S. cerevisiae* minimal centromere (Chapter 2), the organization of the *Dp1187* centromere is quite different both in terms of size and sequence composition, although the presence of an AT-rich domain(s) is a common feature. With the less well-defined centromeres of other higher organisms, except for the presence of AT-rich repetitive DNA, there appears to be relatively little organizational resemblance.

Newt

A number of centromeric repeat families have been identified in different species and subspecies of the amphibian newt *Notophthalmus viridescens* (Diaz *et al.* 1981), the European crested newt *Triturus cristatus* (Baldwin and Macgregor 1985), and the Italian smooth newt *Triturus vulgaris meridionalis* (Barsacchi-Pilone *et al.* 1986; Cremisi *et al.* 1988). For example, one of these repeats is the MspI family (Cremisi *et al.* 1988). This DNA represents about 1% of the genome of *T.v. meridionalis*. It has a monomer of 197 bp, and is organized in large clusters of tandemly arrayed units. The DNA is present at the centromeres of all the chromosomes except for one. It is well conserved in the *T. vulgaris* subspecies, and significantly less well conserved in the related species *T. helveticus*, where it appears to be present at the centromeres of only a few chromosomes, and is not found in other *Triturus* species analysed, such as *T. italicus*, *T.c. carnifex* and *T.a. apuanus*. The 197 bp monomer has an A + T content of 55%, and shows less than 10% variation between different cloned units. Some short direct and inverted repeats are present, and stop codons are found in all six reading frames. Database search reveals a stretch of 13 nucleotides within the MspI sequence which is similar to the core consensus sequence of the centromeric element CDE I of yeast *S. cerevisiae*.

Nematodes

Caenorhabditis elegans is a free-living organism that has a small genome size of 10^8 bp contained in six chromosomes. It is a popular model system for the study of complex biological problems, and is the subject of a complete physical mapping programme (Waterston *et al.* 1993). As discussed in Chapter 4, the chromosomes in this organism are holocentric. The genome of *C. elegans* carries over 100, and possibly as many as 1,000, distinct families of short, interspersed repeats, with each family containing fewer than 100 members (Emmons *et al.* 1980). Physical mapping of five of these repeat families on the different chromosomes indicates that although the distribution of members of each family is relatively even along the chromosomes, members of more than one family tend to cluster in some locations (Naclerio *et al.* 1992). Several other families of repeated DNA sequences that have been characterized include one family that consists of transposable (Tcl) elements (Emmons *et al.* 1983; Liao *et al.* 1983), and a second family that has a conserved inverted repeat structure but does not appear to be an active transposon (Felsenstein and Emmons 1987). Whether any of these sequences have centromeric properties is not clear.

DNA with replication and centromere functions

One of the repeat families that appears to have properties that are pertinent to centromeric roles in *C. elegans* is *CeRep3* (Felsenstein and Emmons 1988). This family consists of a set of repetitive DNA elements about 1 kb in length, dispersed some 50–100 times in the *C. elegans* genome. The sequence has a G + C content of 34%, which is typical for *C. elegans* DNA (36%). The repeat family is well conserved in various wild-type *C. elegans* strains but is absent from related species of nematodes. Interestingly, sequences within the repeats show homology to both the autonomously replicating sequence (*ARS*) and centromeric (*CEN*) consensus sequence of *S. cerevisiae*. When *C. elegans* DNA segments containing this repeat are cloned into appropriate shuttle vectors and introduced into *S. cerevisiae* cells, the nematode segments confer *ARS* function to the plasmids, as judged by an increased frequency of transformation compared with control plasmids without *ARS* function. Some but not all, also confer to the plasmids increased mitotic stability, increased frequency of 2 + :2 − segregation in meiosis, and decreased plasmid copy number. These results therefore indicate possible *ARS* and *CEN* functions in the *CeRep3*-containing *C. elegans* DNA segments.

In addition to *CeRep3*, seven DNA fragments (referred to as *SEG* or *seg*regator sequences) have been isolated from *C. elegans* that enhance the mitotic segregation of autonomously replicating plasmids in *S. cerevisiae* (Stinchcomb *et al.* 1985). Based on this segregation assay, there may be as many as 30 such segregator sequences in the *C. elegans* genome. Two of the seven sequences are single copy DNA, while the other five are repeated in the genome. Cross-hybridization analysis indicates that four of the five repeats belong to the same family. Interestingly, additional members of this repeat family, when tested, do not show *SEG* function. These results indicate that this repeat family is neither indispensable nor sufficient for segregator function.

Although the *CeRep3* repetitive family and *SEG* sequences demonstrate effects that are similar to those of *S. cerevisiae* centromeric DNA when assayed in *S. cerevisiae*, suggesting that they may have replication and centromere functions in *C. elegans*, direct proof of this will have to come from extra-chromosomal DNA transformation of *C. elegans* cells. It is possible that these sequences may affect segregation in *S. cerevisiae* through a mechanism distinct from that of a centromere within the host organism itself.

Chironomus

The isolation and characterization of a 155 bp tandemly repeated sequence that is localized exclusively to all the centromeres in the dipteran insect *Chironomus pallidivittatus* have been reported (Rovira *et al.* 1993). This DNA is detected in closely related species *C. tentans* and *C. thummi thummi*, but not in phylogenetically more remote species. Titration experiments indicate that the sequence corresponds to an average of 50 kb per centromere. A special feature of the 155 bp repeat is a region with two palindromes connected by a region of about 20 bp containing only AT base pairs. This arrangement is reminiscent of the centromere structure of *S. cerevisiae* (Rovira *et al.* 1993; Liao *et al.* 1994).

Arabidopsis thaliana (also mustard, radish, cucumber, and maize)

Arabidopsis thaliana, with a karyotype $2n = 10$ and genome size of 70 Mb, has the lowest DNA content known among higher plant species (Leutwiler *et al.* 1984). Repeated DNA sequences constitute about 25% of the genome of this species, which is also a relatively low proportion by comparison with other plant species (Meyerowitz and Pruitt 1985). One of these repeat sequences is a 180 bp DNA that is present in about 5000 copies and represents 1–1.5% of the *Arabidopsis* genome (Martinez-Zapater *et al.* 1986). The repeating units, which are tandemly organized into arrays longer than 50 kb, have been localized to the centromeric heterochromatin of all the chromosomes (Maluszynska and Heslop-Harrison 1991). Good homology is seen when this repeat is compared with satellite DNA from other species of the same plant family. For example, sequences from *Arabidopsis,* mustard, and radish all show homology, although those of mustard and radish share a significantly greater degree of this homology (Grellet *et al.* 1986). This observation is in agreement with the taxonomic position of these species since radish and mustard are classified in the same tribe *Brassiceae,* whereas *Arabidopsis* is classified in a different tribe *Sysimbrieae.* Significant regions of homology are also found between the repeat and satellite DNA from more distantly related species such as cucumber (see Martinez-Zapater *et al.* 1986) and maize (Peacock *et al.* 1981). The size of the repeat unit is about 180 bp in maize and 360 bp in cucumber.

Another repetitive genomic DNA that maps to the centromere of *A. thaliana* is a DNA that is derived from chromosome 1 (Richards *et al.* 1991). This DNA contains highly reiterated telomere-similar repetitive DNA. The telomere-similar repeats are approximately 500 bp in length and contains a simple sequence region of variable size composed of degenerate telomere repeats. The cloned 500 bp repeat sequence resides next to an AT-rich, dispersed repetitive element that is reiterated approximately five times in the *A. thaliana* genome. This 500 bp repeat has been shown to be closely related in sequence to the abundant 180 bp DNA described above and is proposed to have arisen by insertion of telomeric DNA into the 180 bp repeat sequence (Simoens *et al.* 1988).

In addition to *A. thaliana* and its related species, different families of centromeric repetitive DNA have also been isolated and characterized for other plant species, including *Brachycome dichromosomatica* (an Australian native daisy; Leach *et al.* 1995) and *Crepis capillaris* (Jamilena *et al.* 1995).

Maize B-chromosomes

B-Chromosomes are supernumerary chromosomes causing polymorphisms of chromosomal number in natural population of over a thousand plants and nearly 500 animal species (reviewed by Jones and Rees 1982; Jones and Puertas 1993). Despite their prevalence throughout the plant and animal kingdoms, they do not have any obvious genetic functions, yet are preserved in populations because they have acquired various accumulation mechanisms.

The B-chromosome of maize has been well studied. These chromosomes exert an influence on the primary chromosomes by increasing recombination frequencies (Rhoades 1968) and decreasing overall plant vigor at high copy number (Randolph 1941). The maize B-chromosomes are known to undergo nondisjunction at the second pollen mitosis, producing unequal partitioning of the replicated B-chromosomes into the two sperm involved in double fertilization. This is followed by a preferential fertilization of the egg by the sperm containing the B-chromosomes. Translocation analysis indicates that the portion of the B-chromosome predominantly responsible for this unique mitotic behaviour resides at or near the centromere (Carlson 1986).

A maize B-chromosome-specific centromeric repeat DNA has been described (Alfenito and Birchler 1993). Sequence analysis indicates an A + T nucleotide content of 71%, and the presence of a region containing a DNA motif CCCTAAA or variants (e.g. CCCTAA or CTAAA) that shares between 54 and 69% homology with the telomeres of *Plasmodium falciparum, Arabidopsis thaliana, Plasmodium berghie,* and *Homo sapiens.* However, unlike 'true' telomeric DNA, members of the B-chromosome repeats are neither identical nor tandemly organized.

A portion of this maize B repeat also shows homology to the maize neocentromere (knob). Normally knobs are an apparently inert sequence that have little effect on the maize genome. However, in the presence of an unusually large block of heterochromatin on the long arm of chromosome 10 (*K10*), all knobs throughout the genome are used preferentially as spindle attachment sites during meiosis, i.e. they become neocentromeres. This neocentromere formation leads to the preferential recovery of knobbed chromosomes during megasporogenesis. It is therefore possible that neocentromeres could have similar sequence motifs as normal centromeres, since they can act as a facultative substitute. The sequence homology between the knob and the reported B-specific repeat, coupled with its cytological location, suggests a possible centromeric role for the B repeat sequence.

Neurospora crassa

This is a multicellular filamentous ascomycete which shares the small genome and chromosome size as well as the ease of genetic and biochemical manipulation of yeast. Like *S. pombe* and animal cells, the chromosomes of this organism undergo condensation during mitosis and meiosis. The condensed chromosomes contain heterochromatin-like structures which are associated with the centromere (McClintock 1945; Singleton 1953; Perkins and Barry 1977). A region of 450 kb from the centromere of one of the chromosomes has been cloned into a YAC vector and shown to have three properties that are characteristic of centromeric DNA (Centola and Carbon 1994): (i) the region is recombination deficient, showing only 0.2% of the average recombination frequency of the genome; (ii) DNA from this region is AT rich, with the A + T content being 20% higher than normal genomic DNA; and (iii) the region contains a family of repetitive DNA that is present at or near centromeres on all seven chromosomes of *N. crassa.* In contrast to these properties, the DNA directly flanking both sides of this 450 kb region show higher recombination frequencies and lower A + T contents.

Although the 450 kb region shares the general properties of centromeric DNA, its repetitive DNA sequence appears to be unusual. Unlike centromeric repeats of higher organisms, which are highly redundant and contain many copies of a relatively short repeat sequence, the internal repeats of the 450 kb DNA are relatively long and do not contain short satellite repeat sequences. This structure is akin to the centromere organization in *S. pombe*, where fairly long sequence repeats ($\sim$ 6 kb), in relatively few copies, are present at the centromeres, with DNA sequences highly conserved among copies of these repeats. However, unlike *S. pombe*, the repeats within the 450 kb region are highly divergent and there is no evidence for the large inverted repeat structures seen in *S. pombe*.

Bacteria

Although not a eukaryote, it is of interest to include here a brief discussion on the 'centromere' of bacterial chromosomes to allow comparison. Most bacteria contain a single, circular haploid chromosome. Despite this apparent simplicity they, like all cells, require a good system of ensuring that each daughter cell receives a complete copy of the genome following chromosomal replication. The fact that chromosomal segregation occurs with high fidelity suggests the presence of a specific and efficient partition mechanism. However, structures resembling the mitotic apparatus of higher organisms have not been seen in bacteria, and the nature of the bacterial partition apparatus is just beginning to be understood (reviewed by Wake and Errington 1995). In a prototype bacterial or plasmid partition system, a model composed of four essential components has been proposed (Rothfield 1994) (Fig. 5.6A). In this model, the partition machinery is attached to a specific chromosomal site, the centromere, via a sequence-specific adaptor protein. The adaptor in turn interacts with a transducer, whose role is to attach each chromosome to a cellular attachment site, the anchor. The centromeres have been identified and studied in the circular chromosomes of unit copy plasmids such as F and P1, but not in the bacterial chromosome.

F plasmid

The stable maintenance of low copy number plasmids such as the F plasmid in growing cultures implies that a nonrandom partition mechanism exists which ensures that each daughter cell must receive at least one plasmid copy during cell division. Mini-F plasmids derived from the original F plasmid have been extensively studied for such mechanisms (reviewed by Hiraga 1992). The centromere of mini-F plasmids has been localized to the '*sopC*' region. In this region, there are 12 uninterrupted direct repeats of a 43 bp motif [Fig. 5.6B(-i)]. Within each of the direct repeats, one pair of 7 bp inverted repeats exists. In addition to this *cis*-acting *sopC* centromeric DNA, the participation of two additional *trans*-acting genes (*sopA* and *sopB*) is required for centromere activity.

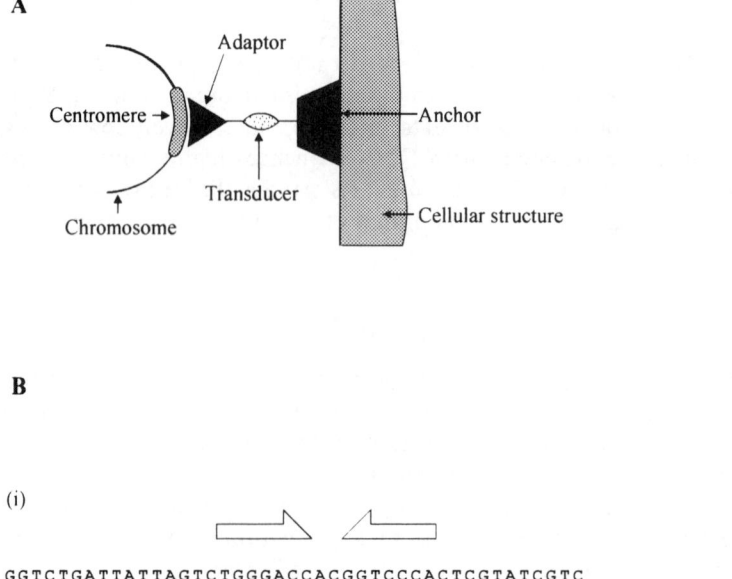

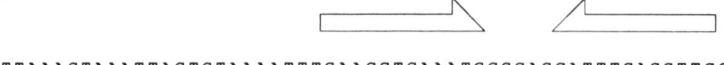

Fig. 5.6 (A) A model for the different components of the bacterial chromosome partition system. (B) Centromere sequences of F and P1 plasmids. (i) A 43 bp repeat motif in the '*sopC*' centromeric region of the F plasmid. A pair of 7 bp inverted repeats (arrows) exists within this motif. (ii) A 60 bp sequence in the *parS* centromeric region of the P1 plasmid, showing a 13 bp palindrome (arrows) and an adjacent AT-rich sequence.

P1 and P7 bacteriophages

The prophage of bacteriophages P1 and P7 are also maintained as unit copy plasmids, and has a loss rate of less than 1 in 10^5 cell divisions. The centromeres of these plasmids have been defined within the '*parS*' locus. This locus lies in a region consisting of a 13 bp palindrome and an adjacent AT-rich sequence [Fig. 5.6B-(ii)]. As with the mini-F plasmids, centromere activity of the *cis*-acting *parS* locus requires the involvement of two additional *trans*-acting genes (*parA* and *parB*). Homologies in deduced polypeptide sequences have been found between *sopA* and *parA*, and between *sopB* and *parB* of the F and P1/P7 systems.

Conclusion

DNA sequence versus conformational requirement of a centromere

In summarizing the information presented in this and Chapters 2 and 3, it is clear that the centromere DNA sequences of all the different eukaryotes studied consistently show little or no homology to one another, which contrasts with the highly conserved sequence nature of another well-studied chromosomal structure, the telomere. Only general similarities can be drawn from the centromeres of the different eukaryotes, such as the abundance of simple or complex repeats (that are more often than not AT rich), and the association of the centromeres with heterochromatin. This apparent lack of phylogenic conservation of the DNA sequence for a structure that performs such universal functions as mitosis and meiosis seriously challenges the dogma that DNA with fundamentally important functions must be well conserved at the nucleotide level throughout evolution. In view of this apparent paradox, it may be necessary to reassess what the functional requirement of a centromere DNA may be. It is conceivable that it is perhaps not a specific DNA sequence, but any one of a variety of sequences that is capable of interacting with the necessary protein components to achieve the correct conformational make-up that is important for centromere function, and that it is this conformational requirement that must be conserved through evolution. If this is true, then it can be further speculated that there may be one or more centromere DNA-binding proteins that are also universally conserved and whose primary role is to recognize this unique conformational requirement and to fold the centromere DNA into the functional higher-order configuration of an active centromere. Already, a number of centromere-binding proteins have been demonstrated to show specific domain homologies across widely separated eukaryotic species (Chapters 2, 3, and 6). Some of these proteins have a proposed role in the structural organization of the centromere and are possible candidates for such conformation-recognizing proteins.

References

Abad, J., Carmena, M., Baars, S., Saunders, R., Glover, D., Ludena, P. *et al.* (1992). Dodeca satellite: a conserved G + C-rich satellite from the centromeric heterochromatin of *Drosophila melanogaster*. Proc. Natl Acad. Sci. USA *89*, 4663–4667.

Agresti, A., Rainaldi, G., Lobbiani, A., Magnani, I., Lernia, R. D., Meneveri, R.*et al.* (1987). Chromosomal location by *in situ* hybridization of the human Sau3A family of DNA repeats. Hum. Genet. *75*, 326–332.

Agresti, A., Meneveri, R., Siccardi, A. G., Marozzi, A., Comeo, G., Gaudi, S. *et al.* (1989). Linkage in human heterochromatin between highly divergent Sau3A repeats and a new family of repeated DNA sequences (HaeIII family). J. Mol. Biol. *205*, 625–631.

Alexandrov, I. A., Akopian, T. A., Vinnik, E. A., Mitkevich, S. P., Kisselev, L. L., and Yurov, Y. B. (1989a). Cloned alpha-satellite fragment—the molecular marker of human chromosome 4: sequence, genomic organization, polymorphism. Cytogenet. Cell Genet. *51*, 949.

Alexandrov, I. A., Akopian, T. A., Vinnik, E. A., Mitkevich, S. P., Kisselev, L. L., and Yurov, Y. B. (1989b). Two alpha satellite domains on human chromosome 18: A novel 18-specific repeated unit. Cytogenet. Cell Genet. *51*, 949.

Alexandrov, I. A., Mashkova, T. D., Akopian, T. A., Medvedev, L. I., Kisselev, L. L., Mitkevich, S. P. *et al.* (1991). Chromosome-specific alpha satellites: two distinct families on human chromosome 18. Genomics *11*, 15–23.

Alfenito, M. and Birchler, J. (1993). Molecular characterization of a maize B chromosome centric sequence. Genetics *135*, 589–597.

Anderson, S., Yates, T., and Cook, J. (1987). Notes on Bolivian mammals IV. The genus *Ctenomys* (Rodentia, Ctenomyidae) in the eastern lowlands. Am. Mus. Novitates *2891*, 1–20.

Baldini, A., Smith, D. I., Rocchi, M., Miller, O. J., and Miller, D. A. (1989). A human alphoid DNA clone from the EcoRI dimeric family: genomic and internal organization and chromosomal assignment. Genomics *5*, 822–828.

Baldini, A., Rocchi, M., Archidiacono, N., Miller, O. J., and Miller, D. A. (1990). A human alpha satellite DNA subset specific for chromosome 12. Am. J. Hum. Genet. *46*, 784–788.

Baldini, A., Archidiacono, N., Carbone, R., Bolino, A., Shridhar, V., Miller, O. J. *et al.* (1992). Isolation and comparative mapping of a human chromosome 20-specific alpha-satellite DNA clone. Cytogenet. Cell Genet. *59*, 12–16.

Baldwin, L. and Macgregor, H. (1985). Centromeric satellite DNA in the newt *Triturus cristatus karelinii* and related species: its distribution and transcription on lampbrush chromosomes. Chromosoma *92*, 100–107.

Barsacchi-Pilone, G., Batistoni, R., Andronico, F., Vitelli, L., and Nardi, I. (1986). Heterochromatic DNA in *Triturus* (Amphibia, Urodela) I. A satellite DNA component of the pericentric C-bands. Chromosoma *93*, 435–446.

Beridze, T. (1986). Satellite DNA. Springer-Verlag, Berlin, 1–149.

Blennow, E., Telenius, H., Vos, D. d., Larsson, C., Henriksson, P., Johansson, O. *et al.* (1994). Tetrasomy 15q: two marker chromosomes with no detectable alpha-satellite DNA. Am. J. Hum. Genet. *54*, 877–883.

Bogenberger, J., Neumaier, P., and Fittler, F. (1985). The Muntjak satellite IA sequence is composed of 31-base-pair internal repeats that are highly homologous to the 31-base-pair subrepeats of the bovine satellite 1.715. Eur. J. Biochem. *148*, 55–59.

Broccoli, D., Miller, O., and Miller, D. (1990). Relationship of mouse minor satellite DNA to centromere activity. Cytogenet. Cell Genet. *54*, 182–186.

Broccoli, D., Trevor, K., Miller, O., and Miller, D. (1991). Isolation of a variant family of mouse minor satellite DNA that hybridizes preferentially to chromosome 4. Genomics *10*, 68–74.

Brown, K. E., Barnett, M. A., Burgtorf, C., Shaw, P., Buckle, V. J., and Brown, W. R. A. (1994). Dissecting the centromere of the human Y chromosome with cloned telomeric DNA. Hum. Mol. Genet. *3*, 1227–1237.

Bukvic, N., Susca, F., Gentile, M., Tangari, E., Ianniruberto, A., and Guanti, G. (1996). An unusual dicentric Y chromosome with a functional centromere with no detectable alpha-satellite. Hum. Genet. *97*, 453–456.

Burgtorf, C. and Bunemann, H. (1993). A telomere-like satellite (GGGTCAT)n comprises 4% of *Drosophila hydei* and is located mainly in centromeric hetero-chromatin of all large acrocentric autosomes. Gene *137*, 287–291.

Cloning and molecular characterization of a novel chromosome specific centromere sequence of Chinese hamster. Nucleic Acid Res. *22*, 3728–3736.

Felsenstein, K. and Emmons, S. (1987). Structure and evolution of a family of repetitive DNA sequences in *Caenorhabditis elegans*. J. Mol. Evol. *25*, 230–240.

Felsenstein, K. and Emmons, S. (1988). Nematode repetitive DNA with ARS and segregation function in *Saccharomyces cerevisia*. Mol. Cell. Biol. *8*, 875–883.

Frommer, M., Prosser, J., Tkachuk, D., Reisner, A. H., and Vincent, P. C. (1982). Simple repeated sequences in human satellite DNA. Nucleic Acids Res. *10*, 547–563.

Gallardo, M. (1991). Karyotypic evolution in *Ctenomys* (Rodentia, Ctenomyidae). J. Mammal. *72*, 11–21.

Garagna, S., Redi, C., Capanna, E., Andayani, N., Alfano, R., Doi, P. *et al.* (1993). Genome distribution, chromosomal allocation and organization of the major and minor satellite DNAs in 11 species and subspecies of the genus Mus. Cytogenet. Cell Genet. *64*, 247–255.

Garagna, S., Broccoli, D., Redi, C., Searle, J., Cooke, H., and Capanna, E. (1995). Robertsonian metacentrics of the house mouse lose telomeric sequences but retain some minor satellite DNA in the pericentromeric area. Chromosoma *103*, 685–692.

Garrido-Ramos, M., Jamilena, M., Lozano, R., Rejon, C. R., and Rejon, M. R. (1994). Cloning and characterization of a fish centromeric satellite DNA. Cytogenet. Cell Genet. *65*, 233–237.

Garrido-Ramos, M., Jamilena, M., Lozano, R., Rejon, C. R., and Rejon, M. R. (1995). The *Eco*RI centromeric satellite DNA of the Sparidae family (Pisces, Perciformes) contains a sequence motive common to other vertebrate centromeric satellite DNAs. Cytogenet. Cell Genet. *71*, 345–351.

Ge, Y., Wagner, M. J., Siciliano, M., and Wells, D. E. (1992). Sequence, higher order repeat structure, and long-range organization of alpha satellite DNA specific to human chromosome 8. Genomics *13*, 585–593.

Gosden, J. R., Mitchell, A. R., Buckland, R. A., Clayton, R. P., and Evans, H. J. (1975). The location of four human satellite DNAs on human chromosomes. Exp. Cell Res. *92*, 148–158.

Gosden, J. R., Lawrie, S. S., and Gosden, C. (1981). Satellite DNA sequences in the human acrocentric chromosomes: information from translocations and heteromorphisms. Am. J. Hum. Genet. *33*, 243–251.

Grady, D., Ratliff, R., Robinson, D., McCanlies, E., Meyne, J., and Moyzis, R. (1992). Highly conserved repetitive DNA sequences are present at human centromeres. Proc. Natl Acad. Sci. USA *89*, 1695–1699.

Graham, G. J., Baro, D. J., Garcia, M. J., and Cummings, M. R. (1985). Molecular organization in the proximal region of human acrocentric chromosomes. Annls NY Acad. Sci. *450*, 55–67.

Greig, G. M., England, S. B., Bedford, H. M., and Willard, H. F. (1989). Chromosome-specific alpha satellite DNA from the centromere of human chromosome 16. Am. J. Hum. Genet. *45*, 862–872.

Greig, G. M. and Willard, H. F. (1992). Beta satellite DNA: characterization and localization of two subfamilies from the distal and proximal short arms of the human acrocentric chromosomes. Genomics *12*, 573–580.

Greig, G. M., Parikh, S., George, J., Powers, V. E., and Willard, H. F. (1991). Molecular cytogenetics of alpha satellite DNA from chromosome 12: fluorescence *in situ* hybridization and description of DNA and array length polymorphisms. Cytogenet. Cell Genet. *56*, 144–148.

Grellet, F., Delcasso, D., Panabieres, F., and Delseny, M. (1986). Organization and evolution of a higher plant alphoid-like satellite DNA sequence. J. Mol. Biol. *187*, 495–507.

Grimaldi, G. and Singer, M. F. (1982). A monkey *Alu* sequence is flanked by 13-base-pair direct repeats of an interrupted α-satellite DNA sequence. Proc. Natl Acad. Sci. USA *79*, 1497–1500.

Grimaldi, G. and Singer, M. F. (1983). Members of the Kpn1 family of long interspersed repeated sequences join and interrupt α-satellite in the monkey genome. Nucleic Acids Res. *11*, 321–338.

Gubbay, J., Collignon, J., Koopman, P., Capel, B., Economou, A., Munsterberg, A. *et al.* (1990). A gene mapping to the sex-determining region of the mouse Y chromosome is a member of a novel family of embryonically expressed genes. Nature *346*, 245–250.

Gupta, R. (1983). Nucleotide sequence of a reiterated rat DNA fragment. FEBS Lett. *164*, 175–180.

Haaf, T. and Willard, H. F. (1992). Organization, polymorphism, and molecular cytogenetics of chromosome-specific alpha-satellite DNA from the centromere of chromosome 2. Genomics *13*, 122–128.

Haaf, T., Warburton, P. E., and Willard, H. F. (1992). Integration of human alpha satellite DNA into simian chromosomes; centromere protein binding and disruption of normal chromosome segregation. Cell *70*, 681–696.

Heartlein, M. W., Knoll, J. H. M., and Latt, S. A. (1988). Chromosome instability associated with human alphoid DNA transfected into the Chinese hamster genome. Mol. Cell. Biol. *8*, 3611–3618.

Higgins, M. J., Wang, H., Shtromas, I., Haliotis, T., Roder, J. C., Holden, J. J. *et* al. (1985). Organization of a repetitive human 1.8kb *Kpn*I sequence localized in the heterochromatin of chromosome 15. Chromosoma (Berl.) *93*, 77–86.

Hiraga, S. (1992). Chromosome and plasmid partition in *Escherichia coli*. Annu. Rev. Biochem. *61*, 283–306.

Holmquist, G. and Dancis, B. (1979). Telomere replication, kinetochore organisers and satellite DNA evolution. Proc. Natl Acad. Sci. USA *76*, 4566–4570.

Horz, W. and Altenburger, W. (1981). Nucleotide sequence of mouse satellite DNA. Nucleic Acids Res. *9*, 683–696.

Hulsebos, T., Schonk, D., Dalen, I. v., Coerwinkel-Driessen, M., Schepens, J., Ropers, H. H. *et al.* (1988). Isolation and characterization of alphoid DNA sequences specific for the pericentric regions of chromosomes 4, 5, 9, and 19. Cytogenet. Cell Genet. *47*, 144–148.

Ikeno, M., Masumoto, H., and Okazaki, T. (1994). Distribution of CENP-B boxes reflected in CREST centromere antigenic sites on long-range α-satellite DNA arrays of human chromosome 21. Hum. Mol. Genet. *3*, 1245–1257.

Jabs, E. W., Wolf, S. F., and Migeon, B. R. (1984). Characterization of a cloned DNA sequence that is present at centromeres of all human autosomes and the X chromosome and shows polymorphic variation. Proc. Natl Acad. Sci. USA *81*, 4884–4888.

Jabs, E. W. and Persico, M. G. (1987). Characterization of human centromeric regions of specific chromosomes by means of alphoid DNA sequences. Am. J. Hum. Genet. *41*, 374–390.

Jackson, M. S., Mole, S. E., and Ponder, B. A. J. (1992). Characterisation of a boundary between satellite III and alphoid sequences on human chromosome 10. Nucleic Acids Res. *20*, 4781–4787.

Jackson, M. S., Slijepcevic, P., and Ponder, B. A. J. (1993). The organisation of repetitive sequences in the pericentromeric region of human chromosome 10. Nucleic Acids Res. *21*, 5865–5874.

Jackson, M. S., See, C. G., Mulligan, L. M., and Lauffart, B. F. (1996). A 9.75-Mb map across the centromere of human chromosome 10. Genomics *33*, 258–270.

Jamilena, M., Garrido-Ramos, M., Rejon, M. R., Rejon, C. R., and Parker, J. (1995). Characterisation of repeated sequences from microdissected B chromosomes of *Crepis capillaris*. Chromosoma *104*, 113–120.

Jantsch, M., Hamilton, B., Mayr, B., and Schweizer, D. (1990). Meiotic chromosome behaviour reflects levels of sequence divergence in *Sus scrofa domestica* satellite DNA. Chromosoma *99*, 330–335.

Johnson, D. H., Kroisel, P. M., Klapper, H. J., and Rosenkranz, W. (1992). Microdissection of a human marker chromosome reveals its origin and a new family of centromeric repetitive DNA. Hum. Mol. Genet. *1*, 741–747.

Jones, R. and Puertas, M. (1993). The B-chromosomes of Rye (*Secale cereale* L.). In: Frontiers in Plant Science Research, eds., Dhir. K.K. Sareen, S. Bhagwati Enterprises, Delhi.

Jones, R. and Rees, H. (1982). B Chromosomes. Academic Press, New York.

Jorgensen, A. L., Bostock, C. J., and Bak, A. L. (1987). Homologous subfamilies of human alphoid repetitive DNA on different nucleolus organizing chromosomes. Proc. Natl Acad. Sci. USA *84*, 1075–1079.

Jorgensen, A. L., Kolvraa, S., Jones, C., and Bak, A. L. (1988). A subfamily of alphoid repetitive DNA shared by the NOR-bearing human chromosomes 14 and 22. Genomics *3*, 100–1009.

Joseph, A., Mitchell, A., and Miller, O. (1989). The organization of the mouse satellite DNA at centromeres. Exp. Cell Res. *183*, 494–500.

Kaelbling, M., Miller, D., and Miller, O. (1984). Restriction enzyme banding of mouse metaphase chromosomes. Chromosoma *90*, 128–132.

Kalitsis, P., Earle, E., Vissel, B., Shaffer, L., and Choo, K. (1993). A chromosome 13-specific human satellite I DNA subfamily with minor presence on chromosome 21: further studies on Robertsonian translocations. Genomics *16*, 104–112.

Kipling, D., Ackford, H., Taylor, B., and Cooke, H. (1991). Mouse minor satellite DNA genetically maps to the centromere and is physically linked to the proximal telomere. Genomics *11*, 235–241.

Kipling, D., Wilson, H., Mitchell, A., Taylor, B., and Cooke, H. (1994). Mouse centromere mapping using oligonucleotide probes that detect variants of the minor satellite. Chromosoma *103*, 46–55.

Kipling, D., Mitchell, A., Masumoto, H., Wilson, H., Nicol, L., and Cooke, H. (1995). CENP-B binds a novel centromeric sequence in the Asian mouse *Mus caroli*. Mol. Cell. Biol. *15*, 4009–4020.

Koide, T., Yoshino, M., Niwa, M., Ishiura, M., Shiroishi, T., and Moriwaki, K. (1992). The amplified long genomic sequence (ALGS) located in the centromeric regions of mouse chromosomes. Genomics *13*, 1186–1191.

Kurnit, D., Shafit, B., and Maio, J. (1973). Multiple satellite deoxyribonucleic acids in the calf and their relation to the sex chromosome. J. Mol. Biol. *81*, 273–284.

Kurnit, D. M., Neve, R. L., Norton, C. C., Bruns, G. A., Ma, N. S., Cox, D. R. *et al.* (1984). Recent evolution of DNA sequence homology in the pericentric regions of human acrocentric chromosomes. Cytogenet. Cell Genet. *38*, 99–105.

Kurnit, D. M., Roy, S., Stewart, G. D., Schwedock, J., Neve, R. L., Bruns, G. A. *et al.* (1986). The 724 family of DNA sequences is interspersed about the pericentromeric regions of human acrocentric chromosomes. Cytogenet. Cell Genet. *43*, 109–116.

Lapeyre, J., Beattie, W., Dugaiczyk, A., Vizard, D., and Becker, F. (1980). *Eco*RI-generated reiterated components of the rat genome I. Sequence of two (92 and 93 bp) related DNA fragments. Gene (Amst.) *10*, 339–346.

Larin, Z., Fricker, M. D., and Tyler-Smith, C. (1994). *De novo* formation of several features of a centromere following introduction of a Y alphoid YAC into mammalian cells. Hum. Mol. Genet. *3*, 689–695.

Le, M., Duricka, D., and Karpen, G. (1995). Islands of complex DNA are widespread in *Drosophila* centric heterochromatin. Genetics *141*, 283–303.

Leach, C., Donald, T., Franks, T., Spiniello, S., Hanrahan, C., and Timmis, J. (1995). Organisation and origin of a B chromosome centromeric sequence from *Brachycome dichromosomatica*. Chromosoma *103*, 708–714.

Lee, C., Ritchie, D., and Lin, C. (1994). A tandemly repetitive, centromeric DNA sequence from the Canadian woodland caribou (*Rangifer tarandus caribou*): its conservation and evolution in several deer species. Chromosome Res. *2*, 293–306.

Lee, C., Li, X., Jabs, E. W., Court, D., and Lin, C. C. (1995). Human gamma X satellite DNA: an X chromosome specific centromeric DNA sequence. Chromosoma *104*, 103–112.

Leutwiler, L., Hough-Evans, B., and Meyerowitz, E. (1984). The DNA of *Arabidopsis thaliana*. Mol. Gen. Genet. *194*, 15–23.

Liao, L., Rosenzweig, B., and Hirsh, D. (1983). Analysis of a transposable element in *Caenorhabditis elegans*. Proc. Natl Acad. Sci. USA *80*, 3585–3589.

Liao, C., Rovira, C., and Edstrom, J.-E. (1994). Constant and variable parts of the 155-bp centromeric repeat in *Camptochironomus*. J. Mol. Evol. *39*, 112–114.

Lin, C., Sasi, R., Fan, Y.-S., and Chen, Z.-Q. (1991). New evidence for tandem chromosome fusions in the karyotypic evolution of Asian muntjacs. Chromosoma *101*, 19–24.

Lin, C. C., Sasi, R., Lee, C., Fan, Y. S., and Court, D. (1993). Isolation and identification of a novel tandemly repeated DNA sequence in the centromeric region of human chromosome 8. Chromosoma *102*, 333–339.

Lohe, A. and Roberts, P. (1988). Evolution of satellite DNA sequences in *Drosophila*. In Heterochromatin: Molecular and Structural Aspects, ed. Verma, R.S. Cambridge University Press, Cambridge, 148–186.

Lohe, A. R., Hilliker, A. J., and Roberts, P. A. (1993). Mapping simple repeated DNA sequences in heterochromatin of *Drosophila melanogaster*. Genetics *134*, 1149–1174.

Lohe, A. R. and Hilliker, A. J. (1995). Return of the H-word (heterochromatin). Curr. Opin. Genet. Dev. *5*, 746–755.

Looijenga, L. H., Smit, V. T., Wessels, J. W., Mollevanger, P., Oosterhuis, J. W., Cornelisse, C. J. *et al.* (1990). Localization and polymorphism of a chromosome 12-specific alpha satellite DNA sequence. Cytogenet. Cell Genet. *53*, 216–8.

Macaya, G., Cortadas, J., and Bernardi, G. (1978). An analysis of the bovine genome by density-gradient centrifugation. Preparation of the dG + dC-rich DNA components. Eur. J. Biochem. *84*, 179–188.

Magnani, I., Sacchi, N., Darfler, M., Nisson, P., Tornaghi, R., and Fuhrman-Conti, A. (1993). Identification of the chromosome 14 origin of a C-negative marker associated with a 14q32 deletion by chromosome painting. Clin. Genet. *43*, 180–185.

Maio, J. J. (1971). DNA strand reassociation and polyribonucleotide binding in the African green monkey, *Cercopithecus aethiops*. J. Mol. Biol. *56*, 579–595.

Maluszynska, J. and Heslop-Harrison, J. (1991). Localization of tandemly repeated DNA sequences in *Arabidopsis thaliana*. Plant J. *1*, 159–166.

Manuelidis, L. (1976). Repeating restriction fragments of human DNA. Nucleic Acids Res. *3*, 3063–3076.

Manuelidis, L. (1978a). Complex and simple sequences in human repeated DNAs. Chromosoma *66*, 1–21.

Manuelidis, L. (1978b). Chromosomal localization of complex and simple repeated human DNAs. Chromosoma *66*, 23–32.

Manuelidis, L. (1981). Consensus sequence of mouse satellite DNA indicates it is derived from tandem 116 base pair repeats. FEBS Lett. *129*, 25–28.

Martinez-Zapater, J., Estelle, M., and Somerville, C. (1986). A highly repeated DNA sequence in *Arabidopsis thaliana*. Mol. Gen. Genet. *204*, 417–423.

Massarini, A., Barros, M., Ortells, M., and Reig, O. (1991). Chromosomal polymorphism and small karyotypic differentiation in a group of *Ctenomys* species from Central Argentina (Rodentia, Octodontidae). Genetica *83*, 131–144.

Masumoto, H., Masukata, H., Muro, Y., Nozaki, N., and Okazaki, T. (1989). A human centromere antigen (CENP-B) interacts with a short specific sequence in alphoid DNA: a human centromeric satellite. J. Cell Biol. *109*, 1963–1973.

McClintock, B. (1945). Neurospora. I. Preliminary observations of the chromosomes of *Neurospora crassa*: update. Am. J. Bot. *32*, 671–678.

McCutchan, T., Hsu, H., Thayer, R. E., and Singer, M. F. (1982). Organization of African green monkey DNA at junctions between alpha satellite and other DNA sequences. J. Mol. Biol. *157*, 195–211.

McDermid, H. E., Duncan, A. M., Higgins, M. J., Hamerton, J. L., Rector, E., Brasch, K. R. *et al.* (1986). Isolation and characterization of an alpha-satellite repeated sequence from human chromosome 22. Chromosoma *94*, 228–234.

Meneveri, R., Agresti, A., Valle, G. D., Talarico, D., Siccardi, A. G., and Ginelli, E. (1985). Identification of a human clustered GC-rich DNA family of repeats (*Sau*3A family). J. Mol. Biol. *186*, 483–490.

Metzdorf, R., Gottert, E., and Blin, N. (1988). A novel centromeric repetitive DNA from human chromosome 22. Chromosoma *97*, 154–158.

Meyerowitz, E. and Pruitt, R. (1985). *Arabidopsis thaliana* and plant molecular genetics. Science *229*, 1214–1218.

Meyne, J., Baker, R., Hobart, H., Hsu, T., Ryder, O., Ward, O. *et al.* (1990). Distribution of non-telomeric sites of the $(TTAGGG)_n$ telomeric sequence in vertebrate chromosomes. Chromosoma *99*, 3–10.

Meyne, J., Goodwin, H. E., and Moyzis, R. K. (1994). Chromosome localization and orientation of the simple sequence repeat of human satellite I DNA. Chromosoma *103*, 99–103.

Miller, J., Hindkjaer, J., and Thomsen, P. (1993). A chromosomal basis for the differential organization of a porcine centromere-specific repeat. Cytogenet. Cell Genet. *62*, 37–41.

Mitchell, A. R., Gosden, J. R., and Miller, D. A. (1985). A cloned sequence, p82H, of the alphoid repeated DNA family found at the centromeres of all human chromosomes. Chromosoma *92*, 369–377.

Moens, P. and Pearlman, R. (1990). Telomere and centromere DNA are associated with the cores of meiotic prophase chromosomes. Chromosoma *100*, 8–14.

Moyzis, R. K., Albright, K. L., Bartholdi, M. F., Cram, L. S., Deaven, L. L., Hildebrand, C. E. *et al.* (1987). Human chromosome specific repetitive DNA sequences. Novel markers for genetic analysis. Chromosoma *95*, 375–386.

Mullenbach, R., Lutz, S., Holzmann, K., Dooley, S., and Blin, N. (1992). A non-alphoid repetitive DNA sequence from human chromosome 21. Hum. Genet. *89*, 519–523.

Muro, Y., Masumoto, H., Yoda, K., Nozaki, N., Ohashi, M., and Okazaki, T. (1992). Centromere protein B assembles human centromeric alpha-satellite DNA at the 17-bp sequence, CENP-B box. J. Cell Biol. *116*, 585–596.

Murphy, T. D. and Karpen, G. H. (1995). Localization of centromere function in a *Drosophila* minichromosome. Cell *82*, 599–609.

Naclerio, G., Cangiano, G., Coulson, A., Levitt, A., Ruvolo, V., and Volpe, A. L. (1992). Molecular and genomic organization of clusters of repetitive DNA sequences in *Caenorhabditis elegans*. J. Mol. Biol. *226*, 159–168.

Narayanswami, S., Doggett, N., Clark, L., Hildebrand, C., Weier, H., and Hamkalo,

B. (1992). Cytological and molecular characterization of centromeres in *Mus domesticus* and *Mus spretus*. Mammal. Genome *2*, 186–194.

Neuer-Nitsche, B., Lu, X., and Werner, D. (1988). Functional role of a highly repetitive DNA sequence in anchorage of the mouse genome. Nucleic Acids Res. *16*, 8531–8539.

O'Hare, K., Alley, M. R., Cullingford, T. E., Driver, A., and Sanderson, M. J. (1991). DNA sequence of the Doc retrotransposon in the white-one mutant of *Drosphila melanogaster* and of secondary insertions in the phenotypically altered derivatives white-honey and white-eosin. Mol. Gen. Genet. *225*, 17–24.

Ohashi, H., Wakui, K., Ogawa, K., Okano, T., Niikawa, N., and Fukushima, Y. (1994). A stable acentric marker chromosome: possible existence of an intercalary ancient centromere at distal 8p. Am. J. Hum. Genet. *55*, 1202–1208.

Ortells, M., Contreras, J., and Reig, O. (1990). New *Ctenomys* karyotypes (Rodentia, Octodontidae) from north-eastern Argentina and from Paraguay confirm the chromosomal multiformity of the genus. Genetica *82*, 189–201.

Ouspenski, I. and Brinkley, B. (1993). Centromeric DNA cloned from functional kinetochore fragments in mitotic cells with unreplicated genomes. J. Cell Sci. *105*, 359–367.

Pardue, M. and Gall, J. (1970). Chromosomal localization of mouse satellite DNA. Science *168*, 1356–1358.

Peacock, W., Dennis, E., Rhoades, M., and Pryor, A. (1981). Highly repeated DNA sequence limited to knob heterochromatin in maize. Proc. Natl Acad. Sci. USA *78*, 4490–4494.

Pech, M., Igo-Kemenes, T., and Zachau, H. (1979). Nucleotide sequence of a highly repetitive component of rat DNA. Nucleic Acids Res. *7*, 417–432.

Perkins, D. and Barry, E. (1977). The cytogenetics of *Neurospora*. Adv. Genet. *19*, 133–285.

Pesce, C., Rossi, M., Muro, A., Reig, O., Zorzo'pulos, J., and Kornblihtt, A. (1994). Binding of nuclear factors to satellite DNA of retroviral origin with marked differences in copy number among the species of the South American rodents of the genus *Ctenomys*. Nucleic Acids Res. *22*, 656–661.

Pietras, D., Bennett, K., Siracusa, L., Woodworth-Gutai, M., Chapman, V., Gross, K. *et al.* (1983). Construction of a small *Mus musculus* repetitive DNA library: identification of a new satellite sequence in *Mus musculus*. Nucleic Acids Res. *11*, 6965–6983.

Polli, E., Ginelli, E., Bianchi, P., and Corneo, G. (1966). Renaturation of calf thymus satellite DNA. J. Mol. Biol. *17*, 305–308.

Porter, S. and Jones, R. (1983). Unusual domains of human alphoid satellite DNA with contiguous non-satellite sequences: sequence analysis of a junction region. Nucleic Acids Res. *11*, 3137–3153.

Prosser, J., Frommer, M., Paul, C., and Vincent, P. C. (1986). Sequence relationships of three human satellite DNAs. J. Mol. Biol. *187*, 145–155.

Randolph, L. (1941). Genetic characteristics of the B-chromosomes in maize. Genetics *26*, 608–631.

Rattner, J. (1991). The structure of the mammalian centromere. BioEssays *13*, 51–56.

Rauch, A., Pfeiffer, R. A., Trautmann, U., Liehr, T., Rott, H. D., and Ulmer, R. A. (1992). A study of ten small supernumerary (marker) chromosomes identified by fluorescence *in situ* hybridization (FISH). Clin. Genet. *42*, 84–90.

Reig, O. (1989). Karyotypic repatterning as a triggering factor in cases of explosive speciation. In Evolutionary Biology of Unstable Populations, ed. Fontdevila, A. Springer-Verlag, Berlin, 246–289.

Reig, O. and Kiblisky, P. (1969). Chromosome multiformity in the genus *Ctenomys* (Rodentia, Octodontidae). Chromosoma *28*, 211–244.

Reig, O., Busch, C., Ortells, M., and Contreras, J. (1990). An overview of evolution, systematics, population biology and speciation in *Ctenomys,* In Evolutionary Biology of Subterranean Mammals, eds. Nevo, E. and Reig, O.A. Alan R. Liss, New York, 71–96.

Reig, O., Ortells, M., Massarini, A., Barros, M., Tiranti, S., and Dyzenchauz, F. (1992). New karyotypes and C-banding patterns of the subterranean rodents of the genus *Ctenomys* (Caviomorpha, Octodontidae) from Argentina. Mammalia *56*, 603–623.

Rhoades, M. (1968). Studies on the cytological basis of crossing over. In Replication and Recombination of Genetics Material, eds. Peacock, W.J. and Brock, R.D. Australian Academy of Science, Canberra.

Richards, E. Goodman, H., and Ausubel, F. (1991). The centromere region of *Arabidopsis thaliana* chromosome 1 contains telomere-similar sequences. Nucleic Acids Res. *19*, 3351–3357.

Roberts, C., Weith, A., Passage, E., Michot, J., Mattei, M., and Bishop, C. (1988). Molecular and cytogenetic evidence for the location of *Tdy* and *Hya* on the mouse Y chromosome short arm. Proc. Natl Acad. Sci. USA *85*, 6446–6449.

Rocchi, M., Baldini, A., Archidiacono, N., Lainwala, S., Miller, O. J., and Miller, D. A. (1990). Chromosome-specific subsets of human alphoid DNA identified by a chromosome 2-derived clone. Genomics *8*, 705–709.

Rocchi, M., Archidiacono, N., Ward, D. C., and Baldini, A. (1991). A human chromosome 9-specific alphoid DNA repeat spatially resolvable from satellite 3 DNA by fluorescent *in situ* hybridization. Genomics *9*, 517–523.

Rosenberg, H., Singer, M., and Rosenberg, M. (1978). Highly reiterated sequences of SIMIANSIMIANSIMIANSIMIANSIMIAN. Science *200*, 394–402.

Rossi, M., Reig, O., and Zorzopulos, J. (1990). Evidence of rolling circle replication in a major satellite DNA from the South American rodents of the genus *Ctenomys*. Mol. Biol. Evol. *7*, 340–350.

Rossi, M., Pesce, C., Reig, P., Kornblihtt, A., and Zorzopulos, J. (1993a). Retroviral-like features in the monomer of the major satellite DNA from the South American rodents of the genus *Ctenomys*. DNA Sequencing *3*, 379–381.

Rossi, M., Reig, O., and Zorzopulos, J. (1993b). A major satellite DNA from the South American rodents of the genus *Ctenomys*: quantitative and qualitative differences in species with different geographic distribution. Z. Saügetierk, *58*, 244–251.

Rossi, M., Redi, C., Viale, G., Massarini, A., and Capanna, E. (1995). Chromosomal distribution of the major satellite DNA of South American rodents of the genus *Ctenomys*. Cytogenet. Cell Genet. *69*, 179–194.

Rothfield, L. (1994). Bacterial chromosome segregation. Cell *77*, 963–966.

Rovira, C., Beermann, W., and Edstrom, J.-E. (1993). A repetitive DNA sequence associated with the centromeres of *Chironomus pallidivittatus*. Nucleic Acids Res. *21*, 1775–1781.

Schildkraut, C., Marmur, J., and Doty, P. (1962). Determination of the base composition of deoxyribonucleic acid from its buoyant density in CsCl. J. Mol. Biol. *4*, 430–443.

Sealy, L., Hartley, J., Donelson, J., Chalkley, R., Hutchison, N., and Hamkalo, B. (1981). Characterization of a highly repetitive sequence DNA family in rat. J. Mol. Biol. *145*, 291–318.

Simoens, C., Gielen, J., VanMontagu, M., and Inze, D. (1988). Characterization of highly repetitive sequences of *Arabidopsis thaliana*. Nucleic Acids Res. *16*, 6753–6766.

Singleton, J. (1953). Chromosome morphology and the chromosome cycle in the ascus of *Neurospora crassa*. Am. J. Bot. *407*, 124–144.

Smith, G. P. (1976). Evolution of repeated DNA sequences by unequal crossover. Science *191*, 528–535.

Stinchcomb, D., Mello, C., and Hirsh, D. (1985). *Caenorhabditis elegans* DNA that directs segregation in yeast cells. Proc. Natl Acad. Sci. USA *82*, 4167–4171.

Tagarro, I., Wiegant, J., Raap, A. K., Gonzalez-Aguilera, J. J., and Fernandez-Peralta, A. M. (1994). Assignment of human satellite I DNA as revealed by fluorescent *in situ* hybridization with oligonucleotides. Hum. Genet. *93*, 125–128.

Taparowsky, E. and Gerbi, S. (1982). Structure of 1.711 g/cm3 bovine satellite DNA: evolutionary relationship to satellite I. Nucleic Acids Res. *10*, 5503–5515.

Thayer, R. E. and Singer, M. F. (1983). Interruption of an α-satellite array by a short member of the *Kpn*1 family of interspersed, highly repeated monkey DNA sequences. Mol. Cell. Biol. *3*, 967–973.

Trowell, H. E., Nagy, A., Vissel, B., and Choo, K. H. A. (1993). Long-range analyses of the centromeric regions of human chromosomes 13, 14 and 21: identification of a narrow domain containing two key centromeric DNA elements. Hum. Mol. Genet. *2*, 1639–1649.

Tyler-Smith, C. (1987). Structure of repeated sequences in the centromeric region of the human Y chromosome. Development *101*, 93–100.

Tyler-Smith, C. and Brown, W. (1987). Structure of the major block of alphoid satellite DNA on the human Y chromosome. J. Mol. Biol. *195*, 457–470.

Tyler-Smith, C. and Willard, H. F. (1993). Mammalian chromosome structure. Curr. Biol. *3*, 390–397.

Tyler-Smith, C., Oakey, R. J., Larin, Z., Fisher, R. B., Crocker, M., Affara, N. A. *et al.* (1993). Localization of DNA sequences required for human centromere function through an analysis of rearranged Y chromosomes. Nature Genet. *5*, 368–375.

Verma, R. A. and Luke, S. (1992). Variations in alphoid DNA sequences escape detection of aneuploidy at interphase by FISH technique. Genomics *14*, 113–116.

Vig, B. and Richards, B. (1992). Formation of primary constrictions and heterochromatin in mouse does not require minor satellite DNA. Exp. Cell Res. *201*, 292–298.

Vig, B. and Zinkowski, R. (1986). Sequence of centromere separation: a mechanism for orderly separation of dicentrics. Cancer Genet. Cytogenet. *22*, 347–359.

Vig, B. K., Latour, D., and Frankovich, J. (1994). Dissociation of minor satellite from the centromere in mouse. J. Cell Sci. *107*, 3091–3095.

Vissel, B. and Choo, K. (1989). Mouse major (gamma) satellite DNA is highly conserved and organized into extremely long tandem arrays: implications for recombination between nonhomologous chromosomes. Genomics *5*, 407–414.

Vissel, B. and Choo, K. H. (1991). Four distinct alpha satellite subfamilies shared by human chromosomes 13, 14 and 21. Nucleic Acids Res. *19*, 271–277.

Vissel, B., Nagy, A., and Choo, K. H. A. (1992). A satellite III sequence shared by human chromosomes 13, 14 and 21, that is contiguous with alpha satellite DNA. Cytogenet. Cell Genet. *61*, 81–86.

Voullaire, L. E., Slater, H. R., Petrovic, V., and Choo, K. H. A. (1993). A functional marker centromere with no detectable alpha-satellite, satellite III, or CENP-B protein: activation of a latent centromere? Am. J. Hum. Genet. *52*, 1153–1163.

Wake, R. and Errington, J. (1995). Chromosome partitioning in bacteria. Annu. Rev. Genet. *29*, 41–67.

Waterston, R., Ainscough, R., Anderson, K., Berks, M., Blair, D.*et al.* (1993). The genome of the nematode *Caenorhabditis elegans*. Cold Spring Harbor Symp. Quant. Biol. *LVIII*, 367–377.

Waye, J. S. and Willard, H. F. (1985). Chromosome-specific alpha satellite DNA: nucleotide sequence analysis of the 2.0 kilobasepair repeat from the human X chromosome. Nucleic Acids Res. *13*, 2731–2743.

Waye, J. S. and Willard, H. F. (1986). Structure, organization, and sequence of alpha satellite DNA from human chromosome 17: evidence for evolution by unequal crossing-over and an ancestral pentamer repeat shared with the human X chromosome. Mol. Cell. Biol. *6*, 3156–3165.

Waye, J. S. and Willard, H. F. (1987). Nucleotide sequence heterogeneity of alpha satellite repetitive DNA: a survey of alphoid sequences from different human chromosomes. Nucleic Acids Res. *15*, 7549–7569.

Waye, J. S. and Willard, H. F. (1989a). Human beta satellite DNA: genomic organization and sequence definition of a class of highly repetitive tandem DNA. Proc. Natl Acad. Sci. USA *86*, 6250–6254.

Waye, J. S. and Willard, H. F. (1989b). Chromosome specificity of satellite DNAs: short- and long-range organization of a diverged dimeric subset of human alpha satellite from chromosome 3. Chromosoma *97*, 475–480.

Waye, J. S., England, S. B., and Willard, H. F. (1987a). Genomic organization of alpha satellite DNA on human chromosome 7: evidence for two distinct alphoid domains on a single chromosome. Mol. Cell. Biol. *7*, 349–356.

Waye, J. S., Durfy, S. J., Pinkel, D., Kenwrick, S., Patterson, M., Davies, K. E. *et al.* (1987b). Chromosome-specific alpha satellite DNA from human chromosome 1: hierarchical structure and genomic organization of a polymorphic domain spanning several hundred kilobase pairs of centromeric DNA. Genomics *1*, 43–51.

Waye, J. S., Creeper, L. A., and Willard, H. F. (1987c). Organization and evolution of alpha satellite DNA from human chromosome 11. Chromosoma *95*, 182–188.

Waye, J. S., Mitchell, A. R., and Willard, H. F. (1988). Organization and genomic distribution of "82H" alpha satellite DNA. Evidence for a low-copy or single-copy alphoid domain located on human chromosome 14. Hum. Genet. *78*, 27–32.

Wevrick, R. and Willard, H. F. (1991). Physical map of the centromeric region of human chromosome 7: relationship between two distinct alpha satellite arrays. Nucleic Acids Res. *19*, 2295–2301.

Wevrick, R., Willard, V. P., and Willard, H. F. (1992). Structure of DNA near long tandem arrays of alpha satellite DNA at the centromeres of human chromosome 7. Genomics *14*, 912–923.

Willard, H. F. (1985). Chromosome-specific organization of human alpha satellite DNA. Am. J. Hum. Genet. *37*, 524–532.

Willard, H. F. (1991). Evolution of alpha satellite. Curr. Opin. Genet. Dev. *1*, 509–514.

Willard, H. F. and Waye, J. S. (1987). Hierarchical order in chromosome-specific human alpha satellite DNA. Trends Genet. *3*, 192–198.

Willard, H. F., Smith, K. D., and Sutherland, J. (1983). Isolation and characterization of a major tandem repeat family from the human X chromosome. Nucleic Acids Res. *11*, 2017–2033.

Witney, F. and Furano, A. (1983). The independent evolution of two closely related satellite DNA elements in rats (*Rattus*). Nucleic Acids Res. *11*, 291–304.

Wolfe, J., Darling, S. M., Erickson, R. P., Craig, I. W., Buckle, V. J., Rigby, P. W. *et al.* (1985). Isolation and characterization of an alphoid centromeric repeat family from the human Y chromosome. J. Mol. Biol. *182*, 477–485.

Wong, A. and Rattner, J. (1988). Sequence organization and cytological localization of the minor satellite of mouse. Nucleic Acids Res. *16*, 11645–11661.

Wong, A., Biddle, F., and Rattner, J. (1990). The chromosomal distribution of the major and minor satellite is not conserved in the genus *Mus*. Chromosoma *99*, 190–195.

Wu, J. S. and Kidd, K. K. (1990). Extensive sequence polymorphisms associated with chromosome 10 alpha satellite DNA and its close linkage to markers from the pericentromeric region. Hum. Genet. *84*, 279–282.

Wu, J. C. and Manuelidis, L. (1980). Sequence definition and organization of a human repeated DNA. J. Mol. Biol. *142*, 363–386.

Yang, T. P., Hansen, S. K., Oishi, K. K., Ryder, O. A., and Hamkalo, B. A. (1982). Characterization of a cloned repetitive DNA sequence concentrated on the human X chromosome. Proc. Natl Acad. Sci. USA *79*, 6593–6597.

Yu, L., Lowensteiner, D., Wong, E., Sawada, I., Mazrimas, J., and Schmid, C. (1986). Localization and characterization of recombinant DNA clones derived from the highly repetitive DNA sequences in the Indian muntjac cells: their presence in the Chinese muntjac. Chromosoma *93*, 521–528.

Yurov, Y. B., Mitkevich, S. P., and Alexandrov, I. A. (1987). Application of cloned satellite DNA sequences to molecular-cytogenetic analysis of constitutive heterochromatin heteromorphisms in man. Hum. Genet. *76*, 157–164.

6

Centromere proteins of higher eukaryotes

In this chapter, the protein components of the higher eukaryotic centromere are described. These protein components can be broadly classified into two groups. Proteins from the first group are present on the centromere throughout the cell cycle and are referred to as constitutive centromere proteins. Members of this group of centromere proteins are CENP-A, CENP-B, CENP-C, and (maybe) CENP-D. The second group of proteins have been referred to as 'passenger' proteins, since these proteins undergo complex relocations to other cellular organelles during the cell cycle, appearing on the centromere only during specific stages of the cycle (Brinkley *et al.* 1992; Earnshaw and Mackay 1994). To date, more than 30 passenger proteins have been described, although not all have been studied to the required degree of depth to be functionally informative. In this chapter, an attempt has been made to subgroup these different proteins according to their most outstanding proposed function, as follows: (i) centromere assembly, (ii) M phase and checkpoint control, (iii) sister chromatid cohesion, (iv) release of sister chromatid cohesion, (v) chromosome movement, (vi) spindle dynamics, (vii) cytokinesis, and (viii) proteins with undetermined centromeric roles (Tables 6.1 and 6.2). It is, however, noteworthy that whilst this subgrouping provides a point of reference for each protein, some of the proteins may have more than one possible role in the cell cycle, and as such the subgrouping should not be considered unequivocal in some cases. The chapter begins with a definition of the CREST autoimmune antisera, since these antisera have played an important part in the study of centromere proteins.

CREST autoimmune antisera

The discovery of centromere-specific autoantibodies in sera of patients with a variant form of the human autoimmune disease scleroderma has been instrumental in the rapid expansion of our knowledge of the protein composition of the human centromere–kinetochore complex (Fritzler and Kinsella 1980; Moroi *et al.* 1980, 1981; Brenner *et al.* 1981). Clinically, scleroderma (also known as progressive systemic sclerosis) is a chronic systemic rheumatic disease that can affect many organ systems including the skin and subcutaneous tissues, the gastrointestinal tract, the heart, the

Table 6.1 Protein components of the higher eukaryotic centromeres

Protein[1]	Mol. wt (kDa)[2]	Centromere sublocalization	Possible functions	Species
Assembly				
CENP-A	17	Kinetochore chromatin (outer kinetochore domain)	Centromere-specific core histone	Humans, cattle, muntjac, hamster, chicken, rat
CENP-B	80	Central domain (heterochromatin)	α-Satellite assembly	A variety of mammals, chicken
CENP-C	140	Inner kinetochore plate	Kinetochore assembly; cell cycle control in G_1	Humans, mouse, monkey, hamster, amphibian
CENP-F (Mitosin) (p330[d])	400 (350) (330)	Outer surface of kinetochore	Nuclear reorganization at G_2; kinetochore maturation and function from G_2 to anaphase	A variety of mammals
HMG-I	10	n.a. (central domain?)	Nucleoprotein complex formation; α-satellite DNA compaction	Wide range of mammals and nonmammals
pJα	10–15	n.a. (central domain?)	α-Satellite assembly?	Primates
AC1 antigen	170	Ring around centromere	Centromere stabilization	Hamster, humans
M phase control				
p34[cdc2](cdc2)	34	Kinetochore	Master control of mitosis progression and exit	All eukaryotes
MPM2	>40 proteins	Inner and outer kinetochore plates	M phase induction and regulation	All eukaryotes

Protein	Size (kDa)	Location	Function	Organism
Checkpoint				
3F3/2	n.a.	Middle kinetochore plate	Tension-sensitive checkpoint for metaphase to anaphase transition	Mammals, *Drosophila*
ZW10	n.a.	Kinetochore	Tension-sensitive regulation of anaphase onset	*Drosophila*, humans
Cohesion				
INCENP$_I$ INCENP$_{II}$	133 145	Pairing domain	Sister chromatid cohesion; cytoplasmic microtubule bundling; cytokinesis	Birds, mammals
CLiPs	50 and others	Inner centromere	Sister chromatid pairing	Muntjacs, humans, mouse
Mei-S332	44	n.a. (pairing domain?)	Meiotic sister centromere cohesion	*Drosophila*
Ord	55	n.a. (pairing domain?)	Sister chromatid cohesion in meiosis and germline mitosis	*Drosophila*
Cor1	30	n.a. (pairing domain?)	Meiotic sister chromatid and sister centromere cohesion	Hamster, rat
Release				
Topo IIα	170	Throughout centromere	Centromere maturation; release of sister chromatid cohesion; many aspects of nucleic acids metabolism	All eukaryotes
PIM	n.a.	n.a. (pairing domain?)	Release of sister chromatid cohesion	*Drosophila*
Movement				
CENP-E (154 antigen)	275 (250)	Periphery of kinetochore	Chromosome movement; microtubule crosslinking; couplers to depolymerizing microtubules	Humans
MCAK	90	Throughout centromere and between kinetochore plates	Microtubule dynamics modulation and anchor onto kinetochore; tubulin subunit exchange between sister centromeres	All eukaryotes

Table 6.1 *contd*

Protein[1]	Mol. wt (kDa)[2]	Centromere sublocalization	Possible functions	Species
Movement *contd*				
Kid	73*	n.a.	Mitotic spindle formation; chromosome movement along microtubule; centromere DNA binding	Humans, monkey
Cytoplasmic dynein	> 1,000	Corona of kinetochore	Chromosome movement; spindle dynamics; many diverse cellular functions	All eukaryotes
Spindle dynamics				
Tubulin	50	On corona	Major component of spindle microtubule	Probably all species
NuMA (Centrophilin)	236*	Kinetochore or corona	Mitotic spindle dynamics; post-mitotic nuclear reformation	Mammals, fish, bird
Cytokinesis				
TD-60	60	Central domain	Cytokinesis	Humans, hamster, cattle, mouse
Others				
CENP-D (RCC1)	47	n.a.	Guanine exchange factor for Ran; nucleosome binding	All eukaryotes

n.a. = data not available.
[1] Not included in this table are a number of the less well-characterized centromere-associated proteins: 4G10, JB, MSA-36, 9H8, 3G3, 3H1, 2D3, 1F1, S5–39, 37A5, and 32–9. Refer to the text for a discussion of these proteins.
[2] Determined by SDS–PAGE (or where indicated by an asterisk, predicted from the open reading frame).

lungs, and the kidneys. Examination of sera from different patients with scleroderma indicates that approximately 30% contain antibody to centromeres. The patients with the anti-centromere antibodies fall into a subgroup which has less extensive involvement of the skin than do those with diffuse scleroderma. This group of patients generally do not have sclerodermatous involvement of the trunk and the proximal portions of the extremities. However, their disease is characterized by prominence of calcinosis, Raynaud's phenomenon, esophageal dysmotility, sclerodactyly, and telangiectasia (a constellation of symptoms commonly referred to as CREST). Although the anti-centromere antibody is present in high frequency in CREST patients, not all patients with this syndrome will necessarily develop such an autoantibody (Moroi *et al.* 1980).

Use of CREST antisera has led to the direct identification of a host of important centromere-associated proteins that have included CENP-A, -B, -C, -D, -F, and others. A point of interest is that CREST antisera routinely exhibit cross-recognition of CENP-A and CENP-B (Earnshaw and Rothfield 1985, Rothfield *et al.* 1987; Weiner *et al.* 1991). The reason for this is not clear but it is not due to significant sequence homology between the two antigens (Palmer *et al.* 1991). CREST antisera have also been shown to cross-react with centromere proteins from a wide range of mammalian species, as well as kinetochores of some higher plants (Mole-Bajer *et al.* 1990; Palevitz 1990; Houben *et al.* 1995).

Centromere assembly

Centromere assembly occurs by the recruitment of *trans*-acting proteins. Seven proteins are thought to have a possible role in this process. Three of these proteins, CENP-A, CENP-B, and CENP-C, are specifically and constitutively present on the centromere, and as such are likely to be critical to both the interphase and mitotic/meiotic centromeres. Whether the spatial distribution and structural organization of these proteins are in any way modified during the different stages of the cell division cycle is presently not known. The other four proteins which may have a role in the assembly, maturation, and/or stabilization of a functional centromere are CENP-F, HMG-I, pJα, and AC1.

CENP-A

Centromere sublocalization and species conservation

Early work using CREST antisera identified CENP-A as a 17 kDa protein (Guldner *et al.* 1984; Earnshaw and Rothfield 1985; Valdivia and Brinkley 1985; McNeilage *et al.* 1986) that is specific for the human centromere (Guldner *et al.* 1984; Earnshaw and Rothfield 1985). Cloned human CENP-A cDNA reveals an open reading frame that encodes a protein of 140 amino

Table 6.2 Mitotic cell cycle distribution of centromeric passenger proteins

Protein[1]	Interphase	Prophase	Prometaphase	Metaphase	Anaphase	Telophase
Assembly						
CENP-F (Mitosin) (p330[d])	Nucleus and centromere (S–G_2 phases)	Nucleus; centromere	Centromere	Centromere	Centromere (anaphase A); spindle midzone	Midbody
HMG-I	Nucleus	n.a.	Chromosome	Centromere; telomere; chromosome	Overall level diminishing	Overall level diminishing
AC1 antigen	Centriole	n.a.	n.a.	Centromere; centriole	n.a.	n.a.
M phase control						
$p34^{cdc2}$(cdc2)	Nucleus; cytoplasm; centrosome (G_2)	Centrosome; nucleus; cytoplasm	Centromere; centrosome; spindle	Centromere; centrosome; spindle	Centrosome; spindle; midzone	Midbody; centrosome; reforming nuclei
MPM2	Nucleus; centriole	Centromere; centrosome; chromosome	Centromere; centrosome; chromosome	Centromere; centrosome; chromosome; spindle	Centromere (anaphase A); centrosome; chromosome; midzone	Centrosome; midbody
Checkpoint						
3F3/2	Nucleus	Centromere; centrosome	Centromere; centrosome	Centrosome	Centrosome	n.a.
ZW10	Nucleus	Nucleus	Centromere	Spindle	Centromere	Midbody
Cohesion						
INCENP$_I$ INCENP$_{II}$	Nucleus	Chromosome	Centromere	Metaphase plate (late metaphase)	Spindle midzone; cell cortex	Midbody; cell cortex
CLiPs	n.a.	Centromere; chromosome	Centromere; chromosome	Centromere; chromosome	Centromere	Centromere

Release							
Topo IIα	Throughout cell and centromere (S–G₂ phases)	Chromosome scaffold; centromere	Chromosome scaffold; centromere	Chromosome scaffold; centromere	Centromere; chromosome scaffold	Chromosome scaffold	n.a.
PIM	Cytoplasm	Throughout cell	Throughout cell	Throughout cell	Throughout cell; centromere	Decreasing level	Absent
Movement							
CENP-E (154 antigen)	Cytoplasm (from late S phase on)	Cytoplasm	Centromere	Centromere	Centromere	Centromere (anaphase A); spindle midzone	Midbody
MCAK	Cytoplasm; nucleus	Cytoplasm	Centromere	Centromere	Centromere	Centromere	Centromere
Kid	Nucleus	Centrosome; throughout nucleus	Centrosome; chromosome	Centromere; chromosome	Centrosome; chromosome	Centromere; centrosome; chromosome	Centrosome
Cytoplasmic dynein	Cytoplasm	Centromere; centrosome	Centromere; spindle pole fibres	Spindle between centromere and pole	Spindle between centromere and pole	Decreasing level in anaphase B	Decreasing level
Spindle dynamics							
Tubulins	Various organelles; absent on centromere	Various organelles; absent on centromere	Various organelles, including centromere	Various organelles, including centromere	Various organelles, including centromere	Various organelles, including centromere	Various organelles; absent on centromere
NuMA (Centrophilin)	Nucleus	Nucleus	Spindle; centromere	Spindle; centromere	Spindle; centromere	Spindle (polar end)	Reforming nuclei
Cytokinesis							
TD-60	Nucleus	Centromere; chromosome	Centromere; chromosome	Centromere; chromosome	Centromere; chromosome	Spindle midzone	Telophase disc; midbody

n.a. = data not available.

[1] Not included in this table are the four constitutive proteins (CENP-A, -B, -C, and -D, which are present at the centromere throughout the cell cycle), the three meiotic centromere proteins (Mei-S332, Ord and Cor1), and the less well-characterized proteins (4G10, JB, MSA-36, 9H8, 3G3, 3H1, 2D3, 1F1, S5–39, 37A5, and 32–9). Refer to the text for a discussion of these proteins.

acids (Sullivan *et al.* 1994). The protein has been shown by indirect immunofluorescence (Kingwell and Rattner 1987) and immunoelectron microscopy (Cooke *et al.* 1990) to be associated with kinetochore chromatin or with chromatin closely apposed to the outermost domain of the kinetochore. In addition to humans, CENP-A has been detected in bovine (Palmer *et al.* 1990) and Indian muntjac (Hadlaczky *et al.* 1986; Kingwell and Rattner 1987), and may also be present in Chinese hamster ovary, chicken, and rat cells (Valdivia and Brinkley 1985; Palmer *et al.* 1987). Comparison of a partial nucleotide sequence of human and bovine CENP-A reveals a high degree of conservation (90% identity at the nucleotide and 95% similarity at the amino acid levels). A protein, CSE4p, which shares homology of a critical domain with CENP-A has also been found in *S. cerevisiae* (see below).

Structural properties

A centromere-specific core histone

CENP-A shares a similar organization with the histone H3 protein (Sullivan *et al.* 1994). Histone H3, like the other core histones, possesses two domains: a flexible and highly basic N-terminal tail that is dispensable for nucleosome assembly (Allen *et al.* 1982) and viability in yeast (Mann and Grunstein 1992), and a globular C-terminal portion that assembles with histone H4 to form the proteinaceous core of the nucleosome (Arents *et al.* 1991; Richmond *et al.* 1984). The N-terminus of CENP-A shares the basic and flexible nature of the histone tail but no amino acid sequence similarity with histone H3 (Fig. 6.1).

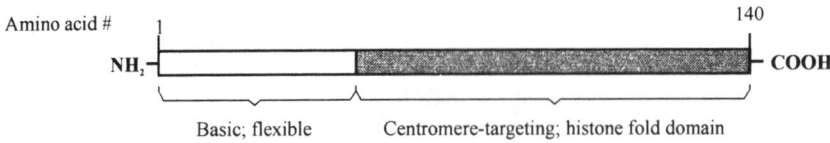

Fig. 6.1 Structural domains of human CENP-A protein, showing the basic and flexible N-terminus that shows no amino acid sequence similarity with histone H3, and the histone fold domain that shares 62% identity with human histone H3 and is responsible for centromere targeting.

The C-terminal domain of histone H3 is folded into an extended dumbbell-shaped structure termed the histone fold domain that typifies all four core histones (Arents *et al.* 1991). The homology between CENP-A and histone H3 begins abruptly at the border between the N-terminal domain and the histone fold domain. The histone fold domain of human CENP-A is 62% identical to that of human histone H3 (Sullivan *et al.* 1994). Since CENP-A is found in association with histone H4 and the other core histones in particles that co-purify with nucleosome core particles (Palmer and Margolis 1985; Palmer *et al.* 1987), it is assumed that CENP-A acts as a histone H3 homologue replacing one or both copies of histone H3 in a certain set of centromeric nucleosomes.

The major function of the core histones is to bind to DNA, folding it across the nucleosome surface. In particular, nucleosomal DNA makes several contacts with histone H3 in its path across the surface of the histone octamer (Mirzabekov *et al.* 1978; Arents and Moudrianakis 1993). The high degree of sequence identity seen between CENP-A and histone H3 suggests that CENP-A nucleosomes closely resemble normal nucleosomes. The sequences that correspond to the positions where DNA enters and exits the nucleosome are highly conserved between CENP-A and histone H3. However, CENP-A is diverged from a conserved region of histone H3 that is found near the nucleosome two-fold axis (Camerini-Otero and Felsenfeld 1977); the positioning of nucleosomes on DNA *in vitro* is facilitated by placing intrinsically bent or flexible DNA near this dyad axis (Constanzo *et al.* 1990; Schrader and Crothers 1990). CENP-A is therefore differentiated from histone H3 in a region that may be involved in nucleosome–DNA recognition.

Centromere-targeting domain

Since CENP-A is detected only at centromeres, it might be expected that some portion of the protein is responsible for this site-specific targeting. Sullivan *et al.* (1994) tested the relative involvement of the basic N-terminal domain and the C-terminal histone fold domain by constructing chimeric molecules composed of CENP-A and histone H3. The expectation was that the N-terminus, being unique, would be required for centromere localization while the conserved histone fold domain would provide the structure needed for nucleosome assembly. Surprisingly, however, the N-terminus was incapable of selectively directing the histone fold domain of H3 to the centromere. Rather, the C-terminal histone fold domain of CENP-A is itself sufficient for selective assembly at the centromere (Fig. 6.1). This selective assembly or interaction could occur either through direct recognition of centromeric DNA, or by binding to one or more proteins that recognize the DNA. The possibility that CENP-B may mediate the localization of CENP-A to the centromere has been tested and excluded (Sullivan *et al.* 1994). Although the core histone-related nature of CENP-A would favour a mechanism of direct centromere DNA binding, direct experimental proof of this is required.

A new histone H3-like class of protein

The sequence arrangement of CENP-A places it into a growing class of proteins referred to as histone H3-like proteins. This class of protein now includes two putative proteins from *C. elegans* (Wilson *et al.* 1994) and a protein (CSE4p) from *S. cerevisiae* (Stoler *et al.* 1995; Chapter 2). Like

CENP-A, a 98 amino acid domain of CSE4p and one of the *C. elegans* homologues show 64% and 52% identity, respectively, with the histone fold domain of histone H3. Several distinguishing features of the histone-like proteins are suggestive of these being a new class of proteins (Basrai and Hieter 1995). For example, the histone H3-like proteins have insertions of 2–3 amino acids at the same relative positions, and there is substitution of a conserved phenylalanine of histone H3 with tryptophan, an amino acid normally absent in all of the core histones.

Role of CENP-A

Studies on CENP-A suggest that modification of chromatin through incorporation of divergent core histone may be an important theme in chromosome structure evolution. The unique structure of CENP-A indicates that the centromere is differentiated from the rest of the chromosome at the most fundamental level of chromatin structure—the nucleosome. The finding that CSE4p, the *S. cerevisiae* homologue of CENP-A, is required for proper chromosome segregation (Stoler *et al.* 1995) implies that the modified centromeric nucleosomes have an important role in centromere function. It is unlikely that the modified nucleosomes are required for the actual assembly of microtubule-dependent motor proteins onto the chromosome, since these interactions can be reconstituted *in vitro* using naked DNA (Hyman *et al.* 1992; Middleton and Carbon 1994). The specific configuration of chromatin at the centromere could be important for insulating the centromere from transcriptional activity or provide a necessary component of the sister chromatid pairing mechanism. Alternatively, the modified centromeric chromatin could provide the chromatin fibre with the mechanical stability necessary to integrate the vast excess of forces (Nicklas 1988) generated at the kinetochore with the chromosome scaffold to effect chromosome movements (Sullivan *et al.* 1994).

The centromere specificity of CENP-A makes it unique amongst somatic histones. In addition, this centromere-specific histone has the unusual property that it is quantitatively retained in chromatin during spermatogenesis in bulls (Palmer *et al.* 1990). During mammalian spermatogenesis, somatic histones are ultimately replaced by cysteine- and arginine-rich sperm-specific basic peptides called protamines, a process resulting in loss of the nucleosomal organization of chromatin and its replacement by a highly compacted structure characteristic of mature sperm (reviewed by Grimes 1986). Although the degree of replacement varies among different species, in most mammals the portion of the genome that remains as nucleohistone is < 2% (O'Brien and Bellvé 1980; Balhorn 1982; Uschewa *et al.* 1982; Moss *et al.* 1989); in humans, somewhat higher levels (~ 15%) of the somatic and testes-specific histones are retained (Gatewood *et al.* 1987; Tamphaichitr *et al.* 1978). It is therefore striking that CENP-A is retained during spermatogenesis in bulls, where replacement of somatic histones is essentially complete. Furthermore, indirect immunofluorescence studies indicate that CENP-A is retained in sperm nuclei in discrete foci, rather than being dispersed throughout the sperm head. This observation, in conjunction with demonstration in rat sperm that centromeric satellite DNA is organized in foci (Moens and Pearlman 1989), suggests continued association of CENP-A with the centromere in the sperm head. Thus, pre-existing CENP-A–centromere interactions are likely to be important in organizing the centromeres of the paternal genome during early embryogenesis.

CENP-B

Centromere sublocalization

Human CENP-B is a relatively abundant protein that is specific for the centromere (Earnshaw and Rothfield 1985). Immunofluorescence analysis reveals a broad and variable staining of the centromere on different human metaphase chromosomes (Earnshaw *et al.* 1987b). This result is quite different from those obtained for most of the other centromere-binding proteins, where two distinct dots of relatively constant intensity are generally observed. By immunoelectron microscopy, > 95% of CENP-B is found to be distributed throughout the centromeric heterochromatin (central domain) beneath the kinetochore (Cooke *et al.* 1990; Saitoh *et al.* 1992). Confirmation of heterochromatin association comes from the demonstration that CENP-B is capable of binding (discussed below) to a subset of the high-abundance centromeric α-satellite DNA.

Structural properties

Human CENP-B is a 599 amino acid polypeptide with a molecular weight of 80 kDa (Earnshaw and Rothfield 1985; Earnshaw *et al.* 1987b; Sullivan and Glass 1991). It is encoded by an intronless gene and is present as a single copy in the genome on chromosome 20p13 (Sugimoto *et al.* 1993; Seki *et al.* 1994). The protein contains two functionally distinct domains that have been well characterized. These are the N-terminal DNA-binding domain and the C-terminal dimerization domain. In addition, a third, internal self-association domain has been described (Fig. 6.2).

DNA-binding (CENP-B box) domain

In agreement with the ultrastructural localization of CENP-B to the heterochromatic central domain of the centromere, CENP-B has been shown to bind directly to human α-satellite DNA (Masumoto *et al.* 1989; Muro *et al.* 1992; Pluta *et al.* 1992; Yoda *et al.* 1992). This binding is achieved through the recognition of a 17 bp sequence [CT(T/A)(C/T)G(T/ G)TGGAAA(C/A)GG(G/A)A], known as the CENP-B box motif, within the 171-bp α-satellite DNA monomer. This DNA-binding/centromere-localization domain is present at the N-terminal end (Sullivan and Glass 1991; Pluta *et al.* 1992; Yoda *et al.* 1992; Sugimoto *et al.* 1994a; Kitagawa *et al.* 1995) in a 125 amino acid region (Yoda *et al.* 1992) that is perfectly conserved between human, mouse, and African green monkey CENP-B (Sullivan and Glass 1991; Yoda *et al.* 1996). Within this 125 amino acid region, four α-helices have been predicted. Deletion of helix 1 or helix 4 results in the abolition of DNA-binding activity, suggesting that a tertiary structure composed of the α-helices is necessary for this activity (Yoda *et al.* 1992).

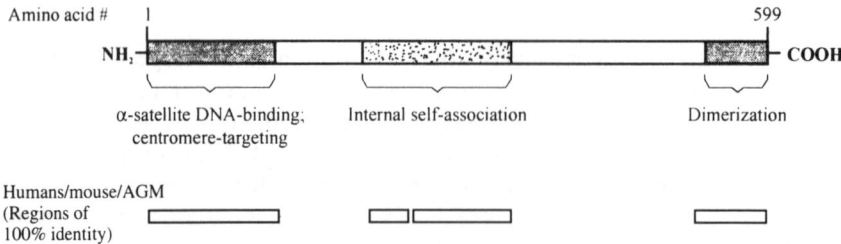

Fig. 6.2 Structural organization of human CENP-B protein, showing the domains for α-satellite DNA-binding/centromere-targeting, internal self-association, and dimerization. The boxes below represent regions that are perfectly conserved between human, mouse, and African green monkey (AGM) CENP-B.

The presence of α-helices suggests that CENP-B might be a member of the HLH family of DNA binding protein. CENP-B, however, differs from the typical HLH family of proteins in a number of significant ways: (i) the DNA-binding domain of CENP-B contains four potential α-helices, whereas HLH proteins have two; (ii) the DNA sequence of the CENP-B binding site, CENP-B box, is longer than that of the HLH protein binding site (Pabo *et al.* 1992); and (iii) the DNA-binding activity of CENP-B has been shown to be independent of dimerization (see below), whereas in HLH proteins the HLH motif and upstream basic region mediates both DNA binding as well as dimerization (Davis *et al.* 1990; Pabo and Sauer 1992). These observations indicate that CENP-B does not belong to the HLH family and that it is a new type of DNA-binding protein (Yoda *et al.* 1992; Kitagawa *et al.* 1995).

Dimerization domain

CENP-B has been shown to form a stable and unique complex *in vitro* as a dimer containing two DNA molecules (Muro *et al.* 1992; Yoda *et al.* 1992). This dimerization activity is located in a 59 amino acid segment in the C-terminal region which is separable from the N-terminal DNA-binding domain (Muro *et al.* 1992; Yoda *et al.* 1992; Sugimoto *et al.* 1994a; Kitagawa *et al.* 1995) and which, like the DNA-binding domain, is totally conserved between the human, mouse, and African green monkey proteins (Sullivan and Glass 1991; Yoda *et al.* 1996). Since the dimerization domain is relatively rich in hydrophobic amino acids, protein–protein interaction may be assumed to be based on hydrophobic association. Computer analysis predicts two potential α-helices which, when either is deleted results in the abolition of dimerization activity (Kitagawa *et al.* 1995).

Internal self-association domain

By electrophoretic mobility shift assay and an assay involving the use of a chemical cross-linking reagent to fix physically associated CENP-B proteins, an internal region of the protein that undergoes self-association has also been identified (Sugimoto *et al.* 1994a). This region includes two separable segments that are proposed to act in concert to produce higher-order protein assembly (Sugimoto *et al.* 1994a). The DNA sequence within these segments is again perfectly conserved in human, mouse, and African green monkey CENP-B (Sullivan and Glass 1991; Yoda *et al.* 1996). Secondary structure prediction indicates that this domain contains a series of alternating β-sheets and α-helices (Sullivan and Glass 1991).

Phosphorylation of CENP-B during mitosis

CENP-B is modified specifically at or just preceding metaphase (Kitagawa *et al.* 1995). This modification has been shown to be due to protein phosphorylation. In agreement with this, CENP-B has a number of phosphorylation target sites for mitogen-activated protein kinase and casein kinase II. The metaphase-specific phosphorylation does not appear to have an effect on the activity of the DNA-binding or dimerization domains. It is possible that such a cell cycle-specific modification is related to other activities of CENP-B, such as interaction with other proteins.

Chromosomal distribution of CENP-B

Genomic abundance

Based on quantitative protein purification and activity recovery, the number of CENP-B molecules per HeLa cell has been estimated to be $\sim$ 20 000 (Muro *et al.* 1992). This value agrees well with that of $\sim$ 375 copies per typical chromatid (or $\sim$ 17 000 per diploid human genome) obtained independently by immunoelectron microscopic analysis (Cooke *et al.* 1990). On different human chromosomes, a wide variation in the level of CENP-B protein has been observed (Earnshaw *et al.* 1987b). An extreme example is the Y chromosome which, despite the presence of α-satellite DNA, consistently lacks the CENP-B boxes, and thus any detectable amounts of CENP-B protein. In addition to Y, unusual chromosomal variants or marker chromosomes that are devoid of both α-satellite DNA and CENP-B proteins have also been described (Chapter 7).

Prevalence of the CENP-B box motif in α-satellite DNA

Choo *et al.* (1991) analysed the sequences of ∼ 300 different α-satellite DNA monomers, representing 33 different subfamilies from all 24 different human chromosomes, and reported a consensus sequence of this DNA (Chapter 5). Given the actual frequency of nucleotides observed at each position of this consensus, the probability of a CENP-B box occurring in a given α-satellite DNA monomer is 0.17%, or approximately once in 600 monomers (Pluta *et al.* 1992). However, this frequency is likely to be a gross under-estimation since there is a general bias for the more heterogeneous α-satellite DNA subfamilies to be sequenced and reported, and also because these hetero-geneous subfamilies tend to have a significantly lower prevalence of the CENP-B box compared to the more abundant and homogeneous subfamilies (e.g. Masumoto *et al.* 1989; Ikeno *et al.* 1994).

Distribution within a centromere and a model for α-satellite DNA compaction

The distribution of CENP-B protein within the centromeres of human chromo-somes 7 and 21 have been determined. As described earlier (Chapter 5), the centromeres of these chromosomes each consists of two distinguishable α-satellite DNA domains (Wevrick and Willard 1989, 1991; Wevrick *et al.* 1992; Trowell *et al.* 1993, Ikeno *et al.* 1994). On these chromosomes, the larger and more homogeneous domains (D7Z1 and D21Z1) have been shown to bind strongly to CENP-B proteins, whereas the smaller and more heterogeneous domains (D7Z2 and α21-II) both show an absence or a greatly reduced level of binding to this protein (Haaf and Ward 1994; Ikeno *et al.* 1994). In agreement with this, sequence determination has demonstrated a high prevalence of the CENP-B box motif amongst the α-satellite DNA monomers of D7Z1 and D21Z1, with this motif being only infrequently present in monomers associated with D7Z2 and α21-II [Fig. 6.3, A (ii) and B (ii)] (Waye *et al.* 1987; Choo *et al.* 1988; Wevrick and Willard 1991; Vissel and Choo 1992; Vissel *et al.* 1992; Ikeno *et al.* 1994).

Fig. 6.3 Distribution of CENP-B on (A) human chromosome 7, and (B) human chromosome 21. (i) Long-range maps of α-satellite domains. (ii) Properties of the different α-satellite DNA domains. (iii) A hypothetical model on how CENP-B proteins (open circles) may participate in folding α-satellite DNA arrays by binding to CENP-B box motifs (solid squares) on α-satellite monomers via their N-terminal centromere-targeting domain and forming homodimers via their C-terminal dimer-ization domain. Open arrowheads indicate the 171 bp monomeric units of α-satellite DNA. (iv) Further compaction of the folded α-satellite DNA by multimerization of homodimers through cross-linking at the internal self-association domain.

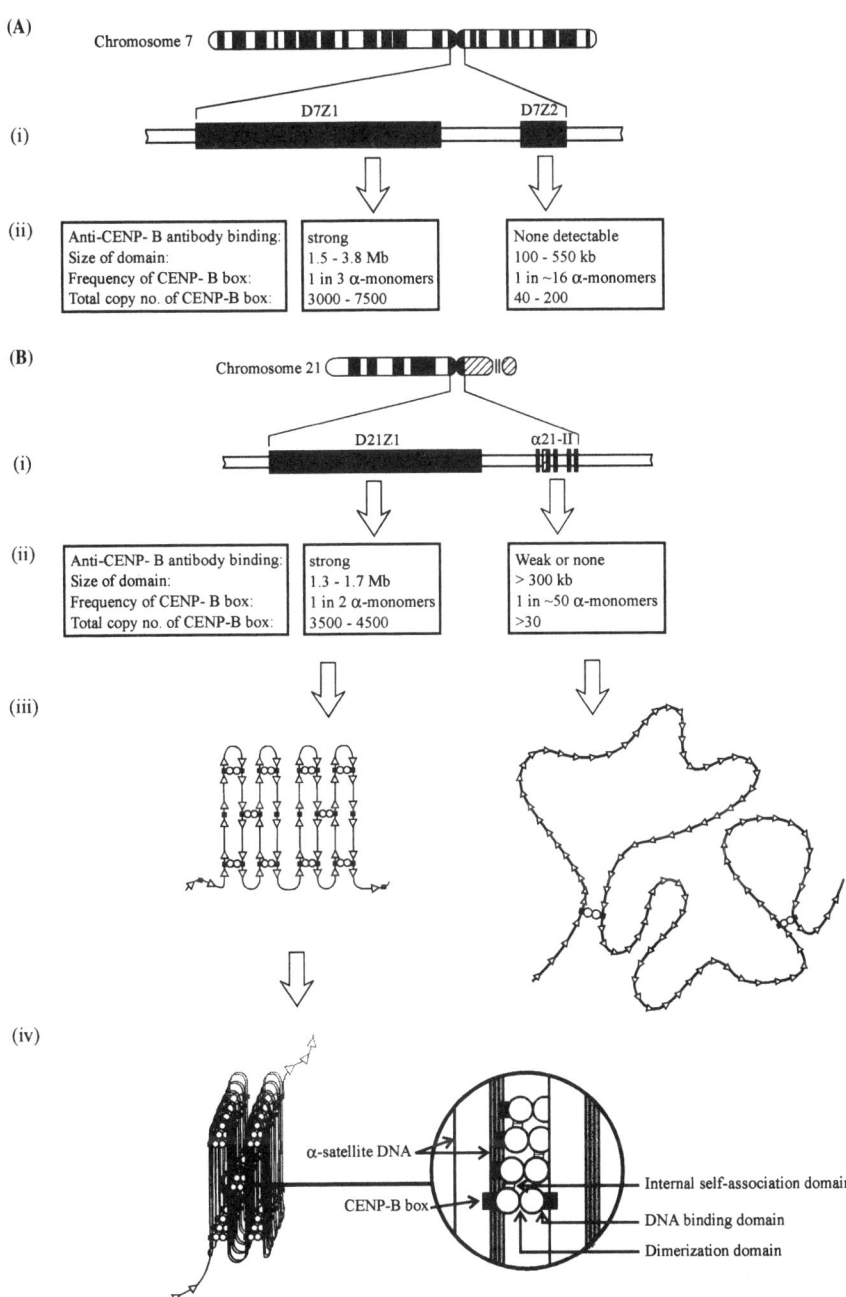

(A)

Chromosome 7

(i) D7Z1 D7Z2

(ii)
Anti-CENP- B antibody binding:	strong	None detectable
Size of domain:	1.5 - 3.8 Mb	100 - 550 kb
Frequency of CENP- B box:	1 in 3 α-monomers	1 in ~16 α-monomers
Total copy no. of CENP-B box:	3000 - 7500	40 - 200

(B)

Chromosome 21

(i) D21Z1 α21-II

(ii)
Anti-CENP- B antibody binding:	strong	Weak or none
Size of domain:	1.3 - 1.7 Mb	> 300 kb
Frequency of CENP- B box:	1 in 2 α-monomers	1 in ~50 α-monomers
Total copy no. of CENP-B box:	3500 - 4500	>30

(iii)

(iv)

α-satellite DNA

CENP-B box

Internal self-association domain

DNA binding domain

Dimerization domain

Fig. 6.3 [B (iii) and B (iv)] depicts a model for the compaction of the CENP-B box-rich D21Z1 domain by CENP-B proteins. The model is based on the proposed DNA-binding, dimerization, and internal self-association properties of the CENP-B protein. Such a model is probably unlikely to be adequate for other CENP-B box-poor α-satellite DNA arrays, such as those present within D7Z2 and α21-II domains or on the Y chromosome. The folding of these CENP-B box-poor arrays may therefore require a different mechanism, possibly involving other CENP-B box-independent α-satellite DNA binding proteins. One such candidate protein is pJα (Gaff *et al.* 1994; see below). This is a novel nuclear protein which, by gel electrophoretic mobility shift assay, has been shown to specifically recognize a non-CENP-B box motif that is found in a substantial proportion of α-satellite monomers Another potential candidate protein is HMG-I (discussed below).

Timing of α-satellite replication and CENP-B duplication

When the relative replication order of the two chromosome 7 α-satellite DNA arrays (D7Z1 and D7Z2) is examined in interphase nuclei, the results indicate that (i) although D7Z1 is replicated prior to D7Z2 in the majority of cells, the replication timing of one array relative to the other is variable, and (ii) the replication of α-satellite arrays on homologous chromosomes is highly asynchronous (Haaf and Ward 1994). These variable and highly asynchronous patterns of replication may reflect the repetitive nature of α-satellite DNA and differences in the relative abundance of origins of replication within individual arrays. The study further shows that fully replicated α-satellite DNA generally displays only a single CENP-B immunofluorescence signal, suggesting that duplication of CENP-B proteins is either temporally and/or mechanistically separated from centromeric DNA replication. The result agrees with an earlier postulation based on immunofluorescent studies that morphological duplication of centromeric proteins occurs during late G_2 phase or mitotic prophase (Brenner *et al.* 1981; Haaf and Schmid 1989). Thus, α-satellite DNA replication and CENP-B duplication are two separable interphase events.

Evolutionary conservation

Comparison of cloned human, mouse, and African green monkey CENP-B gene sequences reveals a high degree of homology between the three species (Sullivan and Glass 1991; Yoda *et al.* 1996). Within the coding region, there are substantial stretches that show 100% identity between these species (Fig. 6.2). Like their human counterpart, the mouse and African green monkey genes are single copy and intronless. Their coding regions show an overall 96% (mouse) and 98% (African green monkey) sequence similarity to the human protein, with the 5′ and 3′ untranslated sequences of the human and mouse mRNA also demonstrating 95% and 83% conservation, respectively (Sullivan and Glass 1991). This level of conservation at the untranslated regions is surprising and may be correlated with post-transcriptional mechanisms of gene regulation (e.g. Caput *et al.* 1986; Mullner and Kühn 1988).

Although the mouse genome does not contain recognizable α-satellite DNA, CENP-B binding is possible because of the presence of the 17 bp consensus CENP-B box motif in the mouse centromeric minor satellite DNA (Chapter 5). The centromere of African green monkey, on the other hand, contains large amounts of α-satellite DNA but has a low (Yoda *et al.* 1996) or no detectable (Goldberg *et al.* 1996) level of binding to the CENP-B protein. Furthermore, direct sequencing of 100 independent African green monkey α-satellite DNA monomers indicates that neither the canonical 17-bp CENP-B box sequences nor the nine core CENP-B recognition bases are present (Goldberg *et al.* 1996; Yoda *et al.* 1996), although a CENP-B box-like sequence with CENP-B affinity has been identified within the genome of this species (Yoda *et al.* 1996). These results reveal a paradox in that CENP-B is highly conserved in human, mouse, and African green monkey genomes, yet its DNA binding site is conserved in humans and mouse but not in African green monkey (Goldberg *et al.* 1996).

In addition to human, mouse, and African green monkey genomes, the CENP-B gene has been detected in the genomes of chicken (Sullivan and Glass 1991), great ape, tupaias (small squirrel-like mammal) (Haaf and Ward 1995), calf, Indian muntjac (Yoda *et al.*) and possibly *Drosophila* (Avides and Sunkel 1994). Evidence also suggests that CENP-B may have functional homologues in the budding yeast *S. cerevisiae* and the fission yeast *S. pombe*. In *S. cerevisiae* (Chapter 2), the centromeric CBF3C protein has been shown to contain a short, 30 amino acid stretch of acidic serine-rich region that is approximately 40% identical to the first acidic block found in CENP-B, whereas in *S. pombe* (Chapter 3), the functionally important centromeric K repeat contains a relatively large number of a TGGAAA motif that resembles part of the human CENP-B box sequence. However, despite this seemingly widespread conservation, fluorescence *in situ* hybridization has failed to detect the CENP-B box motif in lower primate species, rodents other than mouse, as well as in many other mammals examined (Haaf and Ward 1995). Interestingly, a significant homology has been found between CENP-B and two proteins of unrelated functions: (i) the *S. pombe ARS* (or autonomously replicating sequence)-binding protein 1 (Abp1) (25% identity, 50% similarity) that is involved in the regulation of DNA replication (Murakami *et al.* 1996); and (ii) the murine *jerky* protein (28% identity, 50% similarity), deficiency of which causes epileptic seizures (Toth *et al.* 1995). In both cases, the homology is greatest at the N-terminal DNA-binding portion. In addition to these two proteins, CENP-B shows a substantial resemblance to the transposases, Trigger-1 and -2, suggesting that CENP-B may have been derived from a transposase, followed by lateral transfer between diverse species (Smit and Riggs 1996).

Role of CENP-B

Evidence in favour of CENP-B having a role in centromere function

(i)   At the structural level, CENP-B has all the hallmarks of a protein that is involved in the assembly of the large arrays of centromeric α-satellite DNA. The unusually high degree of sequence homology in the coding and noncoding regions of the CENP-B genes in humans and mouse, and the detection of possible homologues of this gene and/or the CENP-B box motif in species as diverse as chicken, tupaias, *Drosophila*, and yeast, are suggestive of conservation of an important function.

(ii)   CENP-B binding is detected at the centromeres formed following transfection of mammalian cells with cloned α-satellite DNA (Haaf *et al.* 1992; Larin *et al.* 1994).

(iii) Microinjection of CREST anti-CENP-B antibodies into both human and mouse cells results in the disruption of centromere assembly during interphase, leading to the inhibition of kinetochore morphogenesis and function in mitosis (Bernat *et al.* 1990; Simerly *et al.* 1990; Bernat *et al.* 1991). However, unequivocal interpretation of these microinjection data is difficult because the CREST antisera used in these experiments contain varying titres of antibodies recognizing other known CENPs and possibly some as yet unidentified centromeric antigens. It is also difficult to separate out any steric effect bound antibodies may have on the structural and functional properties of the centromere in this type of experiments.

Evidence against a direct role of CENP-B in centromere function

(i)   Both the CENP-B protein and its associated CENP-B box DNA motif are not detectable on the Y chromosomes of humans and mouse.

(ii)   Despite the presence of extensive α-satellite arrays, the centromeres of African green monkey show an absence of the consensus CENP-B box motif and bind CENP-B poorly.

(iii) The protein is present at both the active and inactive centromeres of mitotically stable dicentric human chromosomes (Chapter 7).

(iv)   Neither CENP-B nor α-satellite DNA is detectable on, or necessary for the centromeric functions of, some mitotically stable *de novo* human marker chromosomes (Chapter 7).

(v)   Expression of truncated versions of human CENP-B in HeLa cells does not lead to a mitotic or cell cycle arrest phenotype (Pluta *et al.* 1992).

(vi)   Some species of mice (e.g. *M. caroli*) lack detectable CENP-B protein and obviously do not require this component for their centromere function (Kipling *et al.* 1995).

Consensus on the role of CENP-B in centromere function?

Based on existing evidence, it therefore remains difficult to draw any firm conclusion on whether CENP-B has a direct role in centromere function (see Chapter 5 for a related discussion on the role of human α-satellite and mouse minor satellite DNA). It is possible that centromere activity may not be at all dependent on this protein, but that the protein has simply co-evolved with α-satellite in order to deal with the excess amount of what may be 'selfish' or 'junk' satellite DNA. However, equally plausible is the possibility that the protein may have a direct functional role, but that this role is redundant and shared by some other as yet unknown protein components.

CENP-C

Centromere sublocalization

Immunoelectron microscopy has localized the 140 kDa CENP-C protein (Earnshaw and Rothfield 1985) to a narrow band within the inner kinetochore plate, with no significant signals detected on the outer plate, the fibrous corona, or the central domain (Saitoh *et al.* 1992). However, since the presence or absence of the electron-lucent middle kinetochore plate is difficult to judge in these experiments, the possibility that a significant amount of CENP-C may also be present in this middle zone between the inner and outer plates has not been excluded.

CENP-C gene

Although the gene structure for CENP-C has not been described, genomic DNA analysis indicates that, unlike the intronless CENP-B gene, the CENP-C gene contains intron sequences (McKay *et al.* 1994; P. Kalitsis and K.H.A. Choo, unpublished data). *In situ* hybridization mapping has localized the functional human and mouse genes to the 4q12–q13.3 and 5E2–E5 regions, respectively (McKay *et al.* 1994). These sites are in a region of linkage group conservation between the two species (O'Brien *et al.* 1993). Secondary mapping sites are present in humans on chromosome 12q21.2–q21.33 and in mouse on chromosome 2B (McKay *et al.* 1994). This second location in mouse is believed to be a pseudogene because sequence divergence has introduced multiple stop codons into at least one segment of the coding region sequenced. However, the situation in humans is less clear since there is no sequence data and this region is not syntenic in the two species. It is therefore possible that the second locus in humans could be either a pseudogene or a related functional gene.

Protein structure

CENP-C is a 943 amino acid, hydrophilic, and highly basic protein (Saitoh *et al.* 1992). Comparison of human and mouse proteins indicates an overall 53% amino acid identity (66% similarity), with the C-termini being most highly conserved (77% identity over the last 220 residues) (McKay *et al.* 1994). The protein contains two potential phosphorylation sites and four possible nuclear localization signals that are conserved between humans and mouse (Fig. 6.4) (McKay *et al.* 1994). Analysis of CENP-C truncation mutants expressed *in vivo* demonstrates that CENP-C possesses an autonomous centromere-targeting domain situated at the central region of the polypeptide (Yang *et al.* 1996). Furthermore, *in vitro* assays reveals that a region of CENP-C with the ability to bind DNA is also located at the centre of the CENP-C polypeptide, where it overlaps the centromere-targeting domain (Sugimoto *et al.* 1994b; Yang *et al.* 1996). This DNA-binding domain shows functional redundancy, suggesting that it may contain at least two independent DNA-binding sites (Yang *et al.* 1996). The observation that the minimal DNA-binding and centromere-targeting domains of CENP-C co-localize within the same region of the polypeptide suggests that DNA binding could be involved in the targeting of CENP-C to centromeres *in vivo*. However, more direct proof of this is necessary, especially since the putative CENP-C-binding DNA sequence has yet to be identified.

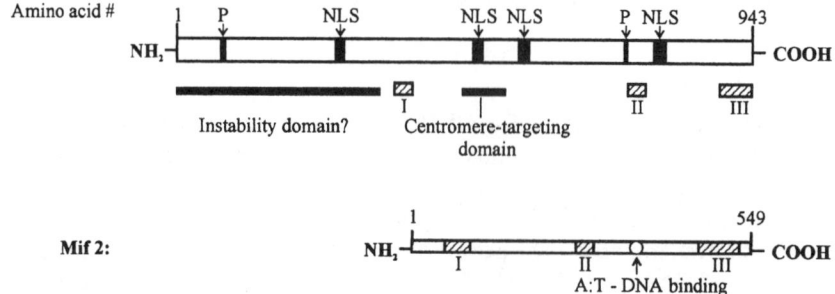

Fig. 6.4 Structural domains of human CENP-C. P and NLS denote the positions of potential phosphorylation sites and nuclear localization signals, respectively. I–III are three regions of substantial homology between human CENP-C and the *S. cerevisiae* Mif2 protein. A circle indicates an AT-rich DNA-binding motif in Mif2 (see Chapter 2) that is not found in CENP-C.

The cellular levels of CENP-C appears to be strictly controlled (Lanini and McKeon 1995). This control has been proposed to be achieved at least in part by the conditional instability of CENP-C. Using deletion mutants, a putative instability sequence has been localized to the first 323 N-terminus amino acids which, when absent results in a gross accumulation of the CENP-C mutants in transfected cells. The absence of this domain does not have a significant effect on CENP-C targeting to centromeres, but in the transfected cells the accumulated CENP-C protein also associates with noncentromeric regions of interphase and mitotic chromatin, with toxic consequences for the cells. Thus, the inherent instability of CENP-C may act as a kinetic proof-reading mechanism to ensure correct kinetochore assembly while eliminating excess build-up to prevent toxicity due to mistargeting. It is unclear how the conditional instability of CENP-C operates, although it is likely that the stability of correctly assembled CENP-C may involve the masking of the N-terminal domain. However, these interpretations have been challenged by results of a separate study in which no specific region of the protein was shown to be associated with increased proteolysis of noncentromeric CENP-C (Yang *et al.* 1996).

Evolutionary conservation

Functional conservation of CENP-C across a number of different species has been demonstrated by the efficient assembly of this protein into kinetochore–centromere complexes when cloned human CENP-C is transfected into cells of monkey, hamster, and amphibian origins (Lanini and McKeon 1995). Furthermore, human CENP-C has been shown to have a significant homology with Mif2, a protein involved in budding yeast chromosome segregation and thought to have a role in kinetochore function (Brown *et al.* 1993; Brown 1995; Meluh and Koshland 1995). Homology between these two proteins is especially prominent within three short stretches of 48 amino acids (region I; 50% similarity), 27 amino acids (region II; 70% similarity), and 52 amino acids (region III; 55% similarity) (Fig. 6.4). The order, spacing and location of these regions are also similar in the two proteins. Significantly, regions II and III coincide with the two longest segments of sequence identity between human and mouse CENP-C. More-over, mutation analysis indicates that these two domains are important to Mif2 function (Chapter 2). Thus, at least some structural and functional components of CENP-C are conserved in widely separated organisms.

Cross-reaction of anti-CENP-C antibodies with CENP-A and CENP-B

Sera from different rabbits immunized with cloned CENP-C fusion proteins all exhibit very weak reaction with proteins co-migrating with CENP-A and CENP-B in immunoblots (Saitoh *et al.* 1992). The appearance of these cross-reacting antibodies is perplexing, given the fact that the amino acid sequences of these proteins contain no regions of identity larger than three residues. Furthermore, the cross-reacting antibodies appear in response to immuniza-tion with two nonoverlapping regions of CENP-C. Rabbit polyclonal antibodies produced to CENP-B fusion proteins, on the other hand, do not recognize CENP-A or CENP-C on immunoblots (Earnshaw *et al.* 1987b). Interestingly, purification of the IgG fraction from sera containing antibodies against CENP-C fusion proteins, or affinity purification using the fusion proteins, both result in the removal of the antibodies cross-reacting with CENP-A and CENP-B (Tomkiel *et al.* 1994). It is possible that these cross-reacting antibodies belong to different immunoglobulin subclasses (Hildebrandt *et al.* 1990), and are thus lost during the IgG purification procedure, although this does not explain why affinity-purified antibodies also lose the CENP-A and CENP-B cross-reactivity.

The origin of the cross-reacting anti-CENP-B and anti-CENP-C antibodies is unclear, although three possibilities appear feasible (Saitoh *et al.* 1992). First, the antibodies may recognize a shared conformational epitope that renatures during immunoblotting but is not apparent from analysis of the primary sequence. Second, immunization with CENP-C may trigger a generalized anticentromere response, causing the rabbits to produce autoantibodies to their own homologues of CENP-A and CENP-B. Third, the rabbits may produce low levels of anti-idiotypic antibodies directed against their primary anti-CENP-C antibodies. Such anti-idiotypic antibodies could recognize sites on CENP-A and CENP-B that interact with CENP-C.

Role of CENP-C

Functional significance

Several lines of evidence suggest that CENP-C has a direct role in kinetochore structure and function: (a) sublocalization of this protein to the inner kinetochore plate indicates that the protein is a key structural component of the kinetochore itself; (b) the protein is found on all human chromosomes and show even immunostaining, as would be expected of a structure that is constant from chromosome to chromosome (Earnshaw *et al.* 1989); (c) it is present at the active but not the inactive centromere of mitotically stable dicentric human chromosomes (Earnshaw *et al.* 1989; Page *et al.* 1995; Sullivan and Schwartz 1995) (Chapter 7); and (d) nuclear microinjection of anti-CENP-C antibodies during interphase causes a disruption in kinetochore assembly and a transient arrest at the following metaphase (discussed below).

Role in kinetochore assembly: Inhibition of CENP-C results in kintochores that are reduced in both number and size

Electron microscopy reveals that kinetochores in anti-CENP-C-arrested metaphase cells have reduced number of trilaminar structures (Tomkiel *et al.* 1994). In addition, the few remaining kinetochores in these cells retain a normal trilaminar morphology but are significantly reduced in diameter. In cells arrested for extended periods, these small kinetochores become disrupted and apparently no longer bind microtubules. This is an unexpected finding, since CENP-C is localized to the inner kinetochore plate, and the majority of microtubules appear to end in the outer plate (Comings and Okada 1971; Witt *et al.* 1980; Reider 1982). One interpretation of this result is that CENP-C could contribute to some aspects of the structural organization of the kinetochore that is necessary for stable binding of other proteins which interact directly with microtubules. Whatever the mechanism, this result suggests that CENP-C plays a critical role in the assembly of kinetochore, the establishment of proper kinetochore size, and the possible stabilization of microtubule attachments.

Role in metaphase checkpoint: Inhibition of CENP-C disrupts mitosis in a cell-cycle dependent manner

Injection of highly purified anti-CENP-C monoclonal antibodies into interphase HeLa nuclei results in a prolonged delay at metaphase (Tomkiel *et al.* 1994). The level of CENP-C at the centromeres of the injected cells is significantly reduced, suggesting that the antibodies interfere with the localization of CENP-C at centromeres during interphase, thus effectively reducing the amount of this protein available for kinetochore assembly. This effect is, however, strictly cell cycle dependent, since introduction of the same antibodies after the initiation of prophase does not disrupt mitosis. Thus, CENP-C may have become resistant to the antibody inhibition during prophase kinetochore assembly, due perhaps to conformational changes or interactions with other centromere proteins, or its binding at centromeres may have been stabilized such that it is less susceptible to displacement by antibodies.

About one-third of the cells that are delayed at metaphase eventually enter cell division. However, chromosome segregation appears to fail in these cells, leading to a wide variety of phenotypes. In some cells, cytokinesis appears overtly normal but produces two daughter cells containing several micronuclei. In other cells, chromosomes that fail to segregate become trapped in the cleavage furrow. The outcome of division in these latter cells seems to depend on the amount of DNA trapped. Some show complete cleavage, with DNA remaining in the midbody. In others, where the majority of the chromosomes remain centrally located, the cleavage furrow forms but subsequently regresses, giving rise to a single daughter cell.

The onset of anaphase has been shown to be delayed by the presence of even a single maloriented chromosome in a mitotic cell (Zirkle 1970; Rieder and Alexander 1989). Such an observation has led to the hypothesis that there is a metaphase checkpoint, or feedback system that monitors the state of spindle assembly. Such a system would not normally allow the metaphase:anaphase transition to occur until all chromosomes have achieved a stable bipolar orientation (Hoyt *et al.* 1991; Li and Murray 1991). The observation that disruption of CENP-C function results in metaphase arrest suggests that proper assembly of kinetochores may be monitored by the cell cycle checkpoint preceding the transition to anaphase. A phosphorylated epitope has recently been described which is differentially expressed at kinetochores in mitotic cells (see '3F3/2 phosphoepitope' below). At metaphase, misaligned chromosomes strongly express this epitope, while chromosomes which are properly congressed at the metaphase plate lack this epitope. Thus, expression of this epitope is a candidate for the inhibitory signal to delay anaphase. It would be interesting to know if kinetochores which are disrupted by anti-CENP-C antibodies continue to express this phosphoepitope.

Possible regulatory role in G_1 phase of the cell cycle

Using HeLa cells synchronized at different cell cycle phases, Knehr *et al.* (1996) investigated the cell cycle-specific expression of CENP-C protein and mRNA. Their results indicate the not unexpected distribution of CENP-C throughout the whole cell cycle but also a somewhat surprising variation in CENP-C level during various cell cycle phases. A parallel increase in the level of both CENP-C protein and mRNA is observed from the S phase through the G_2 phase and mitosis, with the increase peaking at G_1 phase. By setting the mRNA level of G_1 phase at 100%, the abundance of CENP-C transcripts in the S, G_2, and M phase is shown to be 26%, 32%, and 56%, respectively. These results indicate not only an absence of CENP-C degradation at the end of mitosis, but an unexpected increase in CENP-C expression in early G_1 up to the end of this phase, before the protein is partially degraded. Thus, in addition to its role in mitosis, it is possible that CENP-C has a further role in the G_1 phase that may be related to cell cycle control, such as in the regulation of the transition from G_1 to S.

Specific interaction with a nucleolar transcriptional regulator

Using affinity columns coupled with CENP-C, Pluta and Earnshaw (1996) identified two related nucleolar proteins that specifically and reproducibly bound to the carboxyl-terminal third of the CENP-C protein. Microsequencing reveals the two proteins, which appear as a doublet of similar size of ~ 100 kD, as the two highly related nucleolar transcription factors, UBF1 and UBF2 (upstream binding factors 1 & 2; also known as NOR-90 or nucleolus organizer region-90). Double indirect immunofluorescence using monospecific antibodies demonstrates that a subset of CENP-C and these transcription factors is colocalized at nucleoli of interphase HeLa cells, suggesting that the *in vitro* interaction detected by affinity chromatography may reflect an interaction that occurs *in vivo*.

CENP-C is not the only centromere protein that has been found to interact with a non-centromeric protein. CENP-F was also cloned (under the name of mitosin) based on its ability to bind to a portion of the retinoblastoma protein *in vitro* (see CENP-F/mitosin below). In addition to CENP-F, the retinoblastoma protein has also been found to interact with UBF (Shan *et al.*

1992; Cavanaugh *et al.* 1995), providing an indirect link between a centromere protein and a nucleolar transcription factor. In contrast to CENP-F, a direct biochemical link between CENP-C and UBF has been demonstrated.

The biological significance of a biochemical interaction between a kinetochore protein and a nucleolar protein is not understood. It is possible that *in vivo* such an interaction could serve to regulate the function of CENP-C during interphase by sequestering the protein and its bound centromeric DNA, UBF/NOR-90, or both. Alternatively, the association of CENP-C with nucleoli could play an architectural role in the organization of the interphase nucleus. In any case, the observation that only a subset of interphase centromeres is juxtaposed with nucleoli at any point during interphase suggests a possible novel centromere function during this phase.

CENP-F

Centromere sublocalization

Immunostaining of micrococcal nuclease-digested plates of the kinetochore that become devoid of chromosomal DNA but continue to associate with kinetochore microtubules shows that at least a portion of CENP-F is localized to the kinetochore (Rattner *et al.* 1993). In addition, immunoelectron microscopy (Saitoh *et al.* 1992) of the centromere region of Indian muntjac chromosomes reveals that CENP-F antigens are confined to the surface of the kinetochore. These observations suggest that CENP-F is mapped to the outer surface of the outer kinetochore plate.

Structural properties

CENP-F is an ~ 400 kDa protein that has been identified using human CREST autoimmune serum (Rattner *et al.* 1993). This protein, 3,210 amino acids in length (Liao *et al.* 1995), is probably identical to the mitosin and p330[d] proteins (see below). The gene for CENP-F has been localized to human chromosome 1q32-q41 (Testa *et al.* 1994). The primary sequence of CENP-F (Liao *et al.* 1995) shows a consistently low level of homology (<20%) with the rod domains of many cytoskeletal proteins such as myosins, kinesins, lamins, and tropomyosins that probably reflects similarities in secondary structure. CENP-F has four regions that are predicted to form extended coils (Lupas *et al.* 1991) (Fig. 6.5). Within the protein are two pairs of direct repeats, the centrally located pair having a repeating unit of 96 amino acids, while the more C-terminal pair having a longer repeating unit of 178 amino acids. The protein contains 11 leucine heptad repeats that are known to be frequently involved in protein–protein interactions (Landschultz *et al.* 1988; Johnson and McKnight 1989). Four consensus nuclear localization sequences (Chelsky *et al.* 1989; Robbins *et al.* 1991), as well as clusters of consensus phosphorylation sites for cdc2 kinase, MAP kinases or cyclin-dependent kinases, are present at the N- and C-terminal regions. The C-terminus exhibits two additional distinguishing features that include a high proline content and a sequence that fits an NTP-binding motif (Walker *et al.* 1982; Fry *et al.* 1986; Linder *et al.* 1989; Saraste *et al.* 1990).

Cell cycle distribution of CENP-F

The distribution of CENP-F is cell cycle dependent (Rattner *et al.* 1993; Liao *et al.* 1995). Studies in HeLa cells reveal that it is absent or masked in G_1 or shortly after release from the G_1/S boundary. As cells progress through S phase, CENP-F levels gradually increase and reach peak concentrations at G_2 and M. In G_2 cells, the protein is uniformly distributed throughout the nucleus (except for the nucleoli), and is a component of the nuclear matrix. In late G_2 cells, the uniform nuclear distribution gives way to localization around the nuclear boundary and multiple centromeres within the nucleus. Not all centromeres exhibit CENP-F staining, reflecting perhaps the asynchronous maturation of the centromere–kinetochore complex. By prophase, CENP-F is clearly present at all the centromeres as discrete pairs of foci. Since fully formed kinetochores are not detectable until the onset of nuclear envelope breakdown, CENP-F should be considered to be localized at prekinetochores at this stage. The assembly of CENP-F at prekinetochores correlates with increased chromatin condensation and a general reduction in nuclear staining. As cells progress closer to mitosis, the intensity of the nuclear rim staining also diminishes.

In prometaphase and metaphase cells, CENP-F is associated with kinetochores. This association is still detectable at kinetochores after the onset of sister chromatid separation in early anaphase. By late anaphase, CENP-F becomes diffusely distributed throughout the cell except for a distinct, narrow strip of staining at the spindle midzone. The narrow strip of staining becomes concentrated in the intercellular bridge at either side of the midbody due to cleavage furrow formation during cytokinesis. After the completion of mitosis, the majority of CENP-F is degraded and the cells return to the background staining seen in early G_1 cells. During the different stages of mitosis, CENP-F protein is also found throughout the cytoplasm of the cells.

Role of CENP-F

CENP-F possesses two extended coil domains that flank a central core and an ATP-binding motif that may be part of a mechanochemical domain. This general organization is reminiscent of some aspects of the SMC (stability of mini chromosomes) family of proteins that have been implicated in chromatin compaction (Strunnikov *et al.* 1993; Hirano and Mitchison 1994; Saitoh

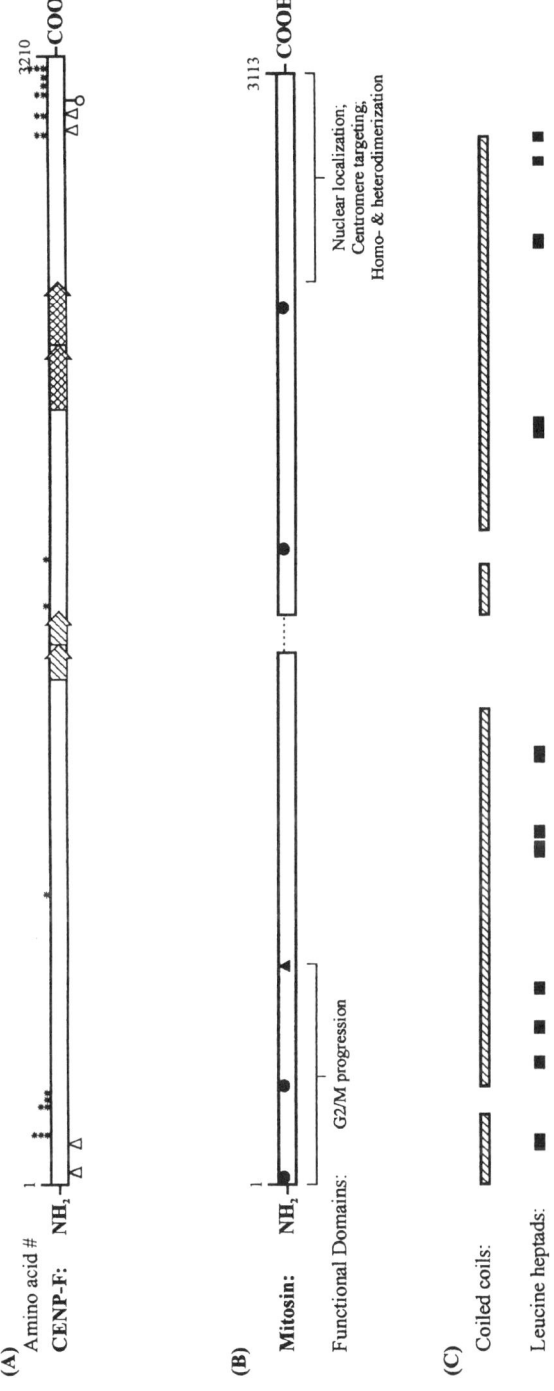

Fig. 6.5 Structural properties of human CENP-F, and comparison with mitosin. (A) CENP-F. Single-hatch and cross-hatch arrows denote two pairs of direct repeats. Open triangle, open circle, and asterisk denote potential nuclear localization, NTP-binding, and phosphorylation sequence motifs, respectively. (B) Mitosin, where deviations from CENP-F are indicated by solid circles (amino acid substitutions), a solid triangle (single amino acid deletion), and dotted line (a 96 amino acid deletion). The demonstrated functional domains for mitosin are shown. (C) Common regions between CENP-F and mitosin predicted to form coiled coil domains and leucine heptad repeats are shown by the hatched and solid boxes, respectively.

et al. 1994; reviewed by Peterson 1994; Saitoh *et al.* 1995), although there are several notable differences that suggest that CENP-F is not closely related to the existing SMC family. Like this family of proteins (Peterson 1994), CENP-F also appears to possess a flexible central region that may allow it to act as a hinge during certain condensation events such as the reorganization of the centro-mere–kinetochore complex during the different stages of the cell cycle. This structural feature of CENP-F, the apparent association of the protein with the kinetochore earlier than any other known transient kinetochore proteins (except TopoIIα; see below), and the timing of its appearance corresponding to the period when the centromere begins its transformation from an interphase structure to a highly organized trilaminar kinetochore structure suggest that the protein may function during the early steps in kinetochore maturation.

In addition to a possible role in kinetochore maturation, CENP-F is the first kinetochore protein that is found to be part of the nuclear matrix, an association that is most prominent at G_2. Since the protein is not readily detected in cells at other stages of interphase, this temporary association with the matrix suggests that it is probably not essential for matrix integrity but may be necessary for nuclear reorganization events that are associated with the onset of cell division. It is noteworthy that CENP-F contains clusters of consensus phosphorylation sites for cell cycle-dependent kinases. In HeLa cells, the cdc–cyclinB1 kinase complex [see p34^{cdc2} protein kinase below] enters the nucleus during late G_2 (Pines and Hunter 1991) and is primarily responsible for initiating mitosis. Phosphorylation of CENP-F by the cdc2–cyclinB1 kinase complex may therefore be part of the mechanism that promotes any function CENP-F may have in nuclear reorganization or the redistribution of the protein between nuclear matrix and kinetochore.

Mitosin

Mitosin and CENP-F are probably products of the same gene

Mitosin (Zhu *et al.* 1995a, 1995b) is a centromere-binding protein which shares most of the characteristics of CENP-F. It is a nuclear phosphoprotein that has an identical cell cycle distribution as that described for CENP-F, is found on the coronal surface of the outer kinetochore plate, and whose gene has been mapped to the same human chromosome region of 1q32–41 as CENP-F. Mitosin is 3113 amino acids in length and is shorter than CENP-F due predominantly to a 96 amino acid deletion in a region corresponding to the central direct repeat domain of the CENP-F sequence (Fig. 6.5). There are also five amino acid substitution differences between the two proteins. It is possible that the 96 amino acid deletion is the consequence of alternative splicing or an error in cloning, and if excluded from the sequence comparison, the oveall identity between mitosin and CENP-F is 99.84%. Thus, mitosin and CENP-F are quite likely to be the products of the same gene.

Structural domains

Experiments involving exogenous expression of truncated mitosins have allowed the definition of functional domains of mitosin. Overexpression of a mutant mitosin protein in which the N-terminal 632 residues have been deleted blocks cell cycle progression mainly at G_2/M, suggesting that elements within this domain may play an important role in mitotic phase progression (Zhu *et al.* 1995a, 1995b) or that improperly assembled structures involving defective mitosin are subjected to checkpoint arrests. Truncation experiments further indicate that the C-terminal 626 residues of mitosin contain centromere targeting and nuclear localization signals, and appear to be essential for the ability of mitosin to bind to itself or other putative mitosin-associated proteins.

Interaction with Rb protein *in vitro*

The mitosin gene was originally isolated by screening an expression library using purified retinoblastoma (Rb) protein as a probe (Zhu *et al.* 1995b). Binding of mitosin to Rb is subsequently shown to specifically involve the C-terminal 211 amino acids of mitosin. This indicates that mitosin interacts at least *in vitro* with Rb protein. Rb is hypophoshorylated during early G_1 phase, hyperphosphorylated during late G_1 through G_2, and undergoes dephosphorylation during M (Buchkovich *et al.* 1989; Chen *et al.* 1989; Ludlow *et al.* 1993). Direct microinjection of unphosphorylated Rb inhibits G_1 progression (Goodrich *et al* 1991), suggesting that the hypophosphorylated form of Rb is active in controlling G_1 entry. It has also been shown that mitosin and hypophosphorylated Rb can both be detected in the chromosome scaffold (Zhu *et al.* 1995b). In addition, there are indications that Rb may have a potential role in G_2 (Karantza *et al.* 1993) and M phase progression (Hirano *et al.* 1988; Booher and Beach 1989; Doonan and Morris 1989; Zhu *et al.* 1995b). However, the biological significance of the interaction between mitosin and Rb is at present unclear.

p330[d]

p330[d], mitosin, and CENP-F are probably the same protein

p330[d] is another example of proteins that have been identified by human autoantibodies (Casiano *et al.* 1993). The protein, which has a molecular weight of 330 kDa, is conserved in a variety of mammalian cells, including human, rat, rat kangaroo, Indian muntjac, and mouse cells. It is located at the external surface of centromeres, and has a cell cycle distribution pattern that is identical to those described for mitosin and CENP-F. Thus, it is likely that p330[d], mitosin, and CENP-F are the same protein with different names.

HMG-I

The high-mobility group (HMG) proteins are abundant, heterogeneous, non-histone components of chromatin. Typically, these proteins have a molecular mass of < 30 kDa, are highly charged, and have a high electrophoretic mobility (thus their name) and solubility in 2–5% trichloroacetic acid (reviewed by Johns 1982). This heterogeneous group of proteins contains the HMG DNA-binding motif that is shared by the abundant non histone components of chromatin and by specific regulators of transcription and cell differentiation. Common properties of HMG proteins include interaction with the minor groove of the DNA helix, binding to irregular DNA structures, and the capacity to modulate DNA structure by bending. DNA bending induced by the HMG domain can facilitate the formation of higher-order nucleoprotein complexes, suggesting that HMG proteins may have an architectural role in assembling such complexes (reviewed by Grosschedl *et al.* 1994).

HMG-I, a member of the HMG proteins, is an abundant protein of ~ 10 kDa that binds to DNA in the minor groove wherever there are runs of six or more A:T base pairs (Solomon *et al.* 1986). This protein (originally named α-protein) was isolated based on its ability to bind *in vitro* to the 172 bp repeat of the African green monkey α-satellite DNA (Strauss and Varshavsky 1984). The protein is found in cultured mammalian cells ranging from humans to murine and in nonmammalian species including *Drosophila* (Alfageme *et al.* 1980; Rodriguez-Alfageme *et al.* 1980; Levinger and Varshavsky 1982a, 1982b; Garreau and Williams 1983; Strauss and Varshavsky 1984; Tharappel and Elgin 1986; Ashley *et al.* 1989; Disney *et al.* 1989).

Cell cycle distribution

The principal location of HMG-I in interphase mouse and human cells is in the nucleus (Disney *et al.* 1989). The protein is not excluded from, nor is it specifically and exclusively localized to, the nucleoli of these cells. During prometaphase, HMG-I is occasionally seen as transverse bands on chromosomes. During metaphase, such transverse bands become prominent. The intercalary chromosome banding patterns are reproducible, and when compared with trypsin–Wright stain or Giemsa-treated human and mouse chromosomes, are shown to correspond to G-band regions of these chromosomes. This nonrandom distribution of the HMG-I protein into specific bands on chromosomes contrasts with the uniform/random distribution of histone proteins on similarly prepared metaphase chromosomes. In addition to G-bands, HMG-I is specifically detected on the centromere and telomere regions of mouse and human metaphase chromosomes. However, the distribution of the protein is different on different chromosomes within the same intact cell, in that it localizes to telomeric and centromeric regions of some chromosomes, and only the telomeric regions or only the centromeric regions of others, with some chromosomes displaying intercalary bands as well. As cells move into anaphase and telophase, the overall HMG-I signal is diminished and becomes more punctate.

Role in the centromere

DNase I footprinting with purified HMG-I protein detects three specific binding sites (I–III) per α-satellite repeat. Within the $(I\text{-}II\text{-}\mathbf{III})_n\text{-}(I\text{-}\mathbf{II}\text{-}III)_{n+1}$ tandem arrangement of the α-satellite DNA, site III (bold) is 145 bp (one core nucleosome length) from site II (bold) on the adjacent α-satellite repeat, while site I lies midway between sites II and III. In the α-nucleosome phasing frame corresponding with this arrangement, sites I–III would be brought into mutual proximity by DNA folding in the nucleosome. In addition to footprinting analysis, an identical phasing with the preferred frame is detected in isolated chromatin by other totally different approaches (Zhang *et al.* 1983; Wu *et al.* 1983), suggesting that HMG-I may function as a nucleosome-positioning or phasing protein (Strauss and Varshavsky 1984). According to this suggestion, HMG-I may underlie both the compactness and transcriptional inactivity of constitutive α-heterochromatin by producing a crystal-like, phased arrangement of α-nucleosomes, which by its very regularity may allow formation of an especially compact and ordered nucleosome packing. Whether HMG-I is sufficient to bring about such transitions, or whether it functions by providing sites for interaction with other proteins such as the centromere-specific core histone CENP-A remains to be determined.

pJα

This is a 10–15 kDa protein identified in HeLa nuclear extracts and shown to bind a significant portion (~14%) of human α-satellite DNA by electrophoretic mobility shift assay (Gaff *et al.* 1994). The protein recognizes a 9-bp motif (GTGAAAAAG) that is located within the same region of the α-satellite DNA monomeric unit that contains the alternative sequence motif for CENP-B box (i.e., pJα and CENP-B are two distinct proteins that recognize alternative nucleotide configurations of the same region of the α-satellite DNA monomer). The two monomeric types containing these alternative nucleotide configurations have been designated as α-satellite prototype-A (containing the binding site for pJα) and prototype-B (containing the binding site for CENP-B) (Romanova *et al.* 1996). Detailed phylogenetic analysis of α-satellite DNA sequences suggests that all α-satellite DNA family members decend from these two prototypes, and that B-type directly decends from A-type and not from some ancestral sequence different from both of them. Furthermore, the analysis shows that selection-driven evolution, rather than random fixation of mutations, formed the distinction between A- and B-types. At present, the function of pJα is obscure, although its α-satellite DNA binding property and its sharing of the same DNA-binding domain as CENP-B suggest a possible role in α-satellite DNA assembly.

AC1 antigen

This is a 170 kDa protein that has been identified by a monoclonal antibody raised against isolated Chinese hamster chromosome scaffolds (Holland *et al.* 1995). The antigen is also detected on human chromosomes, but not on chromosomes of quail, *Drosophila*, wheat, and poppy. Within the Chinese hamster ovary and human metaphase chromosomes, the antigen shows variation in staining pattern at the centromere, appearing as two dots, distorted dots connected by a line, a horseshoe shape encircling the centromere, or occasionally as a complete ring encircling the centromere; no staining of the chromosome arms is observed. The ring seems to be a physically fragile structure, since the incidence of its detection increases with decreasing speed of cytocentrifugation used to spread the cells onto the slides. No AC1 is detected on the centromeres of interphase cells. In addition to the centromeres of metaphase chromosomes, AC1 has been localized to the centrioles of both metaphase and interphase cells. A possible role of this antigen may be to stabilize the centromere, by holding the sister chromatids together until their separation at the metaphase/anaphase transition.

M phase and checkpoint control

The initiation and progression of the mitotic phase is expected to be regulated by many protein factors. Four of these factors are components of the centromere passenger proteins. Two of these, $p34^{cdc2}$ kinase and MPM2 phosphoproteins, play the role of master regulators in many steps of the cell division pathway, whereas the remaining two components, 3F3/2 and ZW10, appear to be checkpoints in metaphase/anaphase transition.

$p34^{cdc2}$ (or cdc2) protein kinase

The cDNA for human $p34^{cdc2}$ protein kinase has been cloned by expressing a human cDNA library in the fission yeast *S. pombe* and selecting for clones that can complement a mutant of the yeast cell cycle control gene *cdc2* (Lee and Nurse 1987). The protein is 297 amino acids long, has a molecular mass of 34 kDa, and contains the consensus ATP-binding and phosphorylation sites that are found in protein kinases (Lee and Nurse 1987). Comparison of human $p34^{cdc2}$ with its *S. pombe cdc2* (Nurse *et al.* 1976; Beach *et al.* 1982; Simanis and Nurse 1986) or *S. cerevisiae CDC28* homologues (Hartwell

1974; Nasmyth and Reed 1980) reveals 63% and 58% identity, respectively, throughout the proteins. In addition to humans and yeasts, a protein kinase that is homologous to $p34^{cdc2}/cdc2/CDC28$ has been found in all eukaryotic species studied (Draetta *et al.* 1987, 1988a, 1988b, 1989; Lee and Nurse 1987; Arion *et al.* 1988; Draetta and Beach 1988; Dunphy *et al.* 1988; Gautier *et al.* 1988; Labbe *et al.* 1988; Lee *et al.* 1988; Rattner *et al.* 1990).

Cell cycle distribution

In human cells, $p34^{cdc2}$ is found in both the cytoplasm and the nucleus. In the cytoplasm, part of the $p34^{cdc2}$ is found at the centrosome, with this localization being most prominent from G_2 through to the completion of telophase (Bailly *et al.* 1989; Riabowol *et al.* 1989). In order to more clearly define the distribution of $p34^{cdc2}$ within the mitotic apparatus, Rattner *et al.* (1990) combined the use of the large chromosomes of Indian muntjac (Brinkley *et al.* 1984; Rattner 1986) and the extraction of cytoplasmic $p34^{cdc2}$ to eliminate background cytoplasmic staining. The study indicates that interphase cells stain positively with anti-$p34^{cdc2}$ antibody, but show no distinct foci of reactivity with the prekinetochores (Earnshaw and Rothfield 1985). During prophase, the centrosomal regions display extensive staining. Coordinated with the increase in detectable $p34^{cdc2}$ in the nucleus and nuclear envelope breakdown in late prophase and prometaphase, $p34^{cdc2}$ is detected in association with the paired kinetochores of all the chromosomes, as well as with a subset of microtubules. Kinetochore staining becomes more defined in metaphase cells, while prominent staining is now seen along the kinetochore microtubules. At the completion of anaphase, $p34^{cdc2}$ is found at each of the poles and on the shortened kinetochore microtubules. Granular staining in the midzone is also observed. During telophase, $p34^{cdc2}$ is found in the reforming nuclei and within the intercellular bridge.

The presence of $p34^{cdc2}$ at the kinetochore of Indian muntjac cells is not due to any transient interaction with tubulin, since chromosomes with kinetochores depleted of tubulin by arresting cells with vinblastine still retain $p34^{cdc2}$ reactivity. In addition to Indian muntjac, the presence of $p34^{cdc2}$ on kinetochore and along kinetochore-to-pole microtubules has also been verified in African green monkey (Rattner *et al.* 1990) and baby hamster kidney cells (Andreassen and Margolis 1994).

Role as a master control enzyme of M phase

The catalytic subunit of $p34^{cdc2}$ associates with its regulatory subunit cyclin B to form a complex referred to as MPF or **m**aturation-**p**romoting **f**actor (Gautier *et al.* 1989). $p34^{cdc2}$ is believed to be universally required as a master control enzyme during mitosis in eukaryotes (Nurse 1990; Norbury and Nurse 1992). Its activity is tightly regulated during the cell cycle and peaks at mitosis (Draetta and Beach 1988; Gautier *et al.* 1989). Exit from mitosis requires the dephosphorylation of p34^{cdc2} (Lorca *et al.* 1992) and the inactivation of p34^{cdc2} by degradation of the cyclin B regulatory subunit (Murray *et al.* 1989; Ghiara *et al.* 1991; Gallant and Nigg 1992). A growing number of key proteins in the mitotic pathway have been shown to be potential or direct phosphorylation substrates for p34^{cdc2}. These proteins include several downstream kinases (e.g. the tyrosine kinases p60^{src} and p150^{abl}, and the serine/threonine-specific casein kinase II), as well as structural proteins that are involved in vital functions, such as nuclear envelope breakdown (e.g. lamin, a membrane-associated intermediate-filament-type karyoskeleton), chromosome condensation (e.g. histones H1 and H3, and certain types of major chromatin-associated non histone proteins of the high-mobility group/HMG family), bipolar spindle formation (e.g. Eg5, a kinesin-related motor protein) (Blangy *et al.* 1995), disassembly of nucleoli (e.g. nucleolin, a major nucleolar protein), mitotic reorganization of microtubules (e.g. microtubule-associated proteins or MAPs), chromosome movement (e.g. various motor molecules, see below), microtubule nucleation (e.g. centrosome-associated factors), microfilament cytoskeleton reorganization (e.g. caldesmon, an actin-binding protein of the actomyosin-associated system), and delay of cytokinesis (e.g. myosin II, another component of the actomyosin-associated system and a regulator of the actin-dependent myosin ATPase) (reviewed by Nigg 1993).

MPM2 phosphoproteins

The MPM2 phosphoproteins are a group of more than 40 phosphoproteins that are recognized by MPM2, a monoclonal antibody that binds an epitope containing phosphothreonine plus additional specific amino acids (Davis *et al.* 1983; Taagepera *et al.* 1994; Westendorf *et al.* 1994). These phosphoproteins tend to be maximally expressed during M phase (Vandre *et al.* 1984). Such an M phase-specific appearance of MPM2 phosphoproteins has been demonstrated in the mitotic cells of a broad range of animal, insect, plant, and fungal species tested, and is likely to be a phenomenon common to all eukaryotic cells (Davis *et al.* 1983; Vandre *et al.* 1986; Engle *et al.* 1988; Kuang *et al.* 1989; Kuriyama *et al.* 1990). In addition to mitotic cells, the phosphoproteins have been detected in various non mitotic or meiotic cells examined (Davis *et al.* 1983; Vandre *et al.* 1986).

Cell cycle distribution

In interphase cells, MPM2 phosphoproteins are present as discrete patches or spots within the nucleus (Vandre *et al.* 1984, 1986; Taagepera *et al.* 1995). The nature of these nuclear patches is not clear, and they are larger and more diffusely staining than are the prekinetochore spots. Within these patches, at least some protein components of the pericentriolar material are recognizable. In prophase, there is a dramatic increase in the level of nuclear MPM2 phosphoproteins. Presence of the phosphoproteins on centromeres and along the chromosome arms is first detected during early prophase. At the time of nuclear envelope breakdown in prometaphase, soluble MPM2 phosphoproteins that are apparently not associated with specific structures become distributed throughout the cytoplasm. Concurrent with this event, a strong presence of the phosphoproteins is seen in the centrosomal region. At metaphase, the phosphoproteins are found, over and above the general cytoplasmic background, at the spindle poles, the centromeres, the chromosome arms, and the spindle itself. Within the centromere, the phosphoproteins are contained in both the outer and the inner dense plates of the kinetochore (Taagepera *et al.* 1995), whereas on the chromosome arms, they are located on the central core or axis of chromosomes and extend along the entire length of sister chromatids (Hirano and Mitchison 1991). The phosphoproteins at the spindle poles persist until the cell re-enters early G_1, but the levels decrease markedly after anaphase. By late anaphase, the phosphoproteins are no longer detected at the centromeres, but begin to appear at the spindle midzone. As the cell progresses into telophase and cytokinesis, the midzone phosphoproteins coalesce on either side of the phase dense matrix of the midbody. As the two daughter cells complete cytokinesis, the intranuclear patches of interphase cells reappear during early G_1.

Role in M phase induction and regulation

The initiation of M phase in the eukaryotic cell cycle is accompanied by a high level of protein phosphorylation (Capony *et al.* 1986; Karsenti *et al.* 1987; Lohka *et al.* 1987; Nishimoto *et al.* 1987b). The presence of MPM2 phosphoproteins throughout mitosis and on important mitotic structures such as centromeres, chromosome axes, the mitotic spindle, and spindle poles, suggests that MPM2 proteins are involved in various aspects of M phase induction and regulation, and that phosphorylation of the MPM2 epitopes is a mechanism for turning on multiple mitotic events in a synchronous fashion. In support of this, phosphorylation of MPM2 epitopes has been shown to be correlated with the activation of the universal M phase inducer, MPF (see p34^{cdc2}) (Kuang *et al.* 1989, 1994), formation of the mitotic spindle (Vandre and Borisy 1989; Centonze and Borisy 1990; Kubiak *et al.* 1992; Masuda *et al.* 1992; Vandre and Burry 1992), and the condensation of interphase chromatin into mitotic chromosomes (Hirano and Mitchison 1991). In addition, several key mitosis-associated proteins, including DNA topoisomerase II (Taagepera *et al.* 1993) (see below), mitogen-activated protein kinase p42^{mapk} (Taagepera *et al.* 1994), *Xenopus* Cdc25 phosphatase (Kuang *et al.* 1994), *Xenopus* Wee1-like kinase (Mueller *et al.* 1995), and the microtubule-associated proteins MAP1 (Vandre *et al.* 1986; Tombes *et al.* 1991) and MAP4 (Tombes *et al.* 1991; Vandre *et al.* 1991), have all been shown to be MPM2 phosphoproteins. As would be expected of phosphoproteins with such diverse functions, microinjection of MPM2 antibodies into cells completely inhibits entry into and exit from mitosis (Davis and Rao 1987; Kuang *et al.* 1989).

Phosphoepitope 3F3/2

3F3/2, a phosphorylated epitope identified by a monoclonal antibody named 3F3/2, is differentially expressed at the kinetochores of chromosomes and shown to play a key role in cell cycle checkpoint control (Gorbsky and Ricketts 1993; Campbell and Gorbsky 1995; Nicklas *et al.* 1995). On SDS–PAGE, the antibody detects a number of protein bands of unknown identity. Immunoelectron microscopy reveals that the phosphoepitope is concentrated in the middle electron-lucent layer of the trilaminar kinetochore structure. The phosphoepitope is widespread and is present in *Drosophila* and a number of mammalian species tested.

Cell cycle distribution

In interphase rat kangaroo cells, 3F3/2 phosphoepitope is not found in the cytoplasm, although irregular patches of it that apparently do not correspond with the positions of interphase kinetochores are detected in the nucleus (Gorbsky and Ricketts 1993). During prophase and early prometaphase, the phosphoepitope is expressed on all the kinetochores and on the mitotic poles. With progression through prometaphase, expression among the kinetochores becomes heterogeneous, with some chromosomes showing strong expression on both kinetochores, others exhibiting weak or no expression on either kinetochore, while in still others, one of the kinetochores shows strong expression while its partner kinetochore shows weak or no expression at all. By late prometaphase, expression of 3F3/2 is lost on most kinetochores. At metaphase, when chromosomes are aligned, the phosphoepitope is no longer detected on kinetochores, and cells enter anaphase shortly thereafter. Absence of this phosphoepitope on kinetochores persists through anaphase, although the mitotic poles remain labeled.

For cells in mid prometaphase, the position of a chromosome within the spindle constellation predicts how the chromosome would express 3F3/2 phosphoepitope. In general, chromosomes that are very near the poles strongly express on both kinetochores. Those that are at the metaphase plate tend to weakly express on one or both kinetochores. Chromosomes situated between the pole and the metaphase plate often exhibit very asymmetric expression on their two kinetochores, generally with the kinetochore facing the metaphase plate showing greater expression. Tracking of chromosome movement indicates that the leading kinetochore, i.e. the one moving toward the metaphase plate, expresses 3F3/2 strongly but the trailing kinetochore expresses the phosphoepitope only weakly. Furthermore, at metaphase, misaligned chromosomes unattached to the mitotic spindle express the phosphoepitope strongly on both kinetochores, even when all the other chromosomes of a cell are assembled at the metaphase plate and lack expression.

Checkpoint control for metaphase to anaphase transition

The transition from metaphase to anaphase is regulated by a checkpoint system that prevents chromosome segregation in anaphase until all the chromosomes have aligned at the metaphase plate (reviewed by Murray 1994; Gorbsky 1995). The cell cycle distribution pattern of 3F3/2 phosphoepitope matches that hypothesized for a checkpoint signal coming from the kinetochores. Support for such a checkpoint role has come from microinjection of 3F3/2 antibody into living mitotic cells (Campbell and Gorbsky 1995). In the injected cells, chromosomes move to the metaphase plate normally but onset of anaphase is delayed for a length of time that is proportional to the concentration of the antibody. In these cells, the normal dephosphorylation of this epitope that occurs as the chromosomes arrive at the metaphase plate is inhibited, and the phosphoepitope remains expressed symmetrically on sister kinetochores throughout the prolonged metaphase. When cells finally enter anaphase, the phosphoepitope becomes dephosphorylated and disappears from the kinetochores. In another study, tension, whether from a micromanipulation needle or from normal mitotic forces, is shown to cause dephosphorylation of the 3F3/2 phosphoepitope (Nicklas *et al.* 1995). In the absence of tension, either naturally or as a result of chromosome detachment by micromanipulation, the epitope becomes phosphorylated and anaphase is delayed indefinitely. It is proposed that tension provides the all-clear signal to the checkpoint by altering the conformation of tension-sensitive kinetochore proteins, such as those recognized by the 3F3/2 antibody, in a way that leads directly to dephosphorylation (Nicklas *et al.* 1995). That dephosphorylation is an essential aspect of passing the metaphase checkpoint has also been documented in fission yeast, *Drosophila*, and mammalian cells (Ohkura and Yanagida 1991; Larsen and Wolniak 1993; Mayer-Jaekel *et al.* 1993; Andreassen and Margolis 1994). It is of interest to note, however, that one study has indicated that rather than kinetochore tension, it is the unattached kinetochores that transmit the 'wait anaphase' signal (Rieder *et al.* 1995).

ZW10

ZW10 is the product of the *l(1)zw10* (abbreviated *zw10*) gene found in *Drosophila*. The protein is essential for accurate mitotic and meiotic chromosome segregation, as null mutations in the *zw10* gene result in very high levels of aneuploidy in a variety of *Drosophila* tissues (Smith *et al.* 1985; Williams *et al.* 1992; Williams and Goldberg 1994; Williams *et al.* 1996). In mutant cells, mitosis and meiosis seem to proceed properly through metaphase; the metaphase spindle is normal in morphology, and all chromosomes congress to the metaphase plate. However, many anaphases are aberrant, with segregating chromosomes that are misoriented or lag at the metaphase plate. In addition, mutations in *zw10* produce precocious sister chromatid separation in colchicine-arrested mitotic cells.

Cell cycle distribution

The cell cycle localization of ZW10 appears to be similar during mitosis, meiosis I, and meiosis II (Williams *et al.* 1992; Williams and Goldberg 1994; Williams *et al.* 1996). The better characterized meiotic cell cycle distribution pattern for ZW10, seen in *Drosophila* spermatocytes (Williams *et al.* 1996), will be described here. In pre-meiotic primary spermatocyte, ZW10 is located inside the nucleus. During meiotic prophase I and early prometaphase I, this intranuclear distribution is uniform and diffuse, with no apparent concentration in any discrete structures. By late prometaphase I, ZW10 begins to localize to specific positions that best reflect the kinetochore sites on the condensed bivalents. In metaphase I, a dramatic change in ZW10 distribution occurs. When the bivalents are situated at the metaphase plate, ZW10 extends in filament-like structures outward from the kinetochores along the long axis of a small subset of spindle microtubules that are presumed to be kinetochore microtubules. From the onset of anaphase through midanaphase I, ZW10 again becomes associated with the kinetochore regions as the homologues separate. During telophase, only a small amount of ZW10 is still found at the kinetochore, while some ZW10 is also observed at the spindle midbody.

The distribution of ZW10 during meiosis II appears quite similar to that seen in meiosis I. The protein is localized to the kinetochore regions of prometaphase II chromosomes. During metaphase II, ZW10 is again found on discrete filaments composing a subset of the kinetochore-to-pole spindle, consistent with localization on kinetochore microtubules. The protein becomes reassociated with the kinetochore regions during anaphase II, and concentrates at the spindle midbody in late telophase.

Role in metaphase–anaphase transition

The unusual distribution pattern of ZW10 at the prometaphase/metaphase/anaphase stages might at first suggest a possible role of the protein in adhesion or separation of sister chromatids, in poleward chromosome movements during anaphase, or in the regulation of entry into anaphase. Several observations suggest that ZW10 is unlikely to be involved in the adhesion or separation of sister chromatids (Williams *et al.* 1996): (a) Unlike *ord* and *mei-S332* mutations (see below), *zw10* mutations do not cause precocious sister chromatid separation during the first meiosis. (b) ZW10 is localized to the kinetochore during meiosis I, when sister centromeres remain attached, as well as during meiosis II, when sister centromeres separate. (c) Mutations in *ord* and *mei-S332* do not affect the ability of ZW10 to localize to the kinetochores, even those of chromatids that were separated precociously during the previous meiotic division.

A role of ZW10 in poleward chromosome movements during anaphase is also believed to be unlikely because of the observation of at least one effect of *zw10* mutations—that is, precocious sister chromatid separation—in cells that are arrested before anaphase by microtubule poisons such as colchicine (Williams *et al.* 1992). These cells do not contain any obvious spindle microtubules upon which ZW10 could influence chromosomal movements. The ZW10 protein would thus need to exhibit two very dissimilar functions if it also played a role in anaphase chromosome migration.

The most plausible role for ZW10 appears to be to act as a component of a feedback control system that renders entrance into anaphase dependent upon bipolar tension exerted across all chromosomes (Niklas *et al.* 1995; Rieder *et al.* 1995). It may be envisioned that ZW10 may either be part of the mechanism that senses tension, not unlike that described for the phosphoepitope 3F3/2, or part of the immediate downstream response to this mechanism specifically leading to anaphase chromatid or chromosome disjunction. This hypothesis is supported by several aspects of the *zw10* mutant phenotype and of the distribution of its gene product (Williams *et al.* 1996): (a) The *zw10* gene acts precisely, at least in a formal sense, as part of a checkpoint that normally prevents sister chromatid separation in the absence of a spindle, as demonstrated by colchicine-treated wild-type cells, in which sister chromatids stay together, and by colchicine-treated *zw10* mutant cells, in which a large percentage of sister chromatids separate. (b) In the absence of colchicine, chromosome missegregation in *zw10* mutant cells occurs if anaphase begins abnormally before all kinetochore-spindle attachments are made. (c) Chromosomes that are or are not under tension display different distribution of ZW10 at metaphase and anaphase. (d) The movements of ZW10 to the presumed kinetochore microtubules during metaphase and back to kinetochore region at anaphase onset might provide a physical basis for the measurement of tension, and indeed might reflect the conveyance of signals between chromosomes and the spindle.

Cohesion

Centromere pairing is required for the bipolar attachment of chromosomes to the mitotic spindle and for chromosome congression (Nicklas 1971, 1974; Rieder 1991) at the metaphase plate. This pairing is achieved by a combination of the action of cohesion proteins and the intertwining of sister chromatid DNA strands to form catenated dimers. Evidence in support of the DNA catenation hypothesis has come from examining the localization and the roles of topoisomerase II (see below) (Wang 1985; Miyazaki and Orr-Weaver 1994). Two groups of proteins that have been postulated to be important in the cohesion of mitotic sister chromatids are INCENPs and CLiPs. In addition, three other proteins, Mei-332, Ord, and Cor1, appear to have a preferential role in meiotic but not mitotic sister chromatid cohesion.

INCENPs (*inner centromere proteins*)

Structural properties

The INCENPs, first identified as a doublet consisting of a shorter INCENP$_I$ and a longer INCENP$_{II}$ polypeptide, are the founder members of the 'chromosome passenger' class of proteins. Although the proteins were initially identified with a monoclonal antibody raised against the bulk proteins of a mitotic chromosome scaffold fraction (Cooke *et al.* 1987), it was their unusual property of relocating onto different structures, including the centromere, during mitosis that has brought attention to them as a new class of proteins (reviewed by Mackay and Earnshaw 1993).

Determination of the primary sequences of INCENP$_I$ and INCENP$_{II}$ reveals that the two proteins are virtually identical, except for a 38 amino acid deletion near the C-terminus of INCENP$_I$ and a single amino acid substitution internally (Fig. 6.6) (Mackay *et al* 1993). The proteins are hydrophilic along their lengths and are highly basic. Both proteins have a region containing heptad repeats that might permit the formation of an α-helical coiled coil domain similar to those of myosins, tropomyosins, and

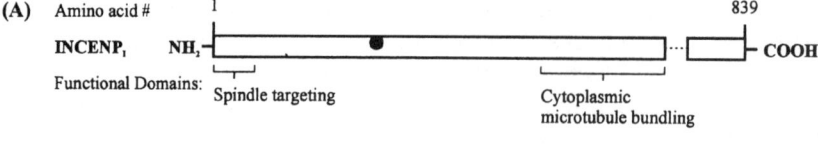

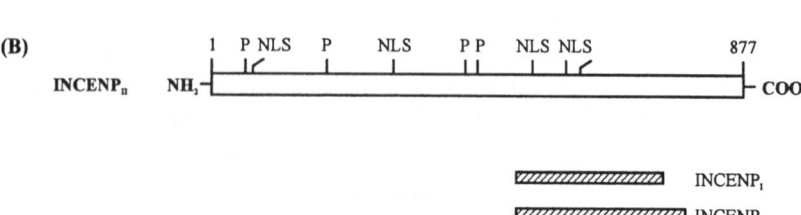

Fig. 6.6 Structures of chicken INCENP$_I$ and INCENP$_{II}$ proteins. (A) INCENP$_I$. Dotted line represents a 38 amino acid deletion and solid circle denotes an amino acid substitution not found in INCENP$_{II}$. (B) INCENP$_{II}$. The positions of potential nuclear localization (NLS), phosphorylation (P) signals, and predicted coiled coil domains (hatched bars), are shown.

keratins (Lupas *et al.* 1991). In addition, several widely spaced partial matches to nuclear localization consensus signals are present, and there are many potential post-translational phosphorylation modification sites for p34^{cdc2} and tyrosine kinase, casein kinase, cAMP-dependent protein kinase, and protein kinase C. The detection of only one copy of the INCENP gene in the genome suggests that the two classes of INCENPs are likely to be derived from the same gene, presumably by alternative splicing of a primary transcript.

Cell cycle distribution

In interphase cells, the INCENPs are confined within the nucleus. During the early stages of mitosis, the protein becomes chromosome bound as part of the chromosome scaffold. As cells progress from prometaphase into metaphase, the INCENPs are concentrated at the centromeric regions of the chromosome. Immunofluorescence analysis of metaphase cells indicates that the proteins can be resolved into two dots on the centromere. These two dots are localized between the sister chromatids, corresponding to the two regions of the proximal arms thought to be the last points of contact between the sister chromatids just before separation at anaphase (Lima de Faria 1955). (It was this striking localization of the proteins between the sister chromatids, apparently on the 'inner side' of the centromere region, that has given rise to their names. Late in metaphase, the proteins detach from the chromosomes, becoming concentrated in linear arrays that transect the metaphase plate (Earnshaw and Cooke 1991) (Fig. 6.7). By early anaphase, they are found in the spindle midzone, intimately associated with the overlapping antiparallel microtubules of the central spindle. In mid anaphase, a portion of the INCENPs also appears at the cell cortex, where the cleavage furrow will later form. In telophase, the proteins are concentrated at each side of the midbody in the intercellular bridge with which they are discarded after cytokinesis.

Evolutionary conservation

Antibodies raised against chicken INCENP$_I$ have been shown to cross-react with human INCENPs. In addition, recombinant chicken INCENP$_I$ when transfected into human and porcine cells behaves like a chromosomal passenger protein through the same stages of cell cycle. These experiments demonstrate that INCENP$_I$ is both structurally and functionally conserved between birds and mammals, and that many or all of the structures with which INCENP$_I$ interacts appear to be conserved between these species (Mackay *et al.* 1993).

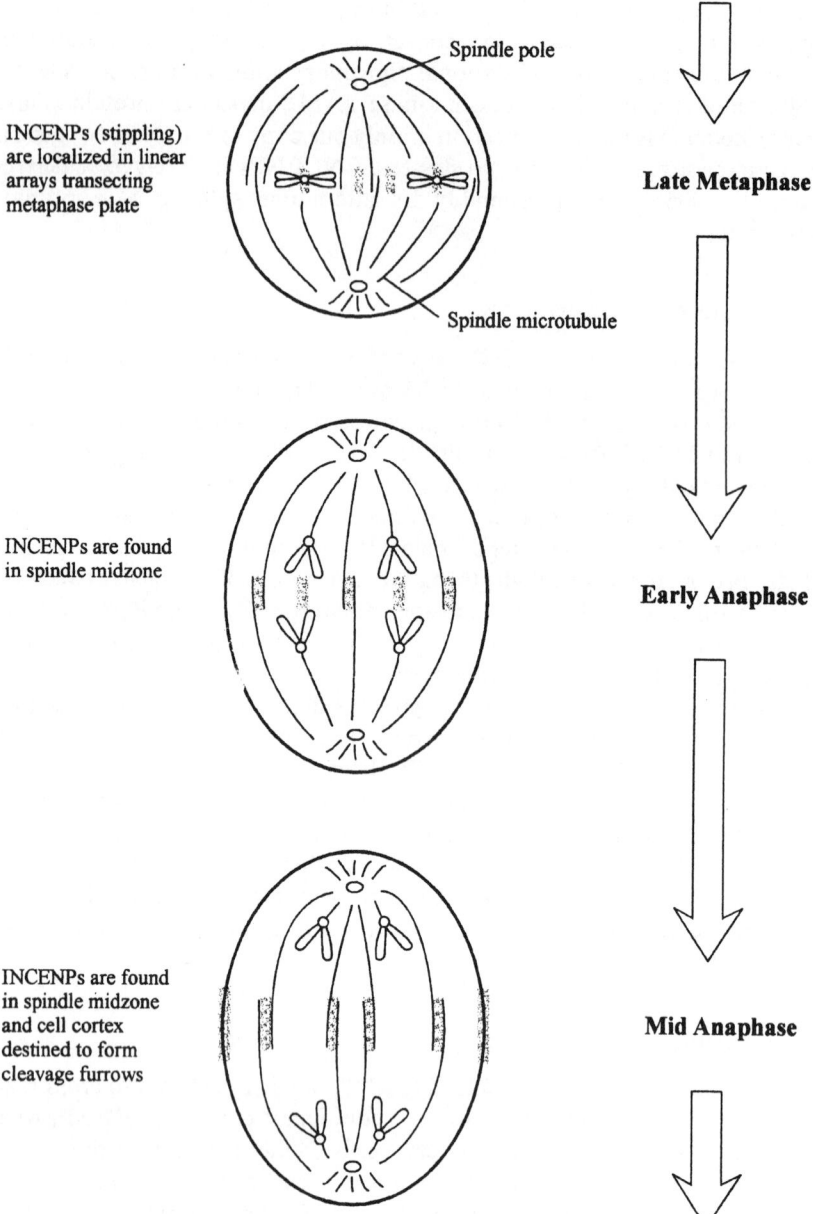

Fig. 6.7 Intracellular distribution of INCENPs during late metaphase, early anaphase, and mid anaphase.

Roles of INCENPs

Sister chromatid cohesion

The localization of the INCENPs at the point of contact between sister chromatid arms suggests that these proteins may be involved in sister chromatid cohesion. This suggestion is strongly supported by the observed disappearance of the proteins from chromosomes during early mitotic anaphase when cohesion is no longer required and sister chromatids begin to separate.

Cytoplasmic microtubule bundling

Another possible role of INCENPs may be in cytoplasmic microtubule bundling. When recombinant $INCENP_I$ proteins are expressed at high levels in interphase cells, it appears in the cytoplasm as well as in the nucleus. In these cells, $INCENP_I$ co-localizes with microtubules, seemingly to bundle them together into coarse, sinus fibrils (Mackay *et al.* 1993). Mutation analysis indicates that the microtubule bundling activity resides within the segment of the $INCENP_I$ polypeptide containing most of the putative coiled coil domain (Fig. 6.6). Deletion of this segment eliminates binding of $INCENP_I$ to cytoplasmic microtubules during interphase, but the distribution of this mutant $INCENP_I$ to chromosomes, the spindle, and the intercellular bridge remains normal (Mackay and Earnshaw 1993; Mackay *et al.* 1993). In a different construct in which the first 42 amino acids from the N-terminus are deleted, the truncated $INCENP_I$ protein becomes completely unable to move from chromosomes to the mitotic spindle, but retains the ability to coat cytoplasmic microtubules when overexpressed in interphase cells. Thus, a region of the protein that is dispensable for association with cytoplasmic microtubules is essential for localization to the mitotic spindle. If bundling of cytoplasmic microtubules reflects INCENP activity at the spindle midzone, it must exist in addition to a separable spindle-binding activity.

Cytokinesis

The concentration of a significant fraction of the INCENPs at the cell cortex in the cleavage furrow during late anaphase and telophase (Earnshaw and Cooke 1991) suggests that the proteins may be functionally important in cytokinesis (for a review on cytokinesis, see Fishkind and Wang 1995). Two lines of evidence (Mackay and Earnshaw 1993) lend support to this suggestion. First, expression of a truncated INCENP lacking a portion of the C-terminal domain is toxic to cells, apparently because it induces frequent failures in cytokinesis. Second, the INCENPs have been shown to be concentrated at the cell cortex where the cleavage furrow will subsequently form before there is any detectable myosin concentration in this area. Thus, INCENP may be one of the earliest known components to become concentrated at the cleavage furrow in mammalian cells, making it a good candidate for being a cleavage stimulus (Rappaport 1986).

CLiPs

CLiPs (or Chromatid Linking Proteins) are a group of antigens identified by autoimmune CREST sera (Rattner *et al.* 1988). These antigens, at least one component of which has a molecular mass of 50 kDa, are conserved in humans, mouse, and Indian and Chinese muntjacs. One subset of CLiPs is detected as a band that extends between the kinetochores of sister chromatids at metaphase, forming a structure which appears to provide a physical link between the two homologous domains that influence disjunction. A second subset of CLiPs is located on the inner surface of the chromatid and is correlated with the close apposition of sister chromatids both at the centromere and the euchromatic arms. Proteins within this subset can be divided into two classes: those on the inner surface of the euchromatic arms that can be disrupted by colcemid treatment, and those found in the primary constriction that are not disrupted by this agent. Following the initiation of anaphase, CLiPs staining disappears from the inner surface of the chromosome, but staining of the kinetochore persists through to telophase. Thus, the CLiPs group of antigens are likely to be involved in the regulation of sister chromatid pairing both at the centromere and at the euchromatic arms, but the persistence of some antigens on kinetochores through to telophase suggests a possible additional role in kinetochore organization and/or chromosomal segregation.

Mei-S332

Genetics of *mei-S332* mutation

Mei-S332 (**mei**otic mutation-**S332**) is a 44 kDa protein that is shown to be required for meiotic sister chromatid cohesion in *Drosophila*. The *mei-S332* mutation (Sandler *et al.* 1968) is unique among known *Drosophila* meiotic mutations because (i) it causes high levels of nondisjunction that appears to occur primarily at the second meiotic division, and (ii) it affects the segregation of all chromosomes in both male and female meiosis and defines a function shared by the two sexes. The latter property was surprising since meiosis proceeds through fundamentally different pathways in *Drosophila melanogaster* males and females, and these pathways were thought to be under separate genetic control. [In *Drosophila* females, meiosis follows a typical pathway in which synapsis and exchange between homologous chromosomes allow their stable orientation toward opposite poles of the spindle. In males, however, recombination, synaptonemal complex, and chiasmata are normally absent, and chromosome pairing is based on the cohesion of specific pairing sites (McKee and Karpen 1990).]

The defect in *mei-S332* males is manifested as premature sister chromatid separation beginning at anaphase I (Fig. 6.8) (Goldstein 1980). By the time the chromosomes recondense during prophase II, most or all pairs of sister chromatids have separated from one another. These precociously separated

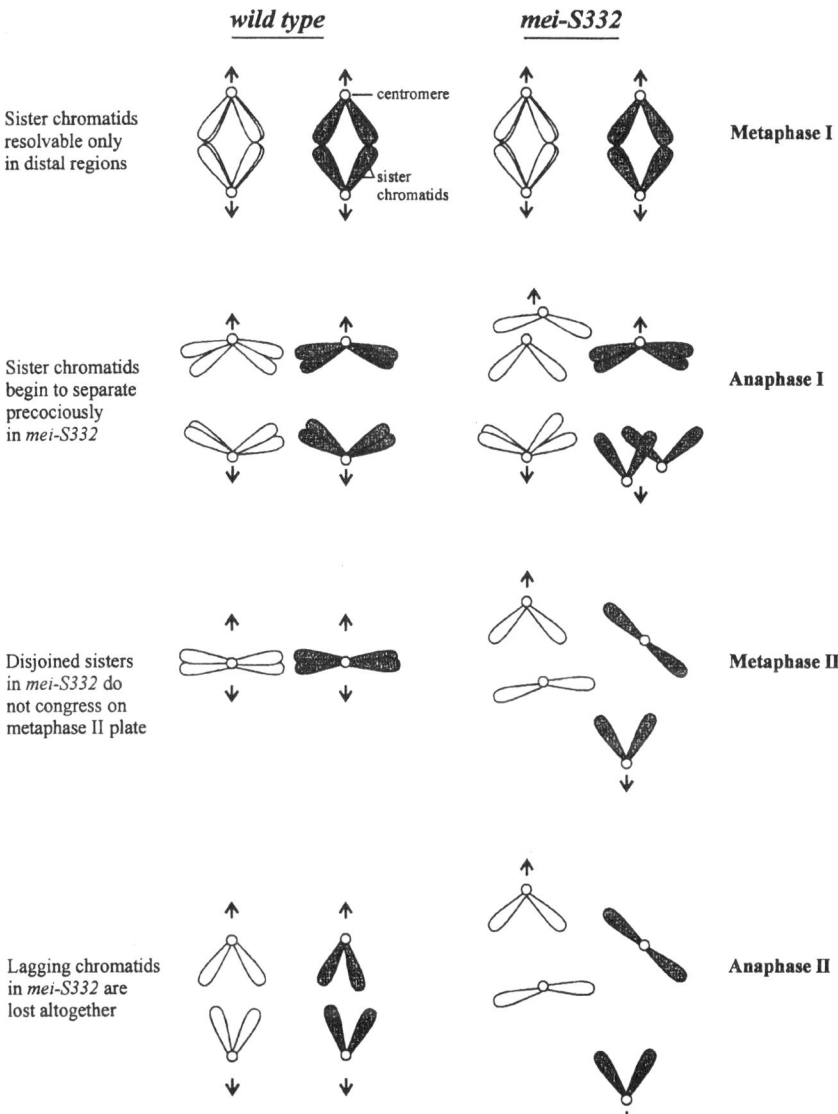

Fig. 6.8 Meiotic segregation of chromosomes in wild-type and *mei-S332 Drosophila* males. The behaviour of two nonhomologous chromosomes (clear or shaded), and the direction of chromosome/chromatid movement (arrow) are shown.

sister chromatids do not congress on the metaphase II plate, but rather move at random with respect to one another. As a result, sometimes one daughter cell receives both sister chromatids, leaving the other with none, or sometimes a chromatid lags at anaphase and is excluded from both daughter nuclei. Thus, a defect in sister chromatid cohesion occurring at anaphase I is manifested genetically as apparent nondisjunction at meiosis II. It is proposed that mei-S332 is required to hold sister centromeres together from the time of kinetochore duplication at metaphase I until the onset of anaphase II. Given that holding sisters together at metaphase I must be a general property of all meiotic systems, the paradoxical effect *mei-S332* mutation has on both sexes can be easily explained.

Meiotic cell cycle distribution

In *Drosophila* spermatocytes, all stages of meiosis are cytologically well resolved and individual chromosome arms and centromere regions are visible. In early prophase I, prior to extensive chromosome condensation, mei-S332 is not detected on chromosomes, but there is a considerable cytoplasmic amount of this protein, possibly localized in some organelles (Kerrebrock *et al.* 1995). As chromosomes condense in prometaphase I and through to metaphase I, the protein is seen on discrete chromosomal sites, whereas the cytoplasmic component diminishes. In anaphase I, mei-S332 is localized to the centromere region (but not the arms) of each chromosome, near the leading edge of the migrating chromosomes. In late anaphase, the protein is present at that part of each chromosome closest to the pole, the centromere region. Association of mei-S332 with the centromeres is seen throughout metaphase II. At anaphase II, as cohesion is lost and the sister chromatids separate, the protein is no longer detectable on the chromosomes. As with the spermatocytes, mei-S332 has also been localized to the centromere region of chromosomes in the oocytes, suggesting that the protein is associated with meiosis in both male and female *Drosophila*.

Role in meiosis and mitosis

The phenotype of the *mei-S332* mutation and the localization of the mei-S332 protein on the centromere region until anaphase II both strongly support a direct role for the protein in meiotic sister centromere cohesion. All the evidence also indicates that the protein has no role in mitosis. Apparent null alleles are fully viable and exhibit normal mitotic chromosome segregation in both genetic and cytological tests, while direct localization experiments have failed to detect mei-S332 on mitotic chromosomes.

Ord

Like mei-332, the 55 kDa ord (for orientation disruptor) protein is another cohesion molecule discovered in *Drosophila*. However, in contrast to the meiotic specificity of mei-332, mutations in the *ord* gene affect several aspects of meiosis as well as mitosis (Mason 1976; Goldstein 1980; Lin and Church 1982; Miyazaki and Orr-Weaver 1992). Meiotically, null alleles of *ord* show missegregation consistent with ratios predicted from premature separation and random segregation of the four sister chromatids through both meiotic divisions, suggesting that ord is necessary to maintain sister chromatid cohesion beginning early in meiosis I when the sister chromatids are held along their lengths. The missegregation phenotype is true for both male and female meiosis. Although no direct localization data are available, it is suggested that the ord protein is likely to be present on both the chromosome arm and the centromere regions (Orr-Weaver 1995).

In addition to the meiotic effect, strong *ord* mutations result in nondisjunction during male germline mitosis (Lin and Church 1982; Miyazaki and Orr-Weaver 1992). However, these same mutations show almost no effect on somatic mitotic divisions, suggesting that the requirement for the gene may be restricted to divisions in the germline (Orr-Weaver 1995).

Cor1

In most sexually reproducing organisms, during early prophase of meiosis each chromosome develops a longitudinal axial core to which the chromatin loops are attached. Subsequently, the cores of each pair of homologous chromosomes become aligned in parallel to form the synaptonemal complex (von Wettstein *et al.* 1984). Each chromosome consists of two sister chromatids, which are attached to their core in a series of loops (Weith and Traut 1980; Heng *et al.* 1994), suggesting that the cores are the source of sister chromatid cohesion during meiotic prophase. Among the protein constituents of the core is a 30 kDa protein, Cor 1, that has been identified in hamster (Dobson *et al.* 1994; Moens and Spyropoulos 1995). A homologue of this protein, referred to as SCP3 (or synaptonemal complex 3), has also been described in rat (Lammers *et al.* 1994).

Meiotic cell cycle distribution

Cor1 is first detected during the leptotene–zygotene stages of meiotic prophase when the cores of chromosomes are still largely unpaired. At pachytene when all chromosomes are fully synapsed, Cor1 distribution coincides with the aligned core. At diplotene, the chromosomes separate (except at the chiasmata) but Cor1 remains present axial to each chromosome. At metaphase I, the protein accumulates at the centromeres, while some remain on either side of the interstitial chiasmata and along the long axes of the chromosomes. When the homologues pull apart at anaphase I, most of the protein has left the chromosomal arms but the centromeres continue to carry an abundance of the protein. The association between centromeres and Cor1 persists till anaphase II, at which time the sister centromeres undergo equatorial segregation and Cor1 lags between the separating centromeres.

Role in meiosis

The above chromosome distribution pattern suggests a role for Cor1 in sister chromatid cohesion throughout meiotic prophase. In addition, its strong presence on the centromeres from metaphase I until the onset of anaphase II suggests an additional role in holding sister centromeres together, thus promoting reductional segregation of sister centromeres in meiosis I. The mechanism that would differentially eliminate axial Cor1 but not centromeric Cor1 remains to be determined. The localization pattern of Cor1 suggests that it may be functionally analogous to the *Drosophila* mei-332 protein, at least after anaphase I (since male *Drosophila melanogaster* lack synaptonemal complexes), but the two proteins share no sequence homology. The Cor1 protein appears to be specific for meiosis but not mitosis, since it is found in meiotically dividing cells but not in somatic nuclei or mitotically active spermatogonia.

Release

The release of cohesion is required for the segregation of sister chromatids to the spindle poles during anaphase. At least two centromere-associated passenger proteins, topoisomerase II and PIM, are thought to regulate this release.

DNA topoisomerase II

Topoisomerase II (Topo II) is a member of the three different families of topoisomerases (reviewed by Roca 1995). It is an essential, highly conserved protein that is found in most, if not all eukaryotic cells (D'Arpa and Liu 1990). Yeast and *Drosophila* have only one form of Topo II, whereas mammalian cells have α and β isoforms that are encoded by different genes. The 170 kDa α isoform is the more abundant protein, constituting 75–80% of total Topo II in proliferating cells, compared with the 180 kDa β isoform (Drake *et al.* 1987, 1988; Chung *et al.* 1989; Woessner *et al.* 1990, 1991). The enzyme works by cleaving and opening one DNA helix transiently, passing a second intact DNA helix through the opening, and then resealing the break (Fig. 6.9; Earnshaw and Heck 1985; Earnshaw *et al.* 1985; Gasser and Laemmli 1986; Roca 1995; Berger *et al.* 1996). Through its ability to alter DNA topology, Topo II has been implicated in a host of cellular processes that require the unwinding and strand separation of double-stranded DNA (Earnshaw and Heck 1988; Smith 1990; Watt and Hickson 1994), including initiation of DNA replication (Nelson *et al.* 1986; Yang *et al.* 1987), DNA repair (Tan *et al.* 1987), recombination (Bae *et al.* 1988), transcription (Cockerill and Garrard 1986; Schröder *et al.* 1987), and chromosome segregation (Rose *et al.* 1990; Klein *et al.* 1992).

Cell cycle distribution

The Topo IIα isoform is undetectable until DNA replication begins in the cells in S phase (Chow and Ross 1987; Woessner *et al.* 1990, 1991; Heck *et al.* 1988). The enzyme increases through S and G_2, and peaks in late G_2/M. After the cells divide and enter G_1, the amount of the enzyme decreases, due to degradation (Heck *et al.* 1988). In contrast, the amount of Topo IIβ isoform remains constant through all phases of the cell cycle (Woessner *et al.* 1991), a pattern of expression suggesting that the two isoforms differ in at least some of their functional roles. Mammalian prophase chromosomes contain an abundance of Topo IIα distributed uniformly throughout their length (Sumner 1996). Chromosomes at a later stage of prophase sometimes show a double line of Topo IIα along their axes, although the chromosomes themselves do not appear to be divided into separate chromatids. At metaphase, the highest concentration of Topo IIα is at the centromeres. This concentration appears to extend throughout the centromere, while the arms display a thin axial distribution (Rattner *et al.* 1996). At anaphase, centromere localization disappears, and only a thin line along the daughter chromosomes remains. The distribution and quantity of this enzyme on chromosomes is therefore consistent with its known functions in chromosome condensation and segregation (discussed below).

When the appearance of Topo IIα on the centromere is compared with that of other transiently associated centromere proteins, Topo IIα is shown to be one of the first proteins to associate with the centromere as it approaches cell division (Rattner *et al.* 1996). In mouse, specific association of Topo IIα with the centromere occurs as early as the late S-G_2 period. Only one other protein, CENP-F, has been shown to associate with the centromere during this pre-mitotic period (Liao *et al.* 1995). The departure of Topo IIα from the centromere during the later stages of cell division appears to coincide with a major movement of other centromeric passenger proteins from this chromosomal region.

Role in chromosome condensation and segregation

During mitosis, Topo II has at least two major chromosomal roles. The first is in the condensation of mitotic chromosomes (Uemura *et al.* 1987; Wood and Earnshaw 1990; Adachi *et al.* 1991; Hirano and Mitchison 1991, 1993; Sumner 1992; Buchenau *et al.* 1993; Gorbsky 1994). The current model of chromatin organization suggests that DNA loops of 50–100 kb are anchored to a network of nonhistone proteins which, in interphase, forms a nuclear matrix and, in mitosis, condenses to form a chromosome scaffold (Laemmli *et al.* 1978). Electron microscopy shows that scaffolds maintain the overall shape of the mitotic chromosome, even after extraction of 95% of the chromosomal proteins and 99% of the chromosomal DNA and RNA (Adolph *et al.* 1977; Paulson and Laemmli 1977; Earnshaw and Laemmli 1983). A major component of the chromosome scaffold is Topo II. In mammalian chromosomes, this protein forms a thin line that extends through the axis or core of each chromatid, presumably corresponding to the chromosome scaffold (Earnshaw and Heck 1985; Earnshaw *et al.* 1985; Gasser *et al.* 1986; Taagepera *et al.* 1993; Giménez-Abián *et al.* 1995; Sumner 1996). Both the Topo IIα and IIβ isoforms are present at the centromeric and axial regions, and both contain the MPM2 phospho-epitope (Taagepera *et al.* 1993). However, despite its localization on centromeres and apparent coincidence with the scaffold distribution, it is not yet possible to conclude unequivocally that Topo II has a structural role in mitotic chromosomes since other studies have indicated that the enzyme may act in a non-structural capacity during the chromosomal events of mitosis, being localized at the sites on the chromosome only where it is required (Hirano and Mitchison 1993; Swedlow *et al.* 1993).

A study of the distribution of Topo IIα in different mammalian cell lines has revealed the following features (Rattner *et al.* 1996): (a) The centromere is distinguished from the remainder of the chromosome by the intensity of its Topo IIα reactivity. (b) The first appearance of Topo IIα at the centromere varies between species but is correlated with the onset of centromeric heterochromatin condensation. (c) The distribution pattern of Topo IIα within the centromere is species- and stage-specific and is conserved only within the kinetochore domain. (d) The Topo IIα inhibitor ICRF-193 prevents the normal accumulation of Topo IIα at the centromere and results in the disruption of chromatin condensation sub-adjacent to the kinetochore as well as the perturbation of kinetochore structure. These features therefore indicate that the distribution of Topo IIα at the centromere is unlike that for the remainder of the chromosome and, together with the observed appearance of Topo IIα on the centromere earlier than most other centromeric passenger proteins, suggest that Topo IIα has a role, whether of a structural or enzymatic nature, for the proper formation and maturation of centromere/kinetochore structure.

The second mitotic role of Topo II relates to sister chromatid disjunction and segregation during anaphase. In this role, Topo II has clearly been demonstrated to be essential for the resolution of interlocking sister chromatid strands that are formed as a result of the replication of topologically fixed DNA (Fig. 6.9; di Nardo *et al.* 1984; Uemura and Yanagida 1986; Uemura *et al.* 1987; Holm *et al.* 1989; Downes *et al.* 1991; Shamu and Murray 1992; Sumner 1992, 1995; Buchenau *et al.* 1993; Clarke *et al.* 1993;

Gorbsky 1994; Ishida *et al.* 1994; Giménez-Abián *et al.* 1995). For example, genetic analysis in yeasts has indicated that Topo II carries out an essential role in the separation of sister chromatids at anaphase (Uemura and Yanagida 1984, 1986; Holm *et al.* 1985). Several lines of evidence suggest that Topo II also plays an essential role in sister chromatid disjunction in higher eukaryotes. For example, teniposide, a potent Topo II inhibitor, blocks disjunction of sister chromatids in the *Xenopus* extract system (Shamu and Murray 1992), and microinjection of anti-Topo II antibodies into *Drosophila* embryos causes a local disruption of anaphase chromatid disjunction (Buchenau *et al.* 1993). The control of Topo II activity appears to be exerted at the level of steric hindrance, since it has been shown that there is no increase in Topo II activity at the metaphase/anaphase boundary (Shamu and Murray 1992).

PIM

In addition to the demonstrated role of Topo II in resolving interlocking sister chromatid DNA strands, interlocking-independent release of sister chromatids at the metaphase/anaphase transition has also been described (Irniger *et al.* 1995; King *et al.* 1995; Tugendreich *et al.* 1995). This interlocking-independent release activity requires the products of two genes identified in *Drosophila: pim* (for **pim**ples) and *thr* (for **thr**ee rows) (Stratmann and Lehner 1996). Since mutants of these two genes show extensive similarities (Stratmann and Lehner 1996), only the properties of the better characterized *pim* gene will be outlined.

   pim expression correlates with mitotic proliferation, with quiescent or endoreduplicating tissues showing no detectable gene activity. During mitotic interphase, PIM accumulates predominantly in the cytoplasm. In prophase and metaphase, it is distributed throughout the cells, whereas in anaphase the protein level decreases rapidly, disappearing totally in telophase. In *pim* mutants, the absence of *pim* function causes a failure of sister chromatids to separate in mitosis and results in a mass of unseparated chromosomes in the equatorial plane of the spindle (Stratmann and Lehner 1996). As a consequent, cytokinesis is also defective. However, cell cycle progression is not blocked by the failure of sister chromatids to separate, and metaphase 'diplochromosomes' with twice the normal number of chromosome arms still connected in the centromeric region are observed in the following mitosis. At later stages, 'quadruplechromosomes' with four times the normal number of arms connected in the centromeric region are formed. Separation of the chromosome arms apparently still occurs in the mutants. These observations suggest that PIM functions at the metaphase/anaphase transition to release the cohesion in the centromeric region. The mitotic degradation after this transition stage might provide one level of regulation preventing a premature separation of sister chromatids during interphase.

Other candidate release proteins

In addition to Topo II and PIM, some proteins of unknown nature must be proteolysed to result in anaphase separation (Holloway *et al.* 1993). These proteins may be directly involved in promoting centromere pairing or in protecting the topological linkages between sister chromatids from being destroyed by Topo II. Other candidate proteins which have been reported to have a possible role in sister chromatid disjunction are the products of the *Drosophila* genes *rough deal* [phenotype: lagging chromosomes and anaphase bridges (Karess and Glover 1989)], *lodestar* [phenotype: chromosome tangling and anaphase bridges (Girdham and Glover 1991)], *fizzy* [phenotype: metaphase arrest with some chromosomes escaping into anaphase (Dawson *et al.* 1993)], and *three row* [phenotype: metaphase arrest (D'Andrea *et al.* 1993; Philp *et al.* 1993)].

A model for higher order centromere assembly, pairing, and release

Fig. 6.9 depicts a model for the higher-order assembly, pairing, and separation of the mitotic centromere–kinetochore complex. In this model, a typical higher eukaryotic centromere is shown as a series of structurally and functionally identical, repeating subunits that are made up of complexes of DNA and kinetochore proteins (Fig. 6.9A) (see Compound centromere in Chapter 4). The portions of the centromere DNA that constitute these subunit structures (designated as the kinetochore-organizer DNA) are separated by linker DNA segments (Zinkowski *et al.* 1991; Brinkley *et al.* 1992). When DNA undergoes replication (Fig. 6.9B), the newly formed sister chromatid strands are inter-locked as a result of the replication of topologically fixed DNA (Sundin and Varshavsky 1980, 1981). During chromosomal condensation in prophase, the sister strands are intertwined at the linker DNA segments to form the pairing domain, while the kinetochore subunit structures face outward to constitute the kinetochore domain that allows spindle microtubule attachment (Fig. 6.9C). This structure is held throughout metaphase until just prior to the onset of anaphase segregation, when the intertwined sister molecules are enzymatically resolved (Fig. 6.9D–H).

Movements

Chromosome movements are in a large part controlled by motor molecules. To date, three proteins (CENP-E, MCAK, and Kid) which belong to the kinesin superfamily of motor molecules have been localized to mammalian centromeres and demonstrated to have a role in chromosome segregation. In addition, a fourth protein (cytoplasmic dynein) belonging to the dynein family of motors has been shown to be associated with the mammalian centromere and involved in chromosome movement.

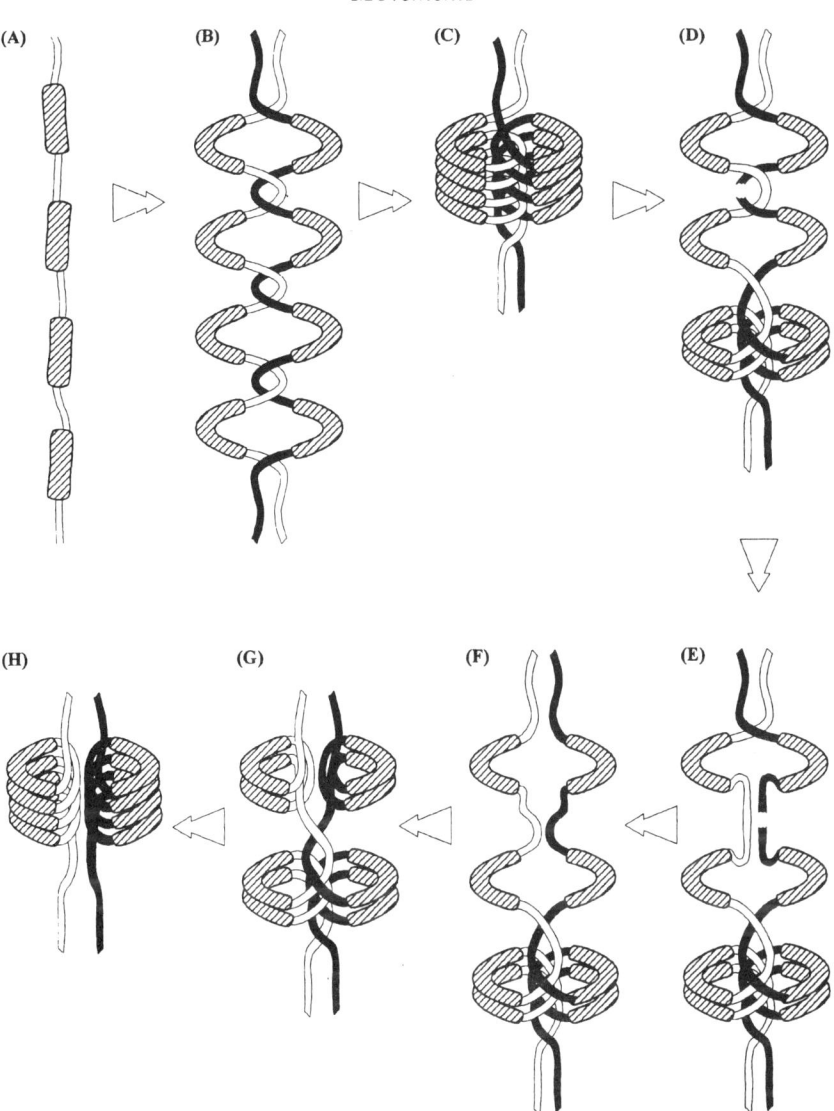

Fig. 6.9 A model for the assembly, pairing, and separation of the higher eukaryotic centromere. (A) A linear centromere molecule at interphase, showing four identical kinetochore subunits formed from an underlying kinetochore-organizer DNA sequence (hatched regions), joined together by linker DNA segments (double-line portions). (B) Newly replicated sister centromeres are intertwined during G_2 phase of the cell cycle. (C) At the onset of mitosis, chromosome condensation and folding bring the kinetochore subunits together in parallel register forming a kinetochore disc at the primary constriction; the linker DNA segments are folded inward to form the centromere-pairing domain. (D–H) Resolution of intertwined sister centromere strands by Topo II during the metaphase/anaphase transition. The enzyme works by cleaving and opening one DNA helix transiently (D), passing a second intact DNA helix through the opening (E), and then resealing the break (F). This process is then repeated until all the catenated kinetochore subunits are resolved (G and H).

The kinesin superfamily of motor molecules

Kinesin was the first member of the kinesin superfamily to be discovered (Brady 1985; Vale *et al.* 1985). This superfamily of molecules, together with the dynein family of related proteins, form the two known classes of microtubule motor proteins (reviewed by Endow 1991; Vale 1992). More than 30 members of the kinesin superfamily of proteins have now been identified (reviewed by Goldstein 1993a). This superfamily of proteins is increasingly being credited for the proper sorting of diverse elements within the cell. The identification of so many different kinesin-related proteins in single organisms has led investigators to suggest that different kinesins are responsible for the transport of different cargoes within the cell, and/or that these different kinesins are subject to differential regulation. Developmental stage-specific and tissue-specific members of this superfamily have been identified (Stewart *et al.* 1991; Aizawa *et al.* 1992). There is also increasing evidence implicating different kinesin-like proteins in neuron-specific (Hall and Hedgecock 1991) or general mitotic/meiotic functions (Saxton *et al.* 1991; Endow 1993; reviewed by Goldstein 1993a).

Mitotic and meiotic functions

Kinesin-like molecules that have been shown by genetic and/or biochemical means to be involved in mitotic and meiotic functions include bimC, KLPA (in *Aspergillus nidulans*; Enos and Morris 1990; O'Connell *et al.* 1993), ncd and nod (in *D. melanogaster*; Endow *et al.* 1990; McDonald and Goldstein 1990; Zhang *et al.* 1990; Endow 1993; Afshar *et al.* 1995; Murphy and Karpen 1995), Kar3p (Chapter 2), CIN8 and KIP1 (in *S. cerevisiae*; Hoyt *et al.* 1992, 1993; Roof *et al.* 1992), cut7 (in *S. pombe*; Hagan and Yanagida 1990, 1992), Eg5 and Kklp1 (in *Xenopus laevis*; LeGuellec *et al.* 1991; Vernos *et al.* 1995), and the three mammalian/higher eukaryotic proteins CENP-E, MCAK, and Kid (described below). Mutations in these genes appear to affect spindle structure and/or the fidelity of chromosome segregation. Immunolocalization studies, where performed, also demonstrate that these kinesin-related proteins associate with the mitotic spindle or structures therein (Hagan and Yanagida 1992; Hatsumi and Endow 1992; Hoyt *et al.* 1992; Theurkauf and Hawley 1992). Interestingly, although these kinesin-like proteins have a role in fundamentally important cellular processes, functional redundancy within the system seems to render some of these genes dispensable (Goldstein 1993b).

Structural and functional domains of kinesin proteins

Kinesins are mechanochemical proteins that are capable of converting the chemical energy derived from the hydrolysis of ATP into a mechanical force that is needed to move their cargo along microtubules, i.e. they are microtubule-activated ATPases (Bloom *et al.* 1988; Scholey *et al.* 1989; Yang *et al.* 1990). Different kinesin proteins have been identified that can move in either direction along a microtubule (McDonald *et al.* 1990; Walker *et al.* 1990); however, it appears that as yet undefined determinants within the motor domain structure may render the directionality uniform for any particular kinesin (Stewart *et al.* 1993). Fig. 6.10 shows the structure of a prototype kinesin protein molecule (Endow 1991; Vale 1992; Goldstein 1993a) but significant variations from this prototype can be found in some members of the superfamily. The prototype structure contains two heavy and two light chains, although a subset of kinesins has been isolated without light chains. The structure can be divided into three domains: a tail domain at the C-terminal region composed of heavy and light chains (Hirokawa *et al.* 1989), a middle stalk domain composed of an α-helical coiled coil primarily formed by the heavy chain (Yang *et al.* 1989; deCuevas *et al.* 1992), and a globular head domain composed of the N-terminal ~ 100 amino acids of the heavy chain (Hirokawa *et al.* 1989; Scholey *et al.* 1989; Yang *et al.* 1989) that is usually 35–45% identical among all kinesin superfamily members. Although the head domain of the majority of kinesins (including CENP-E) has been shown to reside at the N-terminus of the peptide and is responsible for generating the motor activities (i.e. contains the sites for ATP and micro-tubule binding) (Yang *et al.* 1990; Stewart *et al.* 1991), kinesins with the motor domain located at the C-terminus (e.g. Kar3p, ncd, and KLPA) or in the center of the protein (e.g. MCAK) have also been identified.

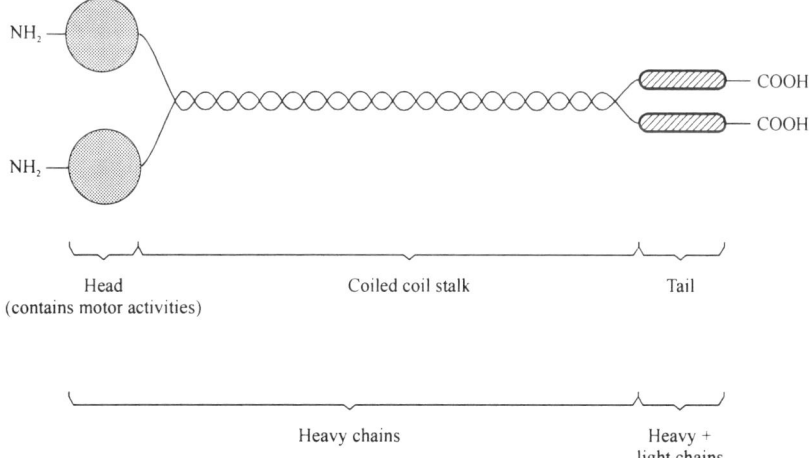

Fig. 6.10 Structural domains of a prototype kinesin motor protein molecule. Each molecule contains two heavy and two light chains. The globular head domains (shaded circles) bind ATP and microtubules, and constitute the motor regions. Attached to the heads is an α-helical coiled coil stalk that extends to two tail structures (hatched rods) composed of a mixture of kinesin light chain and the C-terminal end of kinesin heavy chain.

CENP-E

Centromere sublocalization

CENP-E, a protein originally identified using a monoclonal antibody raised against a chromosome scaffold fraction (Yen *et al.* 1991), is the first member of the kinesin-related motor proteins that has been localized to the mammalian centromere. Immunofluorescence staining of mitotic HeLa chromosomes demonstrates the protein to be present as a characteristic pair of dots on the centromere of each set of paired chromosomes. Unlike the wide variation of centromere staining seen with anti-CENP-B antibodies, the overall CENP-E staining of each chromosome is of similar intensity. When the distances between pairs of dots are compared, CENP-E is found to lie at the periphery of CENP-B, at a position expected for a component of the kinetochore or its associated corona.

Protein structure

CENP-E is a 275 kDa protein that is made up of three principle domains: N- and C-terminal globular domains, each of ~ 500 residues, separated by a central ~ 1700 amino acid-segment that may form a discontinuous α-helix (Fig. 6.11) (Yen *et al.* 1991, 1992). This helical stalk domain contains an extensive, hydrophobic heptad repeat characteristic of coiled coil helices (Lupas *et al.* 1991). The N-terminal 335 amino acids of CENP-E share extensive homology with the motor domain of all known kinesin-like proteins (MacDonald and Golstein 1989; Yang *et al.* 1989; Endow *et al.* 1990; Hagan and Yanagida 1990; Kosik *et al.* 1990; Endow 1991; LeGuellec *et al.* 1991; Otsuka *et al.* 1991; Wright *et al.* 1991; Navone *et al.* 1992). Within this motor region, CENP-E shares virtual sequence identity with the 120 residues that comprise the highly conserved nucleotide- and microtubule-binding sites that are characteristic of the kinesin family of proteins (reviewed by Endow 1991). In sequence and in predicted structure, CENP-E is most like the conventional kinesin heavy chain, although its α-helical stalk segment is nearly four times longer than that of kinesin. Unlike other kinesin-related proteins, sequence comparison reveals an additional 18 amino acid segment near the C-terminus of CENP-E with 78% sequence similarity to sequences in the C-termini of heavy chain kinesins (Yang *et al.* 1989; Kosik *et al.* 1990; Wright *et al.* 1991; Navone *et al.* 1992). As kinesin light chains bind in this globular tail region (Hirokawa *et al.* 1989), this conserved domain may represent a binding site for kinesin light chains or related accessory factors. Two specific N- and C-terminal domains have also been experimentally shown to bind microtubules (Fig. 6.11; discussed below).

Cell cycle distribution

The cell cycle distribution of CENP-E has been studied in detail (Fig. 6.12) (Yen *et al.* 1991, 1992). Cells in the G_1 and early S phases show little

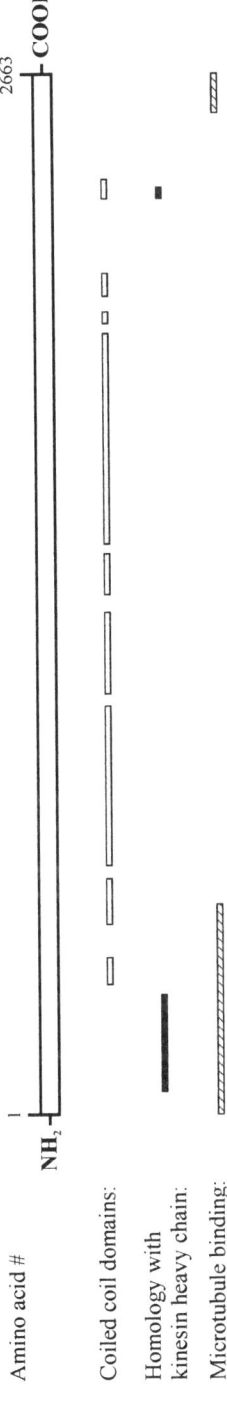

Fig. 6.11 Structural domains of human CENP-E protein. Shaded bars indicate regions predicted to adopt coiled coil conformations. Black bars indicate regions of significant sequence identity to *Drosophila* kinesin heavy chain. Hatched bars denote two microtubule-binding domains.

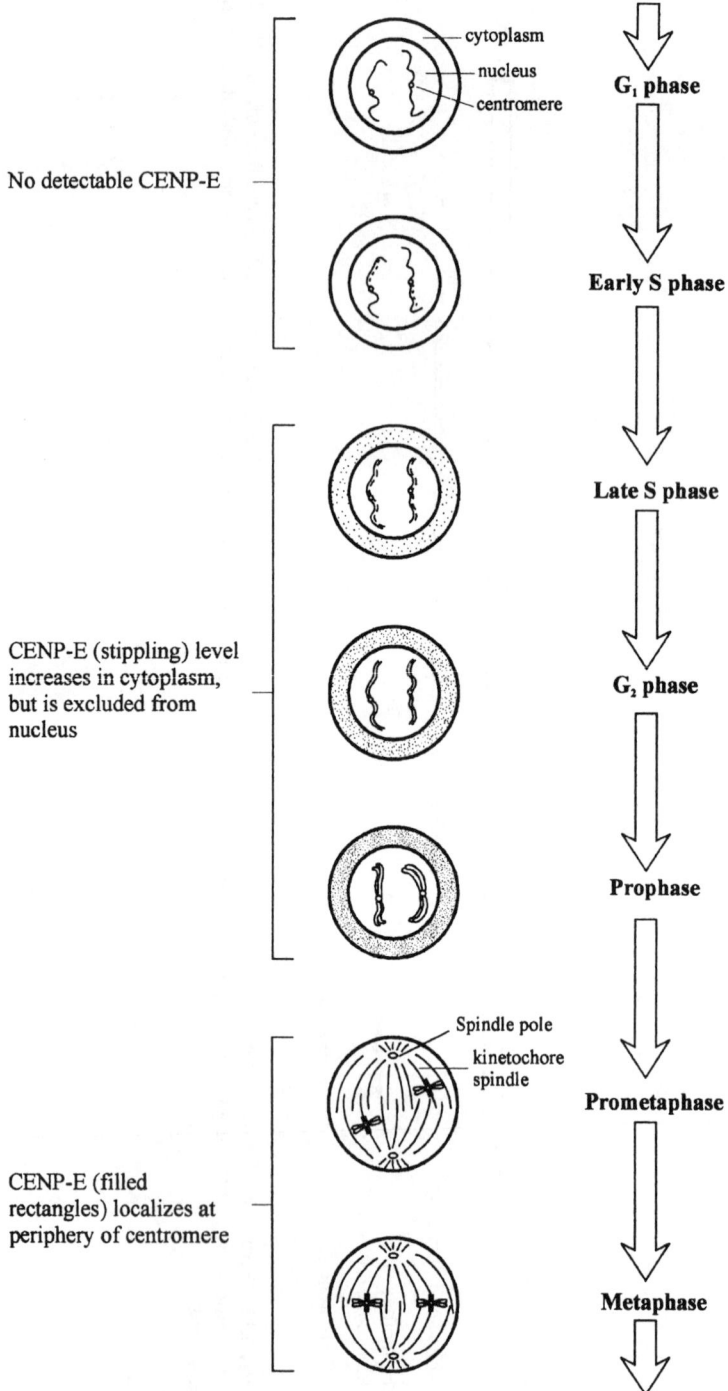

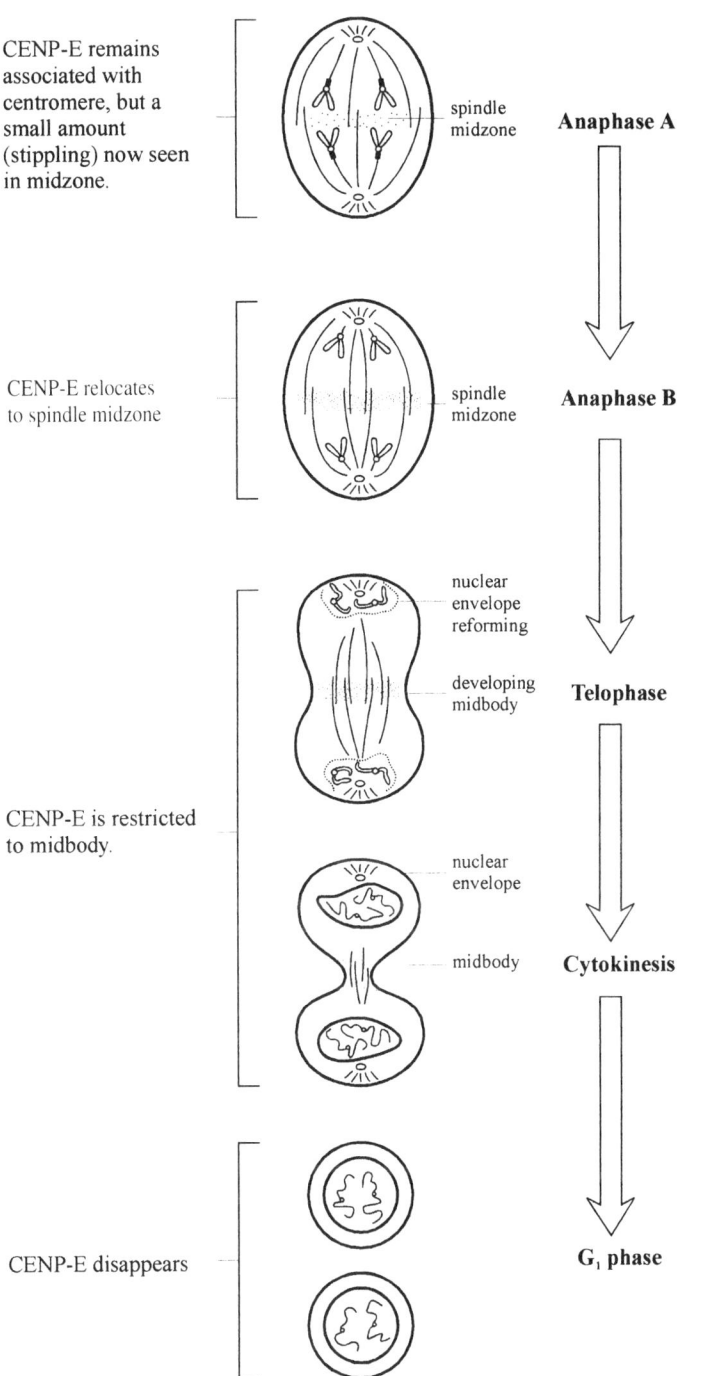

CENP-E remains associated with centromere, but a small amount (stippling) now seen in midzone.

spindle midzone — **Anaphase A**

CENP-E relocates to spindle midzone

spindle midzone — **Anaphase B**

nuclear envelope reforming

developing midbody — **Telophase**

CENP-E is restricted to midbody.

nuclear envelope

midbody — **Cytokinesis**

CENP-E disappears

G₁ phase

Fig. 6.12 Distribution of CENP-E throughout the mammalian mitotic cell division cycle.

detectable CENP-E in either cytoplasmic or nuclear fractions, but cytoplasmic CENP-E levels rise sharply during late S and G_2/M, ultimately reaching the level found in prometaphase cells. At all times CENP-E is excluded from nuclei, with such an exclusion being consistent with the absence of a consensus nuclear targeting signal on the CENP-E polypeptide (Kalderon *et al.* 1984; Lanford and Butel 1984; Robbins *et al.* 1991). Following the appearance of CENP-E at prometaphase, and throughout metaphase, the protein is localized on chromosomes, specifically to the outer surface of the centromeres. When chromosomes begin to separate during early anaphase (anaphase A), the majority of CENP-E remains associated with the chromosomes (Lombillo *et al.* 1995). Although a small amount of CENP-E is visible in the region between the separated chromosomes, indicating that some of the proteins have migrated to the midzone (Lombillo *et al.* 1995), association of the protein with the kinetochore has been shown throughout the entirety of anaphase A (Brown *et al.* 1996). During spindle elongation in anaphase B, CENP-E gradually relocates to the interzonal microtubules. At telophase, the protein is restricted to the midbody and is virtually all lost at the end of mitosis.

Mechanisms for CENP-E accumulation and loss

CENP-E levels increase progressively across the cell cycle, peaking at $\sim 22\,000$ molecules per cell early in mitosis, followed by an abrupt (> 10-fold) loss at the end of mitosis (Brown *et al.* 1994). Pulse-labelling with [^{35}S]methionine reveals that beyond a modest 2-fold increase in synthesis between G_1 and G_2, interphase accumulation results primarily from stabilization of CENP-E during S and G_2. Although moderately unstable during G_1, CENP-E is completely stabilized during S/G_2 and subsequent stages of mitosis through to metaphase. After the anaphase transition, CENP-E is destined for inactivation at the terminal phases of mitosis through pathways that sequester it to the developing midbody and selectively degrade it. However, despite its localization in the midbody, the CENP-E loss that initiates in late anaphase/telophase is independent of cytokinesis, since complete blockage of division (e.g. with cytochalasin) has no effect on CENP-E loss at the M/G_1 transition.

CENP-E polypeptide accumulation and loss are reminiscent of the pattern initially described for cyclins A and B; these cyclins are continuously synthesized between cell divisions and are rapidly degraded near the end of mitosis in invertebrate embryos (Evans *et al.* 1983; Swenson *et al.* 1986; Standart *et al.* 1987). In human cells, cyclin B accumulates sharply in G_2, peaks in early M, and then plummets in abundance by the end of cell division (Pines and Hunter 1989). A comparison of the time of degradation of cyclin A, cyclin B, and CENP-E reveals a different temporal regulation for each during mitosis. The earliest of these to degrade is cyclin A, the loss of which clearly precedes degradation of cyclin B (Luca and Ruderman 1989; Minshull *et al.* 1989, 1990; Lehner and O'Farrell 1990; Pines and Hunter 1990; Hunt *et al.* 1992), while CENP-E is the last of the three proteins to be degraded. Furthermore, cyclin A is degraded before the metaphase/anaphase transition (Minshull *et al.* 1989; Lehner and O'Farrell 1990; Hunt *et al.* 1992; Edgar *et al.* 1994), while cyclin B is degraded after the metaphase/anaphase transition (Holloway *et al.* 1993; Surana *et al.* 1993), and CENP-E is degraded well towards the end of mitosis. The degradation of these three proteins therefore defines a cascade of at least three temporally regulated proteolytic steps during mitosis (Brown *et al.* 1994).

Ubiquitin-mediated proteolysis has been shown to be responsible for the degradation of the cyclins (Glotzer *et al.* 1991; Hershko *et al.* 1991). Since CENP-E contains a sequence highly similar to the 'destruction box' motif proposed to mediate the ubiquitin-dependent proteolysis of cyclin A (Glotzer *et al.* 1991), it is plausible that CENP-E degradation may also be mediated by a ubiquitin-dependent mechanism. Alternatively, since CENP-E contains two regions with strong homology to 'PEST' sequences (which have been shown to signal the rapid degradation of a variety of proteins; Rogers *et al.* 1986; Rechsteiner 1990), CENP-E degradation may be mediated through such motifs.

Role of CENP-E

The importance of CENP-E for centromere function can be inferred from a study examining its distribution on mitotically stable dicentric chromosomes (Sullivan and Schwartz 1995). The results have indicated that the protein is detected only on the active, but not the inactive, centromeres of these chromosomes, suggesting that it is directly associated with centromere function. A number of established or proposed roles for CENP-E are described below.

CENP-E is a minus end-directed motor

In the presence of ATP, partially purified CENP-E has been shown to exhibit motor activity by moving microtubules in a microtubule-gliding assay. In addition, the motor-induced movement has been shown to be minus end-directed (Thrower *et al.* 1995). The CENP-E-associated motor has an apparent native molecular weight of 874 kDa (Thrower *et al.* 1995), which is almost three times larger than that predicted from its amino acid sequence. Other molecular motors, including myosin, kinesin, and dynein, are all complexes composed of one or two heavy chains and a number of low molecular weight proteins. It is therefore possible that CENP-E may also exist in its native state as a dimer and/or complexed with other proteins.

Cross linking of microtubules

Two microtubule-binding domains have been identified on CENP-E protein (Liao *et al.* 1994). The first is found in a N-terminal 540 amino acid segment that shares homology with the motor domain of kinesins (Fig. 6.11). The second domain is located in a C-terminal 99 amino acid segment. The identification of these two biochemically distinct microtubule-binding domains suggests a role for CENP-E as a microtubule crosslinker. Furthermore, it has been shown that the C-terminal domain contains consensus phosphorylation sites (Niggs 1993) for MPF (a cyclin B–cdc2 kinase complex that triggers cell entry into mitosis and whose inactivation is necessary for the transition from metaphase to anaphase; see p34^{cdc2} above) and that phosphorylation of this C-terminal domain directly inhibits its microtubule-binding activity. It is therefore possible that the microtubule crosslinking function of CENP-E is suppressed before anaphase by mitotic kinases such as MPF, which phosphorylates and inhibits the microtubule-binding activity of the C-terminus. The loss of MPF activity at the onset of anaphase (Pines and Hunter 1990) then leads to the dephosphorylation and activation of the C-terminal microtubule-binding domain. This temporally regulated crosslinking of microtubules by CENP-E may help to stabilize the overall structure of the anaphase spindle or serve to push the overlapping set of antiparallel microtubules past each other and elongate the spindle poles (Liao *et al.* 1994).

As couplers to depolymerizing microtubules

One potential mechanism for force generation to move chromosomes is to harness energy released by microtubules in an ATP-independent manner. By demonstrating that CENP-E can act as a molecule that couples its attached kinetochore/chromosome cargo to depolymerizing microtubules (Lombillo *et al.* 1995; reviewed by Desai and Mitchison 1995), it is proposed that, in addition to its role as an ATP-dependent motor, CENP-E is capable of harnessing the energy released by microtubules to move chromosomes in an ATP-independent manner. Also, this coupling activity may allow spindle microtubules to remain kinetochore bound while their lengths change during both prometaphase and anaphase A.

154 Antigen

154 antigen is a 250 kDa protein identified in HeLa cells by a monoclonal antibody raised against mitotic chromosome scaffolds (Compton *et al.* 1991). The antigen first appears at prometaphase when it localizes as a pair of immunofluorescence centromeric dots on each newly condensed chromosome. These dots stain relatively uniformly on all the centromeres, with distance measurement indicating that they are located on sites peripheral to

those of CENP-B, presumably within the kinetochore or corona. At metaphase, the antigen remains localized to the aligned chromosomes. At anaphase, when the chromosomes begin to move toward the poles, the antigen dissociates from the chromosomes and localizes to the spindle midzone. At telophase, the antigen concentrates in the midbody where it is presumably discarded after completion of cytokinesis. Hence, by virtue of the similar protein size and cell cycle distribution, and the fact that 154 antigen and CENP-E were both identified through the use of monoclonal antibodies raised against a chromosome scaffold fraction, it is likely that 154 antigen is the same protein as CENP-E.

MCAK (*m*itotic *c*entromere-*a*ssociated *k*inesin)

Protein structure

MCAK is a 90 kDa protein and the second member of the kinesin superfamily of motor proteins that has been localized to the mammalian centromere. Its cDNA was isolated (Wordeman and Mitchison 1995) using affinity-purified peptide antibodies raised against conserved regions of the kinesin motor domain (Sawin *et al.* 1992). Sequence analysis shows that MCAK has a good homology (40–50% identity, 73% within the motor domain) with another kinesin family member, kif2 (Aizawa *et al.* 1992). Unlike the prototype structure of kinesin-related proteins where the motor domain is displaced either toward the N- or C-terminal end of the protein, the motor domain of MCAK (also kif2) is located centrally within the peptide (Fig. 6.13). In the segment C-terminal to the globular motor domain, there are extended regions of predicted helix domain. Within both the N- and C-terminal nonmotor domains are two regions of predicted coiled coil (Lupas *et al.* 1991) that might be involved in interactions with DNA, light chains, or heavy chains. Putative ATP-binding and nuclear localization sequences (Chelsky *et al.* 1989) are also present.

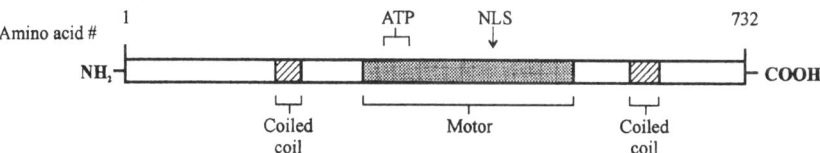

Fig. 6.13 Structural domains of Chinese hamster ovary MCAK protein. The shaded area represents the motor domain. Hatched areas, ATP, and NLS denote predicted regions for coiled coil structures, ATP-binding, and nuclear localization sequences, respectively.

Evolutionary conservation

Monoclonal antibody raised against Chinese hamster MCAK cross-reacts with all organisms so far tested, including *Xenopus laevis, Strongylocentrotus purpuratus, Drosophila melanogaster, Dictyostelium discoideum* amoebae, and *Saccharomyces cerevisiae* (Wordeman and Mitchison 1995). The *Xenopus* homologue of MCAK, designated XKCM1 (for Xenopus kinesin central motor 1), has been isolated and shown to have a similar pattern of subcellular localization as MCAK (Walczak *et al.* 1996). The MCAK protein is thus highly conserved through evolution. On immunoblots, a 90 kDa band is detected in all vertebrates, whereas at least one and sometimes two bands that deviate from 90 kDa are seen in the other organisms.

Cell cycle distribution

During interphase, approximately equal amounts of MCAK are found in the cytoplasm and the nucleus. The prominent targeting of MCAK to the interphase nucleus is consistent with the presence of a consensus nuclear localization sequence within the motor domain of the protein. MCAK associates with the centromere region of mitotic chromosomes at prophase and persists until telophase, after which time the localization disperses. During this association, the protein is found throughout the centromere region and between the kinetochore plates (Wordeman and Mitchison 1995). This localization pattern is in contrast to those of the other two centromere-associated motors, CENP-E and dynein, where the distribution is at the outer surface and corona regions of the kinetochore (Pfarr *et al.* 1990; Steuer *et al.* 1990; Wordeman *et al.* 1991; Yen *et al.* 1991, 1992; Wordeman and Mitchison 1995). The finding of a microtubule-dependent motor such as MCAK throughout the centromere region is surprising since it has been shown that the corona is the site of rapid microtubule-gliding activity in mitotic chromosomes (Reider and Alexander 1990).

Comparison with CENP-E

Both MCAK and CENP-E are kinesin-related proteins that appear on kinetochores at prophase. However, a number of characteristics clearly distinguish the two proteins and indicate that they are likely to have different functions during mitosis. CENP-E is a much larger protein that has an N-terminal motor domain. It accumulates in late S/G_2 phase, associates with kinetochores at prophase, and relocates to the spindle midzone at metaphase. MCAK is a smaller protein with a centrally located motor domain. It is present throughout the cell cycle, associates with centromeres at early prophase (before CENP-E), and remains associated with the centromere until after telophase. CENP-E is a rare protein, whereas MCAK appears abundant. Furthermore, the two proteins have a distinctly different pattern of sublocalization within the centromere region of the mitotic chromosome.

Role of MCAK

Preliminary gliding motility assays indicate that MCAK will bind to microtubules in the absence of ATP. The addition of ATP will cause MCAK to release the microtubule, but this is not accompanied by any sign of gliding motility in either direction (Wordeman and Mitchison 1995). In the light of the peculiar chromosomal location of MCAK (internal to where most of the motor activity is presumed to occur), these results suggest the possibility of a fundamentally different role in chromosome movement than would be expected from a microtubule motor protein. During mitosis, many microtubules are embedded in the kinetochore plate and the chromosomes undergo many reversals in direction (Skibbens *et al.* 1993). The tens of microtubules that are attached to any one kinetochore must undergo coordinated polymerization and depolymerization. During prometaphase, a pair of sister chromosomes moving in one direction involves the coordinate polymerization of microtubules at one kinetochore and depolymerization at the other kinetochore. A motor protein that is more deeply embedded in the kinetochore may be used to anchor the centromere firmly to the dynamic microtubule ends, or it may be involved in modulating microtubule dynamics. It is also possible that MCAK may play a role in shuttling tubulin subunits from one side of the centromere to the sister side during rapid prometaphase movements and reversals in direction (Wordeman and Mitchison 1995).

Kid

Kid (kinesin-like DNA-binding protein) is the third member of the kinesin family of proteins which demonstrates centromere localization (Tokai *et al.* 1996). The protein is found exclusively in the nucleus at interphase. In prophase, a fraction of the protein is concentrated at the centrosomes, although a significant fraction remains throughout the nucleus. At prometaphase and metaphase, the protein is present along the length of chromosomes. At anaphase, Kid moves apart with the separating chromosomes and is enriched at the spindle pole-proximal side of the chromosomes. During this phase, at least some population of the protein is localized at the kinetochore, but the distribution is very diffuse. Co-localization of the protein with microtubules also becomes less obvious. At cytokinesis, Kid further segregates from microtubules, but remains around the pole and eventually becomes localized within the two daughter nuclei.

The cloned human cDNA of Kid (Tokai *et al.* 1996) predicts a gene product of 73 kDa that is related to nod, a *Drosophila* protein that is involved in chromosomal segregation during meiosis (Zhang *et al.* 1990). A sequence similar to the motor domains of kinesin-like proteins is present in the N-terminal half of the protein and the ability of this domain to bind microtubules has also been demonstrated. The C-terminal half contains a putative nuclear localization signal and is able to bind to DNA. Proposed roles for this protein include regulation of chromosomal movement along microtubules during mitosis, binding to centromere DNA sequence, and mitotic spindle formation.

Cytoplasmic dynein

Dynein was first discovered in eukaryotic axonemes as the ATPase required for flagellar and ciliary beating (Gibbons and Rowe 1965). The cytoplasmic form of this protein, isolated two decades later (Paschal and Valice 1987), turns out to be a minus end-directed, microtubule-based motor that is thought to drive the movement of intracellular structures as diverse as chromosomes and cellular organelles such as endosomes, lysosomes, and Golgi apparatus. Although structurally quite distinct, both the axonal and cytoplasmic dyneins appear to be similar to myosin, kinesin, and the superfamilies of myosin- and kinesin-related proteins in the general mechanism for coupling the energy of ATP hydrolysis to the production of mechanical force (Goldstein 1991; Hammer 1991; Pollard *et al.* 1991; Bloom 1992; Endow and Titus 1992; Cheney *et al.* 1993; Goodson and Spudich 1993). In the following discussion, only the properties of cytoplasmic dynein that are pertinent to mitotic and centromere functions will be included; for reviews on axonemal dyneins and other intracellular functions of cytoplasmic dyneins, see Vallee and Shpetner 1990; Holzbaur and Vallee 1994; Holzbaur *et al.* 1994; Schroer 1994.

Protein structure

Cytoplasmic dynein has been studied in organisms such as *D. discoideum, S. cerevisiae, S. pombe, A. nidulans, N. crassa,* chicken, cattle, mouse, rat, and humans (reviewed by Holzbaur and Vallee 1994; Schroer 1994). The structure of the protein is well conserved and, with a molecular mass in excess of 10^6 (Vallee *et al.* 1988), is the largest and most complex of the known motor proteins. All cytoplasmic dyneins that have been examined have two large, force-producing globular heads that are connected by slender stalks to a common base (Vallee *et al.* 1988). This base appears to be as massive as each of the heads (Johnson and Wall 1983; Vallee *et al.* 1988) and is composed of several small globular domains that could represent individual dynein accessory subunits (Sale *et al.* 1985). It is thought that the base of cytoplasmic dynein anchors the molecule to organelles such as kinetochores, while the globular heads, with their variety of protrusions (Goodenough and Heuser 1982, 1984; Amos 1989) interact in an ATP-dependent manner with the microtubules (Vallee *et al.* 1988).

Substructurally, cytoplasmic dynein is a multisubunit complex that can be resolved into a number of polypeptides, generally consisting of two identical heavy chains of greater than 400 kDa, three or four 74 kDa intermediate chains, and numerous light-intermediate chains of approximately 55 kDa (Lye *et al.* 1987; Paschal *et al.* 1987, 1993; Vallee *et al.* 1988; Hughes *et al.* 1993). The heavy chains contain the sites for ATP hydrolysis and microtubule binding, and probably make up the bulk of the globular heads, while the intermediate and light-intermediate chains are believed to lie at the base of the molecule.

Heavy, intermediate, and light-intermediate chains

Cytoplasmic dynein heavy chains from different organisms show an overall highly conserved central-third portion, followed by the C-terminal third, with the N-terminal third of the sequence being the most divergent (Gibbons *et al.* 1992; Koonce *et al.* 1992, Mikami *et al.* 1993; Eshel *et al.* 1993; Li *et al.* 1993; Zhang *et al.* 1993; Plamann *et al.* 1994; Xiang *et al.* 1994). Within the highly conserved central portion are four evenly spaced P-loop elements that are involved in γ-phosphate binding and hydrolysis in ATPases and GTPases (Walker *et al.* 1982), although only the absolutely conserved first element (P-1) from the N-terminal end appears to be functional in ATP hydrolysis (Fig. 6.14) (Lee-Eiford *et al.* 1986; Lye *et al.* 1987; Paschal *et al.* 1987; Euteneuer *et al.* 1988; Gibbons *et al.* 1991; Holzbaur and Vallee 1994; Wilkerson *et al.* 1994). The conserved C-terminal region may represent the large globular head of the enzyme, while the divergent N-terminus may correspond to the tail, a structure presumed to interact with other dynein subunits and cargo. The central domain is flanked by regions predicted to form short α-helical coiled coils (Mikami *et al.* 1993). Such a coiled coil configuration, if actually formed, may comprise a neck region separating the globular head domain from the tail segment.

Comparison of the cytoplasmic dynein intermediate chains from different organisms (Paschal *et al.* 1992; Trivinos-Lagos *et al.* 1993; Vaughan and Vallee 1993) and with the axonemal dynein intermediate chain (Mitchell and Kang 1991) indicates a significantly greater degree of conservation in the C-terminal region compared with the N-terminal portion. The pattern of sequence conservation between the intermediate chains suggests that the C-terminal region may be responsible for a common intermediate chain function, such as association with the heavy chain or another dynein subunit. The N-terminal region, in contrast, may confer functional specificity of what cargo the motor will bind, such as directing cytoplasmic dynein to membrane surfaces or to the kinetochore (Paschal *et al.* 1992).

As with the heavy chain, analysis of the primary sequence of one of the polypeptides of light-intermediate chains of cytoplasmic dynein reveals a P-loop element near the N-terminus which shows homology to the ATPase family (Hughes *et al.* 1993, 1995; Gill *et al.* 1994). A possible role that has

(A) Heavy Chain

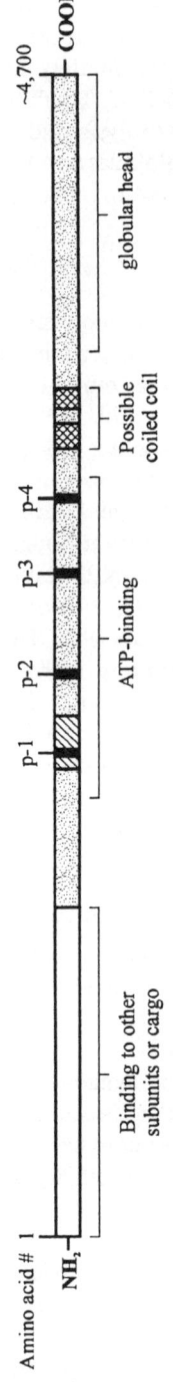

(B) Intermediate Chain

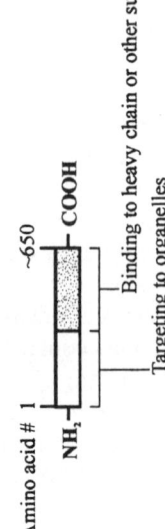

(C) Light intermediate Chain

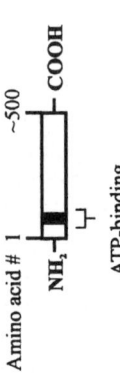

Fig. 6.14 Structural domains of cytoplasmic dynein subunits of (A) heavy, (B) intermediate, and (C) light-intermediate chains. Dark vertical bars, P-loops; cross-hatch areas, possible coiled coils; stippled areas, conserved regions; and single-hatch areas, regions of very high sequence conservation surrounding the first of the four P-loop elements (P-1 to P-4) in the heavy chain.

been suggested for this additional ATPase in the light-intermediate chain of the dynein complex is to serve as a regulator between the cytoplasmic dynein complex and another multiprotein complex, dynactin (see below) (Hughes *et al.* 1995).

Dynactin

Dynactin is a multiprotein complex composed of polypeptides of 150, 135, 62, 50, 45, 37, and 32 kDa (Gill *et al.* 1991; Schroer and Sheetz 1991; Paschal *et al.* 1993). This complex co-purifies with cytoplasmic dyneins and is believed to interact with, as well as have a role in the regulation of the activity of, cytoplasmic dynein (Gill *et al.* 1991; Vaughan and Vallee 1995). To date, the best characterized of the different polypeptides are the 45 and 150 kDa species. The 45 kDa polypeptide is the most abundant component of the dynactin complex (Paschal *et al.* 1993) and appears to be a novel form of actin (Lees-Miller *et al.* 1992; Paschal *et al.* 1993). The 150 kDa polypeptide is well conserved in mammals and birds (Vallee *et al.* 1988; Schroer *et al.* 1989; Steuer *et al.* 1990; Gill *et al.* 1991; Holzbaur *et al.* 1991; Tokito *et al.* 1993). In addition, it shows a good overall homology with the Glued protein (p150^{Glued}) in *Drosophila* (Harte and Kankel 1982; Swaroop *et al.* 1987; Holzbaur *et al.* 1991). Analysis of the predicted amino acid sequence of dynactin/p150^{Glued} reveals an element that is homologous to a microtubule-binding motif found in other microtubule-associated proteins (Berlin *et al.* 1990; Pierre *et al.* 1992). Direct binding of the dynactin/p150^{Glued} polypeptide to microtubules has also been demonstrated (Waterman-Storer *et al.* 1993).

Cell cycle distribution

In interphase cells, cytoplasmic dynein is distributed throughout the cytoplasm and shows a granular pattern superimposed on a uniform background (Fig. 6.15) (Pfarr *et al.* 1990; Steuer *et al.* 1990). No intranuclear protein is observed in these cells. Prophase cells show bright dynein staining at centromeres and centrosomes. Within the centromere, the protein is localized to the fibrous corona on the distal face of the kinetochore plate (Wordeman *et al.* 1991). At prometaphase, increasing dynein staining of fibres emanating from the spindle poles is observed. In cells whose chromosomes have attached to the spindle microtubules, staining becomes concentrated between the poles and the kinetochores. In metaphase cells and anaphase cells, staining of spindle fibres is strong, with much less intense signal in the interzone region between the separating chromosomes. By late anaphase, staining of all mitotic structures decreases, while cells in cytokinesis show an interphase pattern.

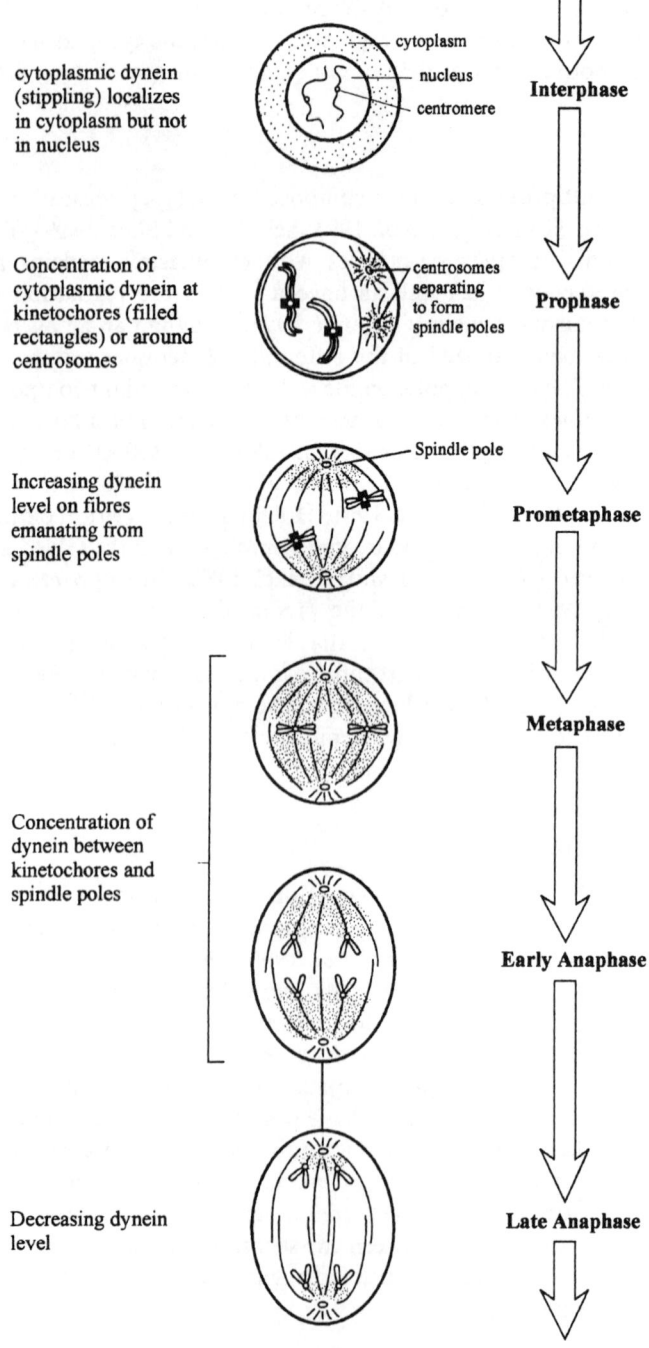

cytoplasmic dynein (stippling) localizes in cytoplasm but not in nucleus

cytoplasm
nucleus
centromere

Interphase

Concentration of cytoplasmic dynein at kinetochores (filled rectangles) or around centrosomes

centrosomes separating to form spindle poles

Prophase

Increasing dynein level on fibres emanating from spindle poles

Spindle pole

Prometaphase

Metaphase

Concentration of dynein between kinetochores and spindle poles

Early Anaphase

Decreasing dynein level

Late Anaphase

Fig. 6.15 Cell cycle distribution of cytoplasmic dynein.

Role in mitosis

In higher eukaryotic prometaphase cells, kinetochores have been shown to attach tangentially to 'astral' (or 'pole'-emanating) microtubules and migrate poleward at rates corresponding to those observed in *in vitro* assays for cytoplasmic dynein force production (Rieder *et al.* 1990). In addition, microtubule-gliding across kinetochores of isolated prometaphase chromosomes has been observed for cytoplasmic dynein (Hyman and Mitchison 1991). These results provide evidence that the protein has a role in chromosome movement during prometaphase. Whether the molecule might also participate in chromosome-to-pole movement during anaphase A is unclear. However, a somewhat different role is revealed when cells are microinjected with antibodies directed against the ATPase domain of cytoplasmic dynein (Vaisberg *et al.* 1993). The injected cells are blocked in prophase, but rather than interfering with poleward chromosome movement, the antibodies prevent spindle formation or, when injected during mitosis, cause the spindle to collapse. These results imply a different role for cytoplasmic dynein involving spindle pole separation.

The cellular role of cytoplasmic dynein has also been studied in lower eukaryotes using mutant strains defective in the heavy chain (Eshel *et al.* 1993; Li *et al.* 1993; Plamann *et al.* 1994; Xiang *et al.* 1994). The results indicate that the protein has a nonessential role in spindle assembly, elongation, and chromosome segregation, but plays a profound role in spindle dynamics and nuclear migration during cell division (Yeh *et al.* 1995). The observation of such 'nonessential' roles raises the possibility that these roles may be functionally redundant and replaceable with one or more other motor proteins (Saunders *et al.* 1993, 1995). A similar functional redundancy is believed to also occur in the kinesin-related motor protein system.

Spindle dynamics modulation

The spindle is a highly dynamic structure whose assembly and disassembly are cell cycle regulated. It is a macromolecule that attaches to and directs the movements of chromosomes during mitosis and meiosis. This macromolecule is most prominently composed of microtubules which, in turn, are composed of tubulin subunits. Some properties of tubulin, and those of a centromere passenger protein, NuMA, which has been proposed to have a possible role in the regulation of spindle dynamics, are described below. In addition to these proteins, a number of the other passenger proteins, including INCENPs and all four motor proteins (CENP-E, MCAK, Kid, and cytoplasmic dynein) discussed earlier, have also been postulated to have a possible direct or secondary role in the modulation of spindle integrity and dynamics.

Tubulin

Tubulins are the building block of microtubules (reviewed by Mandelkow and Mandelkow 1995). Each tubulin molecule is a heterodimer consisting of two closely related and tightly linked globular α-tubulin and β-tubulin polypeptides. Mammalian cells possess multiple forms of these two types of tubulin. The different forms are very similar and will generally co-polymerize into mixed microtubules in the test tube, although they can have distinct locations in the cell and perform subtly different functions. One of the many important functions of microtubules, and thus of tubulin, is to bind to the kinetochore (see for example Brinkley 1990) and, in cooperation with various motors and microtubule-associated proteins, move chromosomes to the poles during mitosis and meiosis.

Immunoelectron microscopic analyses have demonstrated the localization of tubulin to the kinetochore of mammalian chromosomes, in a region corresponding to the corona (Mitchison and Kirschner 1985; Wordeman *et al.* 1991). Very little tubulin is detected on the outer plate itself or the more internal regions of the kinetochore. However, experiments involving cells treated with specific microtubule-depolymerizing drugs indicate that, although the kinetochore stains positively for tubulin before treatment, very little tubulin is detectable at the kinetochore following the treatment. This result suggests that tubulin, rather than being an intrinsic structural component of the kinetochore itself, is present on the corona probably because of the attachment of microtubules to the corona during the period from prometaphase through to anaphase (Mitchison and Kirschner 1985).

NuMA/centrophilin

NuMA (for nuclear protein that associates with the mitotic apparatus) is a high-molecular weight component of the human nuclear matrix that has the unusual property of associating with the microtubules of the spindle apparatus during mitosis (Lydersen and Pettijohn 1980). Over the years, several proteins with physical and cell cycle distribution properties similar to NuMA have been discovered and variously named as SPN (Kallajoki *et al.* 1991), SP-H (Maekawa *et al.* 1991), W1 (Tang *et al.* 1993), and centrophilin (Tousson *et al.* 1991). In particular, the similar identity for two of these proteins, NuMA and centrophilin, has been established by nucleotide sequence comparison (Compton *et al.* 1992; Yang *et al.* 1992).

Protein structure

NuMA is present in all mammalian and at least two nonmammalian cells examined (Price *et al.* 1984; Price and Pettijohn 1986). The predicted 236 kDa human NuMA polypeptide structure is made up of two proline-rich globular domains located at the N- and C-termini, and separated by an $\sim$ 1500 amino acid central α-helical rod domain that contains significant regions of hydrophobic amino acids arranged in a heptad repeat (Compton *et al.* 1992; Yang *et al.* 1992). Although this suggests that NuMA may oligomerize through coiled coil interactions similar to the cytoplasmic intermediate-filament family of proteins, purified NuMA gives no evidence of assembly into filaments or other higher-order structure (Harborth *et al.* 1995). Four predicted p34^{cdc2} phosphorylation sites are located in the C-terminal globular domain. Mutation of a single conserved p34^{cdc2} site alone or in combination with the other sites impair assembly of mitotic spindle and block mitosis, suggesting that the NuMA interaction with the mitotic spindle is controlled by cell cycle-dependent phosphorylation (Compton and Luo 1995).

Cell cycle distribution and role of NuMA/centrophilin

In interphase cells, NuMA is distributed in a punctate fashion throughout the interior of the nucleus, being excluded only from the nucleoli (Lydersen and Pettijohn 1980; Price and Pettijohn 1986; Kallajoki *et al.* 1992; Tang *et al.* 1993). The protein is shown to be a component of nuclear core filaments and is likely to have a structural role in the nucleoskeleton that may be important in interphase nuclear organization and function (Zeng *et al.* 1994). At the onset of mitosis when the nuclear envelope disassembles, NuMA is released from the nuclear compartment and rapidly associates with the microtubules of the mitotic spindle. During the prometaphase and metaphase stages, a proportion of NuMA is detected on the kinetochore or corona region of the centromere (Tousson *et al.* 1991; Compton *et al.* 1992). As the cells progress from prometaphase to metaphase, NuMA progressively accumulates at the polar ends of the mitotic spindle independently of the movements of the chromosomes. NuMA remains associated with the microtubules at the polar ends of the mitotic spindle through anaphase, ultimately releasing from the microtubules in late telophase and re-entering the nucleus of each daughter cell by nuclear pore-dependent import (Compton *et al.* 1992).

In addition to this distribution pattern of the 236 kDa form of NuMA, two smaller isoforms of this protein that are present in the interphase cytosol have been identified (Tang *et al.* 1993, 1994). The three isoforms differ mainly at the C-terminus and are generated by alternative splicing of a common mRNA precursor transcribed from a single gene (Sparks *et al.* 1993). The detailed cell cycle distribution of the smaller isoforms has not been described.

It has been proposed that the role of NuMA/centrophilin is to modulate mitotic spindle dynamics and the re-formation of daughter cell nuclei at the end of mitosis (Compton and Cleveland 1994; Cleveland 1995; Gaglio *et al.* 1995).

Cytokinesis

As described previously, one of the centromere-associated proteins that may have a role in cytokinesis is the INCENPs. In addition to this protein, the TD-60 antigen has also been implicated to have a direct role in this process.

TD-60

Cell cycle distribution

TD-60 (for telophase disc 60 kDa protein) is a mitosis-specific antigen identified by human autoimmune serum (Andreassen *et al.* 1991). Interphase cells demonstrate a low and diffuse amount of this protein. In prophase, the protein appears prominently on the centromere, with the rest of the chromosome showing a weaker presence of the protein. Within the centromere, TD-60 has been shown to co-localize with CENP-B, indicating that the antigen is distributed in the central heterochromatic domain. The antigen remains enriched on the centromere through metaphase and into early anaphase. At mid-anaphase, the antigen leaves the chromosome and is concentrated in a series of narrow bars at the spindle midzone in the region between the two separating chromatid sets (refer to Fig. 6.7). By late anaphase, the protein appears as a series of bars across the full midzone diameter of the cell with the outer edge of the staining closely apposed to the plasma membrane. During telophase, a prominent TD-60-containing equatorial disc (or telophase disc) that fully spans the developing furrow is formed (Fig. 6.16). When the furrow is complete, the presence of TD-60 is restricted to the midbody on the two proximal sides, but not in the centre.

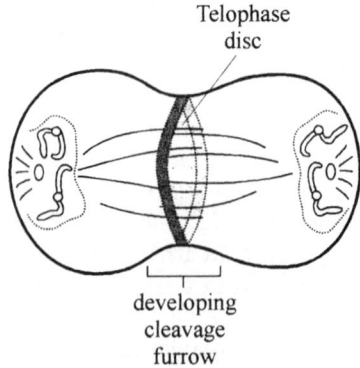

Telophase
disc

developing
cleavage
furrow

Fig. 6.16 A telophase cell showing the position of a TD-60-containing 'telophase disc' (stippled area) that bisects the developing cleavage furrow.

Role in mitosis

A major role of TD-60 is likely to be associated with its involvement as a component of the telophase disc. This disc structure develops initially at the spindle equator or midzone in anaphase, then expands from this site to fully partition the cell at the spindle equator in late anaphase and through telophase. The continuity of the disc, and the distribution of TD-60 throughout this structure, have been unequivocally demonstrated by confocal microscopic analysis and 3D reconstruction of whole cell sections. This disc appears to represent an independent cell organelle, since its integrity is not disturbed by the disassembly of interpolar microtubules or by cell lysis under experimental conditions. Immunocytochemical analysis indicates that the disc appears to contain myosin but not actin. Thus, the position of the telophase disc and the possible presence of cytoplasmic myosin (which may provide the mechanical element for cell cleavage) suggest that the disc is centrally involved in the cytokinetic mechanism. During cytokinesis, the telophase disc may interact with actin in the cell cortex to induce a contraction inward of the disc and cortex toward a central point to produce mitotic cell cleavage (Andreassen *et al.* 1991).

Proteins with an undetermined centromeric role

In addition to the above proteins to which one or more possible functions have been ascribed, at least 12 other proteins that are known to be associated with the centromere, but whose overall properties and/or centromere-related roles are less well understood, have also been reported. The most extensively studied of these proteins is CENP-D. The relevant properties of this and the other proteins are described below.

CENP-D and RCC1

CENP-D and RCC1 are likely to be the same protein

Early work using CREST autoantibodies has identified CENP-D as a 45–50 kDa centromeric antigen (Kingwell and Rattner 1987; Ohtsubo *et al.* 1989; Bischoff *et al.* 1990). Subsequent purification of this antigen and limited sequencing indicate that all the sequenced peptides (205 amino acids) matched exactly to the corresponding regions of a 47 kDa protein known as RCC1 (or regulator of chromosome condensation protein-1) (Bischoff *et al.* 1990), suggesting that CENP-D and RCC1 are likely to be the same protein. In recent years, there has been a deluge of work studying the various structural and functional aspects of the RCC1 protein but, interestingly and somewhat surprisingly, relatively little advance has been made on the specific properties and role of this protein in centromere function.

RCC1

Discovery and evolutionary conservation

RCC1 was first discovered in a temperature-sensitive hamster cell line that carries a single point mutation in the *RCC1* gene (Nishimoto *et al.* 1987a; Uchida *et al.* 1990; reviewed by Dasso 1993). At the restrictive temperature, S phase-arrested mutant cells degrade their RCC1 protein and enter mitosis prematurely, undergoing premature chromosome condensation and nuclear envelope breakdown regardless of the replicative state of the DNA. These early observations indicate that RCC1 is required for the checkpoint control mechanisms that ensure the correct temporal order of events in the cell cycle by detecting unreplicated DNA and blocking premature mitosis when it is present.

Homologues of RCC1 have now been isolated from a wide range of organisms, including humans (Ohtsubo *et al.* 1987, 1989; Bischoff *et al.* 1990), *X. laevis* (Nishitani *et al.* 1990; Dasso *et al.* 1992), *D. melanogaster* (Frasch 1991), *S. pombe* (Matsumoto and Beach 1991), and *S. cerevisiae* (Clark and Sprague 1989; Aebi *et al.* 1990; Clark *et al.* 1991; Fleischmann *et al.* 1991; Forrester *et al.* 1992; Kadowaki *et al.* 1992). A remarkable observation in these studies has been that a wide range of different genetic screens have yielded mutants of RCC1 homologues, and that the phenotypes associated with these mutations are highly pleiotropic. However, despite this, the protein structures of the different homologues are highly conserved (Clarke and Sprague 1989; Aebi *et al.* 1990; Nishitani *et al.* 1990; Uchida *et al.* 1990; Frasch 1991; reviewed by Dasso 1993), and many have been shown to be able to readily complement across species (Nishitani *et al.* 1990; Clark *et al.* 1991; Fleischmann *et al.* 1991; Ohtsubo *et al.* 1991). Furthermore, the homologues are all localized to the nucleus during interphase and all except the *S. cerevisiae* homologue are displaced from the chromosomes during mitosis when expressed in mammalian cells (Ohtsubo *et al.* 1991).

Protein structure

The RCC1 protein structure is unusual in that it is internally divided into a series of seven tandemly arranged peptide segments of 50–60 amino acids in length (Fig. 6.17). With the exception of the fourth repeat, which contains an 18 amino acid insertion, all the peptide segments show a close sequence homology. At the N-terminus of the protein is a basic region of 38 amino acids. A less abundant variant form of the RCC1 protein has also been identified in human cells (Miyabashira *et al.* 1994). This variant, designated RCC1-I and produced by alternative splicing of the *RCC1* gene product (Furuno *et al.* 1991; Miyabashira *et al.* 1994), has a 32 amino acid insertion at the N-terminus basic region. Both RCC1 and RCC1-I are able to complement the mutation in RCC1 mutant cells.

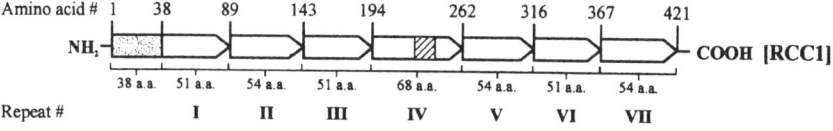

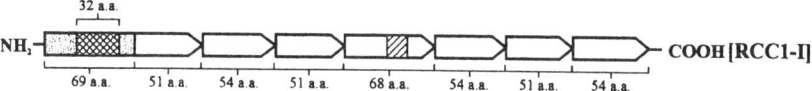

Fig. 6.17 Structural organization of human RCC1 protein and its alternatively spliced product RCC1-I, showing the arrangement of the seven peptide repeats (I–VII). The locations of a 38 residue N-terminus basic region, an 18 residue insertion in the fourth repeat, and a 32 residue insertion at the N-terminus of RCC1-I are indicated by stippled, single-hatch and cross-hatch boxes, respectively.

Interaction with chromatin

RCC1 is a nuclear protein that has been shown to bind to nucleosomes (Ohtsubo *et al.* 1989; Bischoff *et al.* 1990; Frasch 1991). The protein is abundant in tissue culture cells (Bischoff and Ponstingl 1991a) and has been shown to be 2–5% as abundant as histone H1 in *Drosophila* embryos (Frasch 1991) and present in sufficient amounts in *Xenopus* egg nuclei to bind one molecule per nucleosome (Dasso *et al.* 1992). The original belief that the protein binds DNA directly is now considered unlikely, since mutants lacking the ability to bind to naked DNA are still able to associate with chromatin in mammalian cells (Seino *et al.* 1992). Furthermore, studies on the *S. pombe* homologue of RCC1 suggest that the protein binds poorly to DNA on its own, and that its association with chromatin is facilitated through the formation of a high molecular weight complex involving a number of other proteins (Bischoff and Ponstingl 1991a; Lee *et al.* 1993; Saitoh and Dasso 1995). However, despite the abundance of this protein and its association with chromatin, the role of RCC1 in chromatin organization is not certain (Dasso *et al.* 1992).

Role of RCC1

A protein which has been shown to exist as a tight complex with RCC1 is Ran (Bischoff and Ponstingl 1991a). Ran (for *Ras*-related nuclear protein) (originally named TC4; Drivas *et al.* 1990) is a member of the Ras super-family of small GTPases (Bischoff and Ponstingl 1991a). Members of this superfamily are regulated by the binding, hydrolysis, and release of guanyl nucleotides (Takai *et al.* 1992). Unlike other small GTPases, the 25 kDa Ran (reviewed by Moore and Blobel 1994) is primarily localized in the nucleus. It is highly conserved and is expressed in all eukaryotic cells examined to date (Drivas *et al* 1990; Matsumoto and Beach 1991; Belhumeur *et al.* 1993). Like RCC1, the Ran protein is extremely abundant in cells (Bischoff and Ponstingl 1991a).

By acting catalytically as a specific guanine nucleotide exchange factor for Ran (Bischoff and Ponstingl 1991b; Dasso *et al.* 1994), RCC1 interacts enzymatically with Ran to form the core of an evolutionarily conserved GTPase molecular switch (reviewed by Dasso 1993; Moore and Blobel 1994). This switch has been studied in a number of different organisms and has been implicated in diverse cellular processes that include cell cycle control and progression (Nishitani *et al.* 1991; Dasso *et al.* 1992; Ren *et al.* 1993, 1994; Kornbluth *et al.* 1994), DNA replication (Dasso *et al.* 1992), prevention of premature chromosome condensation (Matsumoto and Beach 1991; Nishitani *et al.* 1991), chromatin decondensation after completion of mitosis (Sazer and Nurse 1994), exit from mitosis (Sazer and Nurse 1994), chromosome stability (Clark and Sprague 1989; Matsumoto and Beach 1991), mitotic chromosome segregation (Ouspenski *et al.* 1995), mating in budding yeast (Clark and Sprague 1989), protein import into the nucleus (Melchior *et al.* 1993; Moore and Blobel 1993; Tachibana *et al.* 1994), RNA processing and export from the nucleus (Aebi *et al.* 1990; Forrester *et al.* 1992; Kadowaki *et al.* 1992; Amberg *et al.* 1993; Kadowaki *et al.* 1993), intranuclear trafficking of RNA (Cheng *et al.* 1995), nuclear envelope growth (Dasso *et al.* 1992), and nuclear envelope integrity (Kornbluth *et al.* 1994; Beddow *et al.* 1995; Demeter *et al.* 1995; Dingwall *et al.* 1995). It would seem unlikely that this switch is so multi-functional as to have so many separate roles, but more probable that it has one primary role whose absence wreaks havoc on a whole range of processes indirectly. However, despite intense effort in recent years, no consensus has been reached on what this primary role might be. Two possibilities have been raised (Dasso 1993). In the first, the RCC1–Ran system may act as the first component of a signal transmission pathway, detecting unreplicated DNA and originating the mitosis-inhibiting signal. In the second, the RCC1–Ran complex may be a structural component of the nucleus whose loss leads to a general disruption of nuclear function and an inability to detect unreplicated DNA. However, neither of these models is entirely adequate at present.

Is CENP-D/RCC1 a true centromere-binding protein?

One line of evidence in favour of a centromeric location for CENP-D/RCC1 is that the CREST autoimmune sera used to identify the protein also decorate the kinetochore on chromosomes (Kingwell and Rattner 1987; Bischoff *et al.* 1990). However, since each of these autoimmune sera also recognizes an 18–19 kDa band in immunoblots verifiable as CENP-A (Kingwell and Rattner 1987), cross-reaction with CENP-A antigen could account for part or all of the observed kinetochore staining. In addition to the use of whole autoimmune sera, antibodies affinity purified by specific binding onto CENP-D/RCC1 also result in kinetochore staining (Kingwell and Rattner 1987; Bischoff *et al.* 1990). But once again, these affinity-purified antibodies show cross-reaction with CENP-A in immunoblot analysis, with the converse also true in that antibodies purified for CENP-A will similarly recognize CENP-D/RCC1. In a further study, antibodies raised against synthetic peptides prepared according to the amino acid sequence of the CENP-D/RCC1 protein were shown to stain interphase nuclei strongly and homogeneously, rather than the kinetochores specifically (Ohtsubo *et al.* 1989). This failure to stain kinetochores could arguably be due to different presentation of epitopes on centromere-bound CENP-D/RCC1 proteins not recognizable by the synthetic peptide antibodies. Alternatively, there might be a subpopulation of CENP-D/RCC1 that has undergone post-translational modifications to become centromere specific and recognizable by the autoantibodies, but not by the synthetic peptide antibodies (Earnshaw *et al.* 1987a; Earnshaw and Tomkiel 1992). When all these localization data are taken together, it can be seen that the evidence in favour of CENP-D/RCC1 being a centromere has not been overwhelming. Furthermore, although CENP-D/RCC1 has been implicated in many cellular processes, none has been shown to be directly associated with known centromere functions. Thus, until more definite evidence emerges, the suggestion that CENP-D is a centromere protein should be regarded as tentative. It is also probable that any presence of this protein on the centromere, even if true, may simply reflect a more generalized role of this protein in the multitude of chromosomal or cellular events that the protein is associated with.

4G10 phosphoproteins

Unlike the MPM2 antibody, which detects phosphothreonine-containing antigens, the 4G10 antibody reacts specifically with phosphotyrosine epitopes (Taagepera *et al.* 1995). In interphase cells, 4G10 phosphoproteins are expressed at a low level within the nucleus and cytoplasm, at the centrosome, and along the cell borders. At early prophase, strong expression is detected at centromeres and centrosomes, with maximal expression at centromeres occurring at mid-prometaphase. Within the centromere, the phosphoproteins are found most abundantly on the inner and outer dense plates of the kinetochores but are also present in the corona and the subadjacent heterochromatin. As cells approach metaphase, the phosphoproteins rapidly disappear from the centromeres, until none is detectable during anaphase. In contrast, centrosomes continue to label through anaphase. On immunblots, the 4G10 antibody identifies several protein bands. This, together with the broad localization pattern on centromeres, suggests the likelihood that several centromere/kinetochore proteins contain the 4G10 phosphotyrosine epitopes.

JB antigen

JB antigen is a 38 kDa protein identified by autoantibodies (Kingwell *et al.* 1987). The protein is conserved in mouse, humans, and Indian muntjac, but its apparent absence in rainbow trout suggests that the epitopes recognized by the autoantibodies may be limited in phylogenic distribution. The antigen is undetectable in interphase and prophase cells. Although a low level of it is found on chromosomes in some prometaphase cells, prominent localization of the antigen on chromosomes and centromeres does not occur until metaphase. On the centromere, a broad distribution over the entire structure is observed. At the onset of anaphase, JB antigen leaves the chromosomes and becomes concentrated within the spindle, specifically the spindle midzone. As furrowing proceeds during telophase, staining is excluded from the midbody region but persists in the regions proximal to the midbody. The antigen remains associated with the intercellular bridge of cells that have completed mitosis and have entered the G_1 phase. At the completion of cytokinesis, it is released together with the intercellular bridge remnants into the external cellular milieu.

It is possible that the observed post-anaphase staining of the midzone is not due to the assumed direct association of JB antigen with spindle microtubules, but is the result of interaction with some components of the interzonal cytoplasm or adjacent membranes. Support for the apparent relationship between the antigen and spindle microtubules has come from experiments where cells are treated with colcemid to disrupt the microtubules. In these cells, the lack of JB antigen staining in the intercellular bridge correlates directly with regions of disrupted microtubules (Kingwell *et al.* 1987). Thus, the distribution of this antigen is dependent on the integrity of the underlying microtubules. This relationship suggests that the 38 kDa protein may play a role in the unique interaction between microtubules that occurs within the midzone and intercellular bridge.

MSA-36

MSA-36 is a 36 kDa protein identified by autoantibodies (Rattner *et al.* 1992). The protein is conserved in humans, Rhesus monkey, chimpanzee, African green monkey, Indian muntjac, and mouse. It first appears in interphase cell nuclei during G_2 phase just before the onset of mitosis. In prophase, the protein is distributed in a punctate manner along the condensing chromosomes except at the centromeres where concentration of the protein is particularly prominent. As the cells progress towards metaphase and the degree of chromatin condensation increases, the concentration at the centromeres also increases, while chromosomal level decreases concomitantly. Distribution at the centromere appears as a broad band, suggesting that MSA-36 has an extensive distribution within the centromere. At metaphase, the protein is confined to the metaphase plate. Following the onset of anaphase, chromosomal association disappears and the protein is now associated with the mitotic spindle, particularly in the midzone. This association peaks during early telophase, becoming prominent on either side of the midbody within the intercellular bridge at the completion of telophase. Throughout telophase and cytokinesis, the midbody level decreases with time until it becomes undetectable. Concurrent with this progressive loss of detectable MSA-36 at the intercellular bridge is the appearance of the protein within the reforming nuclei. This appearance is brief and is not seen in subsequent stages of interphase until G_2.

9H8 antigen

9H8 antigen is a 275 kDa phosphorylated protein identified in HeLa and Chinese hamster cells by monoclonal antibodies raised against mitotic chromosome scaffolds and screened for centromere/kinetochore binding (Compton *et al.* 1991). The protein is absent in interphase cells. It first appears during prophase, both at the periphery of the nucleus and in two bright nuclear-associated spots that coincide with the centrosomes. At metaphase, intense concentration is observed over the centromeres of the aligned chromosomes and in the pericentrosomal region. Centromere levels are seen at sites that are peripheral to those for CENP-B. At anaphase, the antigen remains bound to the centromeres and translocates to the poles along with the chromosome. The pericentrosomal distribution of 9H8 antigen remains relatively unchanged at anaphase.

3G3 antigen

Like 9H8 antigen, 3G3 antigen has been identified in HeLa cells by a monoclonal antibody generated against chromosome scaffolds (Compton *et al.* 1991). It has a molecular weight of 205 kDa and does not appear to be phosphorylated. The protein is located in interphase nuclei in a punctate pattern. A similar but slightly more intense distribution pattern is found in prophase cells. During metaphase, the antigen is found on centromeres, at sites peripheral to those of CENP-B, and at the pericentrosomal regions. The pericentrosomal localization remains unchanged at anaphase, while most of the chromosome-bound antigen dissociates from the chromosomes and remains at the midzone. A small amount of the protein is also seen between the midzone and the segregating chromosomes. At telophase, the antigen that accumulates at the midzone becomes restricted to the midbody.

3H1 antigen

3H1 antigen is another protein identified in HeLa and CHO cells by an anti-chromosomal scaffolds monoclonal antibody (Compton *et al.* 1991). It has a molecular weight of 205 kDa and, like 3G3 antigen, does not appear to be phosphorylated. In interphase cells, the protein is found not in the nucleus but in a rather unusual region adjacent to it, where the centrioles are located. At prophase, the antigen is present diffusely over the nuclear envelope. At metaphase, it is redistributed onto pericentrosomal materials and the outer edge of the centromeres, at the position of the kinetochore or its associated corona. At anaphase, the antigen remains attached both to the pericentrosomal regions and to the centromere domains of the migrating chromosomes.

2D3 antigen

The 205 kDa 2D3 antigen is another protein that has been detected in HeLa and CHO cells by anti-chromosomal scaffold monoclonal antibodies (Compton *et al.* 1991). In interphase and prophase cells, these antibodies give punctate nuclear staining. During metaphase, only centrosomes and the pericentrosomal regions are identified. Just after anaphase A, when chromosomes begin to segregate, a small amount of 2D3 antigen is found at the spindle midzone. An unusual property of the anti-2D3 antibodies is that they intensely decorate the kinetochore or corona region of mitotic chromosomes isolated from cells treated with a microtubule-destabilizing drug, but they fail to stain the centromeres of normally cycling mitotic cells. However, by analogy with the route through which other passenger proteins transit from chromosomes onto the midzone, it is possible that at least a portion of 2D3 antigen is present on the centromeres of mitotically cycling cells, but that antibody binding is blocked when the antigen is chromosome associated. Such antibody binding may become observable only when the antigen is released at the anaphase transition, or when its epitope is presented in a favourable way on chromosomes of cells treated with colcemid.

1F1 antigen

1F1 antigen resembles 2D3 antigen in both its 205 kDa size and its odd behaviour of being recognizable on the centromeres of chromosomes isolated from mitotically arrested but not normally cycling cells (Compton *et al.* 1991). However, it differs from 2D3 in several respects. First, it is not detected in CHO cells. Second, it shows intense, instead of punctate, nuclear staining during interphase and prophase. Third, while the antigen undergoes the same redistribution to the pericentrosomal regions as 2D3 antigen during anaphase, it is not detectable in the midzone. Thus, if 1F1 antigen is present at the centromeres of normal cells, its epitope must be blocked from antibody binding throughout its attachment to, and following its dissociation from, the chromosomes. Alternatively, the observed centromere association may be an artefact of a disruption of normal cell structure caused by the drug-induced mitotic arrest.

S5-39 antigen

S5-39 antigen is a 59 kDa protein that has been identified by a polyclonal antibody produced against *Drosophila* embryo proteins that bind to microtubule affinity columns (Kellogg *et al.* 1989). The protein appears to be absent in interphase cells. At prophase, the antigen is found diffusely and irregularly around each centrosome. Presence of the antigen at centrosomal regions persists through metaphase but, in addition, it is detected on the centromeres at the metaphase plate. At anaphase, the antigen remains localized at both the centrosomal regions and the centromeres of the separating chromosomes, but by telophase, only centrosomal localization is observed.

37A5 antigen

37A5 antigen has been identified in CHO cells using a monoclonal antibody raised against a bovine nuclear fraction enriched in several kinetochore polypeptides (Pankov *et al.* 1990). On SDS–PAGE, this antibody recognizes two polypeptides of 140 and 155 kDa. The antigen is present in interphase and prophase cell nuclei. During metaphase, the antigen is located on the kinetochore region of chromosomes. At anaphase, the antigen leaves the kinetochore and appears on the spindle midzone, and by telophase, it is found on both sides of the midbody in the intercellular bridge.

32-9 antigen

32-9 antigen is a 40 kDa protein identified by a monoclonal antibody raised against a tubulin–centromere protein complex produced by crosslinking tubulin to isolated CHO chromosomes (Balczon and Accavitti 1990). The antigen is conserved in HeLa, rat, and Indian muntjac cells. In interphase cells, it is localized in a punctate manner along both cytoplasmic and spindle microtubules. The antigen is detected in the centromeres of metaphase-arrested cells but no information is available on the distribution of this antigen in other stages, nor is it known if the antigen is associated with centromeres in normal, non-drug-arrested metaphase cells.

Conclusion

In this chapter, the properties of a host of 37 proteins that are at one time or another associated with the higher eukaryotic centromere have been described. About two-thirds of these proteins (Tables 6.1 and 6.2) have been characterized sufficiently for one or more proposed or tested functions to be ascribed to them. Some of these functions have been recognized through direct *in vitro* or *in vivo* assays, whilst others have been inferred from mutant phenotypes or from cell cycle and/or subcellular distribution patterns.

Collectively, these proteins cover a diverse array of possible functions that are not always and necessarily associated with those that are classically known for the centromere. The establishment of the roles of these different proteins, especially those in relation to the centromere, has been a major challenge, since functional assays for both the higher eukaryotic centromere and its associated mitotic and meiotic structures are still rather poorly developed. The study of the passenger group of proteins, which represents by far the majority of the number of proteins that interact with the centromere, has the added complication that these proteins transit through different organelles at different stages of the cell cycle. This behaviour suggests the possibility that individual passenger proteins may have multiple temporally regulated functions, each of which may be specific for the organelle that the protein transiently associates with. Alternatively, some of the passenger proteins may only have one primary role but undergo the observed complex cell cycle redistribution to reach a specific destination to execute their functions, or simply to be disposed of. Whatever the roles of these proteins may be, it is clear that the centromere should no longer be regarded as a structure that fulfils only the basic functions of sister chromatid pairing and microtubule attachment. Indeed, increasing evidence now indicates that the centromere has more far-reaching roles that are likely to impinge upon many essential activities of the mitotic and meiotic cell division cycles.

References

Adachi, Y., Luke, M., and Laemmli, U. (1991). Chromosome assembly *in vitro*: topoisomerase II is required for condensation. Cell *64*, 137–148.

Adolph, K., Cheng, S., Paulson, J., and Laemmli, U. (1977). Isolation of a protein scaffold from mitotic HeLa cell chromosomes. Proc. Natl Acad. Sci. USA *11*, 4937–4941.

Aebi, M., Clark, M. W., Vijayraghavan, U., and Abelson, J. (1990). A yeast mutant, *PRP20*, altered in mRNA metabolism and maintenance of the nuclear structure, is defective in a gene homologous to the human *RCC1*, which is involved in the control of chromosome condensation. Mol. Gen. Genet. *224*, 72–80.

Afshar, K., Barton, N. R., Hawley, R. S., and Goldstein, L. S. B. (1995). DNA binding and meiotic chromosomal localization of the *Drosophila* nod kinesin-like protein. Cell *81*, 129–138.

Aizawa, H. Y., Sekine, R., Takemura, Z., Zhang, M., Nangaku, M., and Hirokawa, N. (1992). Kinesin family in murine central nervous system. J. Cell Biol. *119*, 1287–1296.

Alfageme, C., Rudkin, G., and Cohen, L. (1980). Isolation, properties and cellular distribution of D1, a chromosomal protein of *Drosophila*. Chromosoma *78*, 1–31.

Allen, J., Harborne, N., Rau, D. C., and Gould, H. (1982). Participation of core histone tails in the stabilization of the chromatin solenoid. J. Cell Biol. *93*, 285–297.

Amberg, D. C., Fleischmann, M., Stagljar, I., Cole, C. N., and Aebi, M. (1993). Nuclear PRP20 protein is required for mRNA export. EMBO J. *12*, 233–241.

Amos, L. (1989). Brain dynein crossbridges microtubules into bundles. J. Cell Sci. *93*, 19–28.

Andreassen, P., Palmer, D., Wener, M., and Margolis, R. (1991). Telophase disc: a new mammalian mitotic organelle that bisects telophase cells with a possible function in cytokinesis. J. Cell Biol. *99*, 523–534.

Andreassen, P. R. and Margolis, R. L. (1994). Microtubule dependency of p34^{cdc2} inactivation and mitotic exit in mammalian cells. J. Cell Biol. *127*, 789–802.

Arents, G. and Moudrianakis, E. N. (1993). Topography of the histone octamer surface: repeating structural motifs utilized in the docking of nucleosomal DNA. Proc. Natl Acad. Sci. USA *90*, 10489–10493.

Arents, G., Burlingame, R. W., Wang, B. C., Love, W. E., and Moudrianakis, E. N. (1991). The nucleosomal core histone octamer at 3.1Å resolution: a tripartite protein assembly and a left-handed superhelix. Proc. Natl Acad. Sci. USA *88*, 10148–10152.

Arion, D., Meijer, L., Brizuela, L., and Beach, D. (1988). *cdc2* is a component of the M phase-specific histone H1 kinase: evidence for identity with MPF. Cell *55*, 371–378.

Ashley, C., Pendleton, C., Jennings, W., Saxena, A., and Glover, C. (1989). Isolation and sequencing of cDNA clones encoding *Drosophila* chromosomal protein D1. J. Biol. Chem. *264*, 8394–8401.

Avides, M. C. and Sunkel, C. E. (1994). Isolation of chromosome-associated proteins from *Drosophila melanogaster* that bind a human centromeric DNA sequence. J. Cell Biol. *127*, 1159–1171.

Bae, Y.-S., Kawasaki, I., Ikeda, H., and Liu, L. (1988). Illegitimate recombination mediated by calf thymus DNA topoisomerase II *in vitro*. Proc. Natl Acad. Sci. USA *85*, 2076–2080.

Bailly, E., Doree, M., Nurse, P., and Bornens, M. (1989). p34^{cdc2} is located in both nucleus and cytoplasm; part is centrosomally associated at G_2/M and enters vesicles at anaphase. EMBO J. *8*, 3985–3995.

Balczon, R. and Accavitti, M. (1990). Identification of a 40×10^3 M_r centromere-associated protein in cultured mammalian cells. J. Cell Sci. *97*, 705–713.

Balhorn, R. (1982). A model for the structure of chromatin in mammalian sperm. J. Cell Biol. *93*, 298–305.

Basrai, M. A. and Hieter, P. (1995). Is there a unique form of chromatin at the *Saccharomyces cerevisiae* centromeres? BioEssays *17*, 669–672.

Beach, D. H., Durkacz, B., and Nurse, P. M. (1982). Functionally homologous cell cycle control genes in budding and fission yeast. Nature *300*, 706–709.

Beddow, A. L., Richards, S. A., Orem, N. R., and Macara, I. G. (1995). The Ran/TC4 GTPase-binding domain: identification by expression cloning and characterization of a conserved sequence motif. Proc. Natl Acad. Sci. USA *92*, 3328–3332.

Belhumeur, P., Lee, A., Tam, R., DiPaolo, T., Fortin, N., and Clark, M. W. (1993). *GSP1* and *GST2* genetic suppressors of the *prp20-1* mutant in *Saccharomyces cerevisiae*; GTP-binding proteins involved in the maintenance of nuclear organization. Mol. Cell. Biol. *13*, 2152–2161.

Berger, J., Gamblin, S., Harrison, S., and Wang, J. (1996). Structure and mechanism of DNA topoisomerase II. Nature *379*, 225–232.

Berlin, V., Styles, C., and Fink, G. (1990). BIK1, a protein required for microtubule function during mating and mitosis in *Saccharomyces cerevisiae*, colocalizes with tubulin. J. Cell Biol. *111*, 2573–2586.

Bernat, R. L., Borisy, G. G., Rothfield, N. F., and Earnshaw, W. C. (1990). Injection of anticentromere antibodies in interphase disrupts events required for chromosome movement in mitosis. J. Cell Biol. *111*, 1519–1533.

Bernat, R. L., Delannoy, M. R., Rothfield, N. F., and Earnshaw, W. C. (1991). Disruption of centromere assembly during interphase inhibits kinetochore morphogenesis and function in mitosis. Cell *66*, 1229–1238.

Bischoff, F. and Ponstingl, H. (1991a). Mitotic regulator protein RCC1 is complexed with a nuclear ras-related polypeptide. Proc. Natl Acad. Sci. USA *88*, 10830–10834.

Bischoff, F. and Ponstingl, H. (1991b). Catalysis of guanine nucleotide exchange on Ran by the mitotic regulator RCC1. Nature *3554*, 80–82.

Bischoff, F., Maier, G., Tilz, G., and Ponstingl, H. (1990). A 47-kDa human nuclear protein recognized by antikinetochore autoimmune sera is homologous with the protein encoded by *RCC1*, a gene implicated in onset of chromosome condensation. Proc. Natl Acad. Sci. USA *87*, 8617–8621.

Blangy, A., Lane, H., d'Herin, P., Harper, M., Kress, M., and Nigg, E. (1995). Phosphorylation by p34^{cdc2} regulates spindle association of human Eg5, a kinesin-related motor essential for bipolar spindle formation *in vivo*. Cell *83*, 1159–1169.

Bloom, G. S., Wagner, M. C., Pfister, K. K., and Brady, S. T. (1988). Native structure and physical properties of bovine brain kinesin and identification of the ATP-binding subunit polypeptide. Biochemistry *27*, 3409–3416.

Bloom, G. (1992). Motor proteins for cytoplasmic microtubules. Curr. Opin. Cell Biol. *4*, 66–73.

Booher, R. and Beach, D. (1989). Involvement of a type 1 protein phosphatase encoded by *bws1+* in fission yeast mitotic control. Cell *57*, 1009–1016.

Brady, S. T. (1985). A novel brain ATPase with properties expected for the fast axonal transport motor. Nature (Lond.) *317*, 73–75.

Brenner, S., Pepper, D., Berns, M. W., Tan, E., and Brinkley, B. R. (1981). Kinetochore structure, duplication and distribution in mammalian cells: analysis by human autoantibodies from scleroderma patients. J. Cell Biol. *91*, 95–102.

Brinkley, B. R., (1990). Centromeres and kinetochores: integrated domains on eukaryotic chromosomes. Curr. Opin. Cell Biol. *2*, 446–452.

Brinkley, B. R., Valdivia, M. M., Tousson, A., and Brenner, S. L. (1984). Compound kinetochores of the Indian muntjac: evolution by linear fusion of unit kinetochores. Chromosoma (Berl.) *91*, 1–11.

Brinkley, B. R., Ouspenski, I., and Zinkowski, R. P. (1992). Structure and molecular organization of the centromere–kinetochore complex. Trends Cell Biol. *2*, 14–21.

Brown, K. D., Coulson, R. M. R., Yen, T. J., and Cleveland, D. W. (1994). Cyclin-like accumulation and loss of the putative kinetochore motor CENP-E results from coupling continuous synthesis with specific degradation at the end of mitosis. J. Cell Biol. *125*, 1303–1312.

Brown, K., Wood, K., and Cleveland, D. (1996). The kinesin-like protein CENP-E is kinetochore-associated throughout poleward chromosome segregation during anaphase-A. J. Cell Sci. 109, 961-969.

Brown, M. (1995). Sequence similarities between the yeast chromosome segregation protein Mif2 and the mammalian centromere protein CENP-C. Gene *160*, 111–116.

Brown, M. T., Goetsch, L., and Hartwell, L. H. (1993). MIF2 is required for mitotic spindle integrity during anaphase spindle elongation in *Saccharomyces cerevisiae*. J. Cell Biol. *123*, 387–403.

Buchenau, P., Saumweber, H., and Arndt-Jovin, D. (1993). Consequences of topoisomerase II inhibition in early embryogenesis of *Drosophila* revealed by *in vivo* confocal laser scanning microscopy. J. Cell Sci. *104*, 1175–1185.

Buchkovich, K., Duffy, L. A., and Harlow, E. (1989). The retinoblastoma protein is phosphorylated during specific phases of the cell cycle. Cell *58*, 1097–1105.

Campbell, M. S. and Gorbsky, G. J. (1995). Microinjection of mitotic cells with the 3F3/2 anti-phosphoepitope antibody delays the onset of anaphase. J. Cell Biol. *129*, 1195–1204.

Camerini-Otero, R., and Felsenfeld, G. (1977). Histone H3 disulfide dimers and nucleosome structure. Proc. Natl Acad. Sci. USA *74*, 5519–5523.

Capony, J. P., Picard, A., Peaucellier, G., Labbe, J. C., and Doree, M. (1986). Changes in the activity of the maturation-promoting factor during meiotic maturation and following activation of amphibian and starfish oocytes: their correlations with protein phosphorylation. Dev. Biol. *117*, 1–12.

Caput, D., Beutler, B., Hartog, K., Thayer, R., Brown-Shimer, S., and Cerami, A. (1986). Identification of a common nucleotide sequence in the 3'-untranslated region of mRNA molecules specifying inflammatory mediators. Proc. Natl Acad. Sci. USA *83*, 1670–1674.

Casiano, C. A., Landberg, G., Ochs, R. L., and Tan, E. M. (1993). Autoantibodies to a novel cell cycle-regulated protein that accumulates in the nuclear matrix during S phase and is localized in the kinetochores and spindle midzone during mitosis. J. Cell Sci. *106*, 1045–1056.

Cavanaugh, A., Hempel, W., Taylor, L., Rogalsky, V., Todorov, G., and Rothblum, L. (1995). Activity of RNA polymerase I transcription factor UBF blocked by Rb gene product. Nature 374, 177-180.

Centonze, V. E. and Borisy, G. G. (1990). Nucleation of microtubules from mitotic centrosomes is modulated by a phosphorylated epitope. J. Cell Sci. *95*, 405–411.

Chelsky, D., Ralph, R., and Jonak, G. (1989). Sequence requirements for synthetic peptide-mediated translocation to the nucleus. Mol. Cell. Biol. *9*, 2487–2492.

Chen, P. L., Scully, P., Shew, J. Y., Wang, J. Y. J., and Lee, W. H. (1989). Phosphorylation of the retinoblastoma gene product is modulated during the cell cycle and cellular differentiation. Cell *58*, 1193–1198.

Cheney, R., Riley, M., and Mooseker, M. (1993). Phylogenetic analysis of the myosin super-family. Cell Motil. Cytoskeleton *24*, 215–223.

Cheng, Y., Dahlberg, J. E., and Lund, E. (1995). Diverse effects of the guanine nucleotide exchange factor RCC1 on RNA transport. Science *267*, 1807–1810.

Choo, K. H., Vissel, B., Brown, R., Filby, R. G., and Earle, E. (1988). Homologous alpha satellite sequences on human acrocentric chromosomes with selectivity for chromosomes 13, 14 and 21: implications for recombination between nonhomologues and Robertsonian translocations. Nucleic Acids Res. *16*, 1273–1284.

Choo, K. H., Vissel, B., Nagy, A., Earle, E., and Kalitsis, P. (1991). A survey of the genomic distribution of alpha satellite DNA on all human chromosomes, and derivation of a new consensus sequence. Nucleic Acids Res. *19*, 1179–1182.

Chow, K.-C. and Ross, W. (1987). Topoisomerse-specific drug sensitivity in relation to cell cycle progression. Mol. Cell. Biol. *2*, 3119–3123.

Chung, T., Drake, F., Tan, K., Per, S., Crooke, S., and Mirabelli, C. (1989). Characterization and immunological identification of cDNA clones encoding two human topoisomerase II isozymes. Proc. Natl Acad. Sci. USA *86*, 9431–9435.

Clark, K. and Sprague, G. (1989). Yeast pheromone response pathway: characterization of a suppressor that restores mating to receptorless mutants. Mol. Cell. Biol. *9*, 2682–2694.

Clark, K. L., Ohtsubo, M., Nishimoto, T., Goebl, M., and Sprague, G. F. Jr,(1991). The yeast SRM1 protein and human RCC1 protein share analogous function. Cell Regulat. *2*, 781–792.

Clarke, D., Johnson, R., and Downes, C. (1993). Topoisomerase II inhibition prevents anaphase chromatid segregation in mammalian cells independently of the generation of DNA strand breaks. J. Cell Biol. *105*, 563–569.

Cleveland, D. W. (1995). NuMA: a protein involved in nuclear structure, spindle assembly, and nuclear re-formation. Trends Cell Biol. *5*, 60–65.

Cockerill, P. and Garrard, W. (1986). Chromosomal loop anchorage of the kappa immunoglobulin gene occurs next to the enhancer in a region containing topoisomerase II sites. Cell *44*, 273–282.

Comings, D. E. and Okada, T. A. (1971). Fine structure of kinetochore in Indian Muntjac. Exp. Cell Res. *67*, 97–110.

Compton, D. A. and Cleveland, D. W. (1994). NuMA, a nuclear protein involved in mitosis and nuclear reformation. Curr. Opin. Cell Biol. *6*, 343–346.

Compton, D. A. and Luo, C. (1995). Mutation of the predicted p34^{cdc2} phosphorylation sites in NuMA impair the assembly of the mitotic spindle and block mitosis. J. Cell Sci. *108*, 621–633.

Compton, D., Yen, T., and Cleveland, D. (1991). Identification of novel centromere/kinetochore-associated proteins using monoclonal antibodies generated against human mitotic chromosome scaffolds. J. Cell Biol. *112*, 1083–1097.

Compton, D. A., Szilak, I., and Cleveland, D. W. (1992). Primary structure of NuMA, an intranuclear protein that defines a novel pathway for segregation of proteins at mitosis. J. Cell Biol. *116*, 1395–1408.

Constanzo, G., Dimauro, E., Salina, G., and Negri, R. (1990). Attraction phasing and neighbor effects of histone octamers on curved DNA. J. Mol. Biol. *216*, 363–374.

Cooke, C. A., Heck, M. M. S., and Earnshaw, W. C. (1987). The inner centromere protein (INCENP) antigens: movement from inner centromere to midbody during mitosis. J. Cell Biol. *105*, 2053–2067.

Cooke, C. A., Bernat, R. L., and Earnshaw, W. C. (1990). CENP-B: a major human centromere protein located beneath the kinetochore. J. Cell Biol. *110*, 1475–1488.

D'Andrea, R. J., Stratmann, R., Lehner, C. F., John, U. P., and Saint, R. (1993). The *three rows* gene of *Drosophila melanogaster* encodes a novel protein that is required for chromosome disjunction during mitosis. Mol. Biol. Cell. *4*, 1161–1174.

D'Arpa, P. and Liu, L. (1990). The Eukaryotic Nucleus: Molecular Biochemistry and Macromolecular Assemblies, eds. Strauss, P.R. and Wilson, S.H. Caldwell JNJ, Telford, 623–638.

Dasso, M. (1993). RCC1 in the cell cycle: the regulator of chromosome condensation takes on new roles. Trends Biochem. Sci. *18*, 96–101.

Dasso, M., Nishitani, H., Kornbluth, S., Nishimoto, T., and Newport, J. W. (1992). RCC1, a regulator of mitosis, is essential for DNA replication. Mol. Cell. Biol. *12*, 3337–3345.

Dasso, M., Seki, T., Azuma, Y., Ohba, T., and Nishimoto, T. (1994). A mutant form of the Ran/TC4 protein disrupts nuclear function in *Xenopus laevis* egg extracts by inhibiting the RCC1 protein, a regulator of chromosome condensation. EMBO J. *13*, 5732–5744.

Davis, F. M. and Rao, P. N. (1987). Antibodies to mitosis-specific phosphoproteins. In Molecular Regulation of Nuclear Events in Mitosis and Meiosis, eds. Schlegel, R.A., Halleck, M.S., Rao, P.N. Academic Press, New York, 259–291.

Davis, F. M., Tsao, T. Y., Fowler, S. K., and Rao, P. N. (1983). Monoclonal antibodies to mitotic cells. Proc Natl Acad Sci USA *80*, 2926–2930.

Davis, R. L., Feng-Cheng, P., Lasser, A. B., and Weintraub, H. (1990). The MyoD DNA binding domain contains a recognition code for muscle-specific gene activation. Cell *60*, 733–746.

Dawson, I. A., Roth, S., Akam, M., and Artavanis-Tsakonas, S. (1993). Mutations of the *fizzy* locus cause metaphase arrest in *Drosophila melanogaster* embryos. Development *117*, 359–376.

DeCuevas, M., Tao, T., and Goldstein, L. S. B. (1992). Evidence that the stalk of *Drosophila* kinesin heavy chain is an alpha-helical coiled coil. J. Cell Biol. *116*, 957–965.

Demeter, J., Morphew, M., and Sazer, S. (1995). A mutation in the RCC1-related protein pim1 results in nuclear envelope framentation in fission yeast. Proc. Natl Acad. Sci. USA *92*, 1436–1440.

Desai, A. and Mitchison, T. J. (1995). A new role for motor proteins as couplers to depolymerizing microtubules. J. Cell Biol. *128*, 1–4.

Dillman, J., Dabney, L., and Pfister, K. (1996). Cytoplasmic dynein is associated with slow axonal transport. Proc. Natl Acad. Sci. USA *93*, 141–144.

DiNardo, S., Voelkel, K., and Sternglanz, R. (1984). DNA topoisomerase II mutant of *Saccharomyces cerevisiae*: topoisomerase II is required for segregation of daughter molecules at the termination of DNA replication. Proc. Natl Acad. Sci. USA *81*, 2616–2620.

Dingwall, C., Kandels-Lewis, S., and Se'raphin, B. (1995). A family of Ran binding proteins that includes nucleoporins. Proc. Natl Acad. Sci. USA *92*, 7525–7529.

Disney, J., Johnson, K., Magnuson, N., Sylvester, S., and Reeves, R. (1989). High-mobility group protein HMG-I localizes to G/Q- and C-bands of human and mouse chromosomes. J. Cell Biol. *109*, 1975–1982.

Dobson, M., Pearlman, R., Karaiskakis, A., Spyropoulos, B., and Moens, P. (1994). Synaptonemal complex proteins: occurrence, epitope mapping and chromosome disjunction. J. Cell Sci. *107*, 2749–2760.

Doonan, H. J. and Morris, N. R. (1989). The *bimG* gene of *Aspergillus nidulans*, which is required for completion of anaphase, encodes a homolog of mammalian phosphoprotein phosphatase 1. Cell *57*, 987–996.

Downes, C., Mullinger, A., and Johnson, R. (1991). Inhibitors of DNA topoisomerase II prevent chromatid separation in mammalian cells but do not prevent exit from mitosis. Proc. Natl Acad. Sci. USA *88*, 8895–8899.

Draetta, G. and Beach, D. (1988). Activation of *cdc2* protein kinase during mitosis in human cells: cell cycle-dependent phosphorylation and subunit rearrangement. Cell *54*, 17–26.

Draetta, G., Brizuela, L., Potashkin, J., and Beach, D. (1987). Identification of p34 and p13, human homologs of the cell cycle regulators of fission yeast encoded by *cdc2* and *suc1*. Cell *50*, 319–325.

Draetta, G., Piwnica-Worms, H., Morrison, D., Druker, B., Roberts, T., and Beach, D. (1988). cdc2 is a major cell-cycle regulated tyrosine kinase substrate. Nature *336*, 738–743.

Draetta, G., Luca, F., Westendorf, J., Brizuela, L., Ruderman, J., and Beach, D. (1989). cdc2 is complexed with both cyclin A and B: evidence for inactivation of MPF by proteolysis. Cell *56*, 829–838.

Drake, F., Zimmerman, J., McCabe, F., Bartus, H., Per, S., Sullivan, D. *et al.* (1987). Purification of topoisomerase II from amsacrine-resistant P388 leukemia cells. Evidence for two forms of the enzyme. J. Biol. Chem. *262*, 16739–16747.

Drake, F., Hofmann, C., Bartus, H., Mattern, M., Crooke, S., and Mirabelli, C. (1989). Biochemical and pharmacological properties of p170 and p180 forms of topoisomerase II. Biochemistry *28*, 8154–8160.

Drivas, G. T., Shih, A., Coutavas, E., Rush, M. G., and D'Eustachio, P. (1990). Characterization of four novel ras-like genes expressed in a human teratocarcinoma cell line. Mol. Cell. Biol. *10*, 1793–1798.

Dunphy, W. G., Brizuela, L., Beach, D., and Newport, J. (1988). The Xenopus *cdc2* protein is a component of MPF, a cytoplasmic regulator of mitosis. Cell *54*, 423–431.

Earnshaw, W. C. and Cooke, C. A. (1991). Analysis of the distribution of the INCENPs throughout mitosis reveals the existence of a pathway of structural changes in the chromosomes during metaphase and early events in cleavage furrow formation. J. Cell Sci. *98*, 443–461.

Earnshaw, W. C. and Heck, M. (1985). Localization of topoisomerase II in mitotic chromosomes. J. Cell. Biol. *100*, 1716–1725.

Earnshaw, W. C. and Heck, M. (1988). Cell biology of topoisomerase II. Cancer Cells 6, 279–288.

Earnshaw, W. C. and Laemmli, U. (1983). Architecture of metaphase chromosomes and chromosome scaffolds. J. Cell Biol. 96, 84–93.

Earnshaw, W. and MacKay, A. (1994). Role of nonhistone proteins in the chromosomal events of mitosis. FASEB J. 8, 947–956.

Earnshaw, W. C. and Rothfield, N. (1985). Identification of a family of human centromere proteins using autoimmune sera from patients with scleroderma. Chromosoma 91, 313–321.

Earnshaw, W. C. and Tomkiel, J. (1992). Centromere and kinetochore structure. Curr. Opin. Cell Biol. 4, 86–93.

Earnshaw, W. C., Halligan, N., Cooke, C. A., and Rothfield, N. (1984). The kinetochore is part of the metaphase chromosome scaffold. J. Cell Biol. 98, 352–357.

Earnshaw, W. C., Bordwell, C., Marino, C., and Rothfield, N. (1985). Three human chromosomal autoantigens are recognized by sera from patients with anti-centromere antibodies. J. Clin. Invest. 77, 426–429.

Earnshaw, W. C., Halligan, B., Cooke, C. A., Heck, M., and Liu, L. (1985). Topoisomerase II is a structural component of mitotic chromosome scaffolds. J. Cell Biol. 100, 1706–1715.

Earnshaw, W. C., Machlin, P. S., Bordwell, B. J., Rothfield, N. F., and Cleveland, D. W. (1987a). Analysis of anticentromere autoantibodies using cloned autoantigen CENP-B. Proc. Natl Acad. Sci. USA 84, 4979–4983.

Earnshaw, W. C., Sullivan, K. F., Machlin, P. S., Kaiser, D. A., Pollard, T. D., Rothfield, N. F. et al. (1987b). Molecular cloning of cDNA for CENP-B, the major human centromere autoantigen. J. Cell Biol. 104, 817–829.

Earnshaw, W. C., Ratrie, H., and Stetten, G. (1989). Visualization of centromere proteins CENP-B and CENP-C on a stable dicentric chromosome in cytological spreads. Chromosoma (Berl.) 98, 1–12.

Edgar, B., Sprenger, A. F., Duronio, R. J., Leopold, P., and O'Farrell, P. H. (1994). Distinct molecular mechanisms regulate cell cycle timing at successive stages of *Drosophila* embryogenesis. Genes Dev. 8, 440–452.

Endow, S. A. (1991). The emerging kinesin family of microtubule motor proteins. Trends Biochem. Sci. 16, 221–225.

Endow, S. A. (1993). Chromosome distribution, molecular motors and the claret protein. Trends Genet. 9, 52–55.

Endow, S. A. and Titus, M. (1992). Genetic approaches to molecular motors. Annu. Rev. Cell Biol. 8, 29–66.

Endow, S. A., Henikoff, S., and Soler-Niedzeila, L. (1990). Mediation of meiotic and early mitotic chromosome segregation in *Drosophila* by a protein related to kinesin. Nature 345, 81–83.

Engle, D. B., Doonan, J. H., and Morris, N. R. (1988). Cell-cycle modulation of MPM-2-specific spindle pole body phosphorylation in *Aspergillus nidulans*. Cell Motil. Cytol. 10, 432–437.

Enos, A. P. and Morris, N. R. (1990). Mutation of a gene encoding a kinesin-like protein blocks nuclear division in *A. nidulans*. Cell 60, 1019–27.

Eshel, D., Urrestarauzu, L., Vissers, S., Jauniaux, J., Vliet-Reedijk, J. v., Planta, R. J. et al. (1993). Cytoplasmic dynein is required for normal nuclear segregation in yeast. Proc. Natl Acad. Sci. USA 90, 11172–11176.

Euteneuer, U., Koonce, M., Pfister, K., and Schilwa, M. (1988). An ATPase with properties expected for the organelle motor of the giant amoeba *Reticulomyxa*. Nature 332, 176–178.

Evans, T., Rosenthal, E. T., Youngblom, J., Distel, D., and Hunt, T. (1983). Cyclin: a protein specified by maternal mRNA in sea urchin eggs that is destroyed at each cleavage division. Cell *33*, 389–396.

Fishkind, D. J. and Wang, Y. (1995). New horizons for cytokinesis. Curr. Opin. Cell Biol. *7*, 23–31.

Fleischmann, M., Clark, M. W., Forrester, W., Wickens, M., Nishimoto, T., and Aebi, M. (1991). Analysis of yeast *prp20* mutations and functional complementation by the human homologue *RCC1*, a protein involved in the control of chromosome condensation. Mol. Gen. Genet. *227*, 417–423.

Forrester, W., Stutz, F., Rosbash, M., and Wickens, M. (1992). Defects in mRNA 3'-end formation, transcription initiation, and mRNA transport associated with the yeast mutation *prp20*: possible coupling of mRNA processing and chromatin structure. Genes Dev. *6*, 1914–1926.

Frasch, M. (1991). The maternally expressed *Drosophila* gene encoding the chromatin-binding protein Bj1 is a homolog of the vertebrae gene regulator of chromatin condensation, *RCC1*. EMBO J. *10*, 1225–1236.

Fritzler, M. J. and Kinsella, T. D. (1980). The CREST syndrome: a distinct serologic entity with anticentromere antibodies. Am. J. Med. *69*, 520–526.

Fry, D. C., Kuby, S. A., and Mildvan, A. S. (1986). ATP-binding site of adenylate kinase: mechanistic implications of its homology with *ras*-encoded p21, F_1ATPase, and other nucleotide-binding proteins. Proc. Natl Acad. Sci. USA *83*, 907–911.

Furuno, N., Nakagawa, K., Eguchi, U., Ohtubo, M., Nishimoto, T., and Soeda, E. (1991). Complete nucleotide sequence of the human RCC1 gene involved in coupling between DNA replication and mitosis. Genomics *11*, 459–461.

Gaff, C., duSart, D., Kalitsis, P., Iannello, R., Nagy, A., and Choo, K. H. A. (1994). A novel nuclear protein binds centromeric alpha satellite DNA. Hum. Mol. Genet. *3*, 711–716.

Gaglio, T., Saredi, A., and Compton, D. A. (1995). NuMA is required for the organization of microtubules into aster-like mitotic arrays. J. Cell Biol. *131*, 693–708.

Gallant, P. and Nigg, E. A. (1992). Cyclin B2 undergoes cell cycle-dependent nuclear translocation, and when expressed as a non-destructible mutant, causes mitotic arrest in HeLa cells. J. Cell Biol. *117*, 213–224.

Garreau, H. and Williams, J. (1983). Two nuclear DNA-binding proteins of *Dictyostelium discoideum* with a high affinity for poly(dA)-poly(dT). Nucleic Acids Res. *11*, 8473–8484.

Gasser, S. and Laemmli, U. (1986). The organization of chromatin loops: characterization of a scaffold attachment site. EMBO J. *5*, 511–518.

Gasser, S., Laroche, T., Falquet, J., BoyDeLaTour, E., and Laemmli, U. (1986). Metaphase chromosome structure. Involvements of topoisomerase II. J. Mol. Biol. *188*, 613–629.

Gatewood, J. M., Cook, G. R., Balhorn, R., Bradbury, E. M., and Schmid, C. W. (1987). Sequence-specific packaging of DNA in human sperm chromatin. Science *236*, 962–964.

Gautier, J., Norbury, C., Lohka, M., Nurse, P., and Maller, J. (1988). Purified maturation-promoting factor contains the product of a *Xenopus* homolog of the fission yeast cell cycle control gene $cdc2^+$. Cell *54*, 433–439.

Gautier, J., Matsukawa, T., Nurse, P., and Maller, J. (1989). Dephosphorylation and activation of *Xenopus* p34^{cdc2} protein kinase during the cell cycle. Nature (Lond.) *339*, 626–629.

Ghiara, J. B., Richardson, H. E., Sugimoto, K., Henze, M., Lew, D. J., Witenberg, C. *et al.* (1991). A cyclin B homolog in S. *cerevisiae*: chronic activation of the Cdc28 protein kinase by cyclin prevents exit from mitosis. Cell *65*, 163–174.

Gibbons, I. and Rowe, A. (1965). Dynein: a protein with adenosine triphosphatase activity from cilia. Science *149*, 424.

Gibbons, I. R., Gibbons, B. H., Mocz, G., and Asai, D. J. (1991). Multiple nucleotide-binding sites in the sequence of dynein beta heavy chain. Nature *352*, 640–643.

Gibbons, I. R., Asai, D. J., Tang, W. J., and Gibbons, B. H. (1992). A cytoplasmic dynein heavy chain in sea urchin embryos. Biol. Cell *76*, 303–309.

Gill, S. R., Schroer, T. A., Szilak, I., Steuer, E. R., Sheetz, M. P., and Cleveland, D. W. (1991). Dynactin, a conserved, ubiquitously expressed component of an activator of vesicle motility mediated by cytoplasmic dynein. J. Cell Biol. *115*, 1639–1650.

Gill, S. R., Cleveland, D. W., and Schroer, T. A. (1994). Characterization of DLC-A and DLC-B, two families of cytoplasmic dynein light chain subunits. Mol. Biol. Cell *5*, 645–654.

Gimenez-Abian, J., Clarke, D., Mullinger, A., Downes, C., and Johnson, R. (1995). A postprophase topoisomerase II-dependent chromatid core separation step in the formation of metaphase chromosomes. J. Cell Biol. *131*, 7–17.

Girdham, C. H. and Glover, D. M. (1991). Chromosome tangling and breakage at anaphase result from mutations in *lodestar*, a *Drosophila* gene encoding a putative nucleoside triphosphate-binding protein. Genes Dev. *5*, 1786–1799.

Glotzer, M., Murray, A. W., and Kirschner, M. W. (1991). Cyclin is degraded by the ubiquitin pathway. Nature (Lond.) *349*, 132–138.

Goldberg, I., Sawhney, H., Pluta, A., Warburton, P., and Earnshaw, W. (1996). Surprising deficiency of CENP-B binding sites in African green monkey α-satellite DNA: Implications for CENP-B function at centromeres. Mol. Cell. Biol. *16*, 5156–5168.

Goldstein, L. S. (1980). Mechanisms of chromosome orientation revealed by two meiotic mutants in *Drosophila melanogaster*. Chromosoma *78*, 79–111.

Goldstein, L. S. (1991). The kinesin superfamily: tails of functional redundancy. Trends Cell Biol. *1*, 93–98.

Goldstein, L. S. (1993a). With apologies to Scheherazade: tails of 1001 kinesin motors. Annu. Rev. Genet. *27*, 319–351.

Goldstein, L. S. (1993b). Functional redundancy in mitotic force generation. J. Cell Biol. *120*, 1–3.

Goodenough, U. and Heuser, J. (1982). The sub-structure of the outer dynein arm. J. Cell Biol. *95*, 798–815.

Goodenough, U. and Heuser, J. (1984). Structural comparison of purified proteins with *in situ* dynein arms. J. Mol. Biol. *180*, 1083–1118.

Goodrich, D. W., Wang, N. P., Qian, Y. W., Lee, E. Y. H. P., and Lee, W. H. (1991). The retinoblastoma gene product regulates progression through the G1 phase of the cell cycle. Cell *67*, 293–302.

Goodson, H. and Spudich, J. (1993). Molecular evolution of the myosin family: relationships derived from comparisons of amino acid sequences. Proc. Natl Acad. Sci. USA *90*, 659–663.

Gorbsky, G. (1994). Cell cycle progression and chromosome segregation in mammalian cells cultured in the presence of the topoisomerase II inhibitors ICRF-187[(+)-1,2-bis(3,5-dioxopiperazinyl-1-yl)propane: ADR-529] and ICRF-159 (Razoxane). Cancer Res. *54*, 1042–1048.

Gorbsky, G. (1995). Kinetochores, microtubules and the metaphase checkpoint. Trends Cell Biol. *5*, 143–148.

Gorbsky, G. J. and Ricketts, W. A. (1993). Differential expression of a phosphoepitope at the kinetochores of moving chromosomes. J. Cell Biol. *122*, 1311–1321.

Grimes, S. R. (1986). Nuclear proteins in spermatogenesis. Comp. Biochem. Physiol. *83B*, 495–500.

Grosschedl, R., Giese, K., and Pagel, J. (1994). HMG domain proteins: architectural elements in the assembly of nucleoprotein structures. Trends Genet. *10*, 94–100.

Guldner, H. H., Lakomek, H. J., and Bautz, F. A. (1984). Human anti-centromere sera recognize a 19.5 kd nonhistone chromosomal protein from HeLa cells. Clin. Exp. Immunol. *58*, 13–20.

Haaf, T. and Schmid, M. (1989). Centromeric association and non-random distribution of centromeres in human tumour cells. Hum. Genet. *81*, 137–143.

Haaf, T. and Ward, D. C. (1994). Structural analysis of alpha-satellite DNA and centromere proteins using extended chromatin and chromosomes. Hum. Mol. Genet. *3*, 697–709.

Haaf, T. and Ward, D. C. (1995). Rabl orientation of CENP-B box sequences in *Tupaia belangeri* fibroblasts. Cytogenet. Cell Genet. *70*, 258–262.

Haaf, T., Warburton, P. E., and Willard, H. F. (1992). Integration of human alpha satellite DNA into simian chromosomes; centromere protein binding and disruption of normal chromosome segregation. Cell *70*, 681–696.

Hadlaczky, G., Went, M., and Ringertz, N. (1986). Direct evidence for the non-random localization of mammalian chromosomes in the interphase nucleus. Exp. Cell Res. *167*, 1–15.

Hagan, I. and Yanagida, M. (1990). Novel potential mitotic motor protein encoded by the fission yeast *cut7$^+$* gene. Nature *347*, 563–66.

Hagan, I. and Yanagida, M. (1992). Kinesin-related cut7 protein associates with mitotic and meiotic spindles in fission yeast. Nature (Lond.) *356*, 74–76.

Hall, D. H. and Hedgecock, E. M. (1991). Kinesin-related gene *unc-104* is required for axonal transport of synaptic vesicles in *C. elegans*. Cell *65*, 837–847.

Hammer, J. (1991). Novel myosin. Trends Cell Biol. *1*, 50–56.

Harborth, J., Weber, K., and Osborn, M. (1995). Epitope mapping and direct visualization of the parallel, in-register arrangement of the double-stranded coiled-coil in the NuMA protein. EMBO J. *14*, 2447–2460.

Harte, P. J. and Kankel, D. R. (1982). Genetic analysis of mutations at the *glued* locus and interacting loci in *Drosophila melanogaster*. Genetics *101*, 477–501.

Hartwell, L. H. (1974). *Saccharomyces cerevisiae* cell cycle. Bacteriol Rev. *38*, 164–198.

Hatsumi, M. and Endow, S. A. (1992). The *Drosophila* ncd microtubule motor protein is spindle-associated in meiotic and mitotic cells. J. Cell Sci. *103*, 1013–1020.

Heck, M., Hittelman, W., and Earnshaw, W. C. (1988). Differential expression of DNA topoisomerases I and II during the eukaryotic cell cycle. Proc. Natl Acad. Sci. USA *85*, 1086–1090.

Heng, H., Tsui, L.-C., and Moens, P. (1994). Organization of heterologous DNA inserts on the mouse meiotic chromosome core . Chromosoma *103*, 401–407.

Hershko, A., Ganoth, D., Pehrson, J., Palazzo, R. E., and Cohen, L. H. (1991). Methylated ubiquitin inhibits cyclin degradation in clam embryo extracts. J. Biol. Chem. *266*, 16376–16379.

Hildebrandt, S., Weiner, E., Sene'cal, J., Noell, S., Daniels, L., Earnshaw, W. C., and Rothfield, N. F. (1990). The IgG, IgM, and IgA isotypes of antitopoisomerase I and anticentromere autoantibodies. Arthritis Rheumatism *33*, 724–727.

Hirano, T. and Mitchison, T. J. (1991). Cell cycle control of higher-order chromatin assembly around naked DNA *in vitro*. J. Cell Biol. *115*, 1479–1489.

Hirano, T. and Mitchison, T. J. (1993). Topoisomerase II does not play a scaffolding role in the organization of mitotic chromosomes assembled in *Xenopus* egg extracts. J. Cell Biol. *120*, 601–612.

Hirano, T. and Mitchison, T. J. (1994). A heterodimeric coiled-coil protein required for mitotic chromosome condensation *in vitro*. Cell *79*, 449–458.

Hirano, T., Hiraoka, Y., and Yanagida, M. (1988). A temperature-sensitive mutation of the *Schizosaccharomyces pombe* gene *nuc2* + that encodes a nuclear scaffold-like protein blocks spindle-elongation in mitotic anaphase. J. Cell Biol. *106*, 1171–1183.

Hirokawa, N., Pfister, K. K., Yorifuji, H., Wagner, M. C., Brady, S. T., and Bloom, G. S. (1989). Submolecular domains of bovine brain kinesin identified by electron microscopy and monoclonal antibody decoration. Cell *56*, 867–78.

Holland, K., Kereso, J., Zakany, J., Pravnovskzy, T., Monostori, E., Belyaer, N. *et al.* (1995). A tightly bound chromosome antigen is detected by monoclonal antibodies in a ring-like structure on human centromeres. Chromosoma *103*, 559–566.

Holloway, S. L., Glotzer, M., King, R. W., and Murray, A. W. (1993). Anaphase is initiated by proteolysis rather than by the inactivation of maturation promoting factor. Cell *73*, 1393–1402.

Holm, C., Goto, T., Wang, J., and Botstein, D. (1985). DNA topoisomerase II is required at the time of mitosis in yeast. Cell *41*, 553–563.

Holm, C., Stearns, T., and Botstein, D. (1989). DNA topoisomerase II must act at mitosis to prevent nondisjunction and chromosome breakage. Mol. Cell. Biol. *9*, 159–168.

Holzbaur, E. and Vallee, R. (1994). Dyneins: molecular structure and cellular function. Annu. Rev. Cell Biol. *10*, 339–372.

Holzbaur, E., Hammarback, J., Paschal, B., Kravit, N., Pfister, K., and Vallee, R. (1991). Homology of a 150K cytoplasmic dynein-associate polypeptide with the *Drosophila* gene *glued*. Nature *351*, 579–583.

Holzbaur, E., Mikami, A., Paschal, B., and Vallee, R. (1994). Molecular characterization of cytoplasmic dynein. In Microtubules, eds. Hyams, J. and Lloyd, C. Wiley-Liss, New York, 251–268.

Houben, A., Guttenbach, M., KreB, W., Pich, U., Schubert, I., and Schmid, M. (1995). Immunostaining and interphase arrangement of field bean kinetochores. Chromosome Res. *3*, 27–31.

Hoyt, M. A., Totis, L., and Roberts, B. T. (1991). *S. cerevisiae* genes required for cell cycle arrest in response to loss of microtubule function. Cell *66*, 507–517.

Hoyt, M. A., He, L., Loo, K. K., and Saunders, W. S. (1992). Two *Saccharomyces cerevisiae* kinesin-related gene products required for mitotic spindle assembly. J. Cell Biol. *118*, 109–120.

Hoyt, M. A., He, A. L., Totis, L., and Saunders, W. S. (1993). Loss of function of *Saccharomyces cerevisiae* kinesin-related CIN8 and KIP1 is suppressed by KAR3 motor domain mutations. Genetics *135*, 35–44.

Hughes, S., Herskovits, J., Vaughan, K., and Vallee, R. (1993). Cloning and characterization of cytoplasmic dynein 53/55 and 57/59 subunits. Mol. Cell. Biol. *4*, 47a.

Hughes, S., Vaughan, K., Herskovits, J., and Vallee, R. (1995). Molecular analysis of a cytoplasmic dynein light intermediate chain reveals homology to a family of ATPases. J. Cell Sci. *108*, 17–24.

Hunt, T., Luca, F. C., and Ruderman, J. V. (1992). The requirements for protein synthesis and degradation and the control of destruction of cyclins A and B in the meiotic and mitotic cell cycles of the clam embryo. J. Cell Biol. *116*, 707–724.

Hyman, A. A. and Mitchison, T. J. (1991). Two different microtubule-based motor activities with opposite polarities in kinetochores. Nature (Lond.) *351*, 206–211.

Hyman, A. A., Middleton, K., Centola, M., Mitchison, T. J., and Carbon, J. (1992). Microtubule-motor activity of a yeast centromere-binding protein complex. Nature (Lond.) *359*, 533–535.

Ikeno, M., Masumoto, H., and Okazaki, T. (1994). Distribution of CENP-B boxes reflected in CREST centromere antigenic sites on long-range α-satellite DNA arrays of human chromosome 21. Hum. Mol. Genet. *3*, 1245–1257.

Irniger, S., Piatti, S., Michaelis, C., and Nasmyth, K. (1995). Genes involved in sister chromatid separation are needed for B-type cyclin proteolysis in budding yeast. Cell *81*, 269–278.

Ishida, R., Sato, M., Narita, T., Utsumi, K., Nishimoto, T., Morita, T. *et al.* (1994). Inhibition of topoisomerase II by ICRF-193 induces polyploidization by uncoupling chromosome dynamics from other cell cycle events. J. Cell Biol. *126*, 1341–1351.

Johns, E. W. (1982). In The HMG Chromosomal Proteins. Academic Press, New York.

Johnson, K. and Wall, J. (1983). Structure and molecular weight of the hynein ATPase. J. Cell Biol. *96*, 669–678.

Johnson, P. F. and McKnight, S. L. (1989). Eukaryotic transcriptional regulatory proteins. Annu. Rev. Biochem. *58*, 799–839.

Kadowaki, T., Zhao, Y., and Tartakoff, A. (1992). A conditional yeast mutant deficient in mRNA transport from nucleus to cytoplasm. Proc. Natl Acad. Sci. USA *89*, 2312–2316.

Kadowaki, T., Goldfarb, D., Spitz, L. M., Tartakoff, A. M., and Ohno, M. (1993). Regulation of RNA processing and transport by a nuclear guanine nucleotide release protein and members of the Ras superfamily. EMBO J. *12*, 2929–2937.

Kalderon, D., Roberts, B. L., Richardson, W. D., and Smith, A. E. (1984). A short amino acid sequence able to specify nuclear location. Cell *39*, 499–509.

Kallajoki, M., Weber, K., and Osborn, M. (1991). A 210 kDa nuclear matrix protein is a functional part of the mitotic spindle; a micro-injection study using SPN monoclonal antibodies. EMBO J. *10*, 3351–3362.

Kallajoki, M., Weber, K., and Osborn, M. (1992). Ability to organize microtubules in taxol-treated mitotic PtK2 cells goes with the SPN antigen and not with the centrosome. J. Cell Sci. *102*, 91–102.

Karantza, V., Maroo, A., Fay, D., and Sedivy, J. M. (1993). Overproduction of Rb protein after the G_1/S boundary causes G_2 arrest. Mol. Cell. Biol. *13*, 6640–6652.

Karess, R. and Glover, D. M. (1989). *Rough-deal:* a gene required for proper mitotic segregation in *Drosophila*. J. Cell Biol. *109*, 2951–2961.

Karsenti, E., Bravo, R., and Kirschner, M. (1987). Phosphorylation changes associated with the early cell cycle in *Xenopus* eggs. Dev. Biol. *119*, 442–453.

Kellogg, D., Field, C., and Alberts, B. (1989). Identification of microtubule-associated proteins in the centrosome, spindle, and kinetochore of the early *Drosophila* embryo. J. Cell Biol. *109*, 2977–2991.

Kerrebrock, A., Moore, D., Wu, J., and Orr-Weaver, T. (1995). Mei-S332, a *Drosophila* protein required for sister-chromatid cohesion, can localize to meiotic centromere regions. Cell *83*, 247–256.

King, R., Peters, J.-M., Tugendreich, S., Rolfe, M., Hieter, P., and Kirschner, M. (1995). A 20S complex containing CDC27 and CDC16 catalyzes the mitosis-specific conjugation of ubiquitin to cyclin B. Cell *81*, 279–288.

Kingwell, B. and Rattner, J. (1987). Mammalian kinetochore/centromere composition: a 50 kDa antigen is present in the mammalian kinetochore/centromere. Chromosoma *95*, 403–407.

Kingwell, B., Fritzler, M., Decoteau, J., and Rattner, J. (1987). Identification and characterization of a protein associated with the stembody using autoimmune sera from patients with systemic sclerosis. Cell Motil. Cytoskeleton *8*, 360–367.

Kipling, D., Mitchell, A. R., Masumoto, H., Wilson, H. E., Nicol, L., and Cooke, H. J. (1995). CENP-B binds a novel centromeric sequence in the Asian mouse *Mus caroli*. Mol. Cell. Biol. *15*, 4009–4020.

Kitagawa, K., Masumoto, H., Ikeda, M., and Okazaki, T. (1995). Analysis of protein–DNA and protein–protein interactions of centromere protein B (CENP-B) and properties of the DNA–CENP-B complex in the cell cycle. Mol. Cell. Biol. *15*, 1602–1612.

Klein, F., Laroche, T., Cardenas, M., Hofmann, J.-X., Schweizer, D., and Gasser, S. (1992). Localization of RAP1 and topoisomerase II in nuclei and meiotic chromosomes of yeast. J. Cell Biol. *117*, 935–948.

Knehr, M., Poppe, M., Schroeter, D., Eickelbaum, W., Finze, E., Kiesewetter, U., *et al.* (1996). Cellular expression of human centromere protein C demonstrates a cyclic behavior with highest abundance in the G_1 phase. Proc Natl Acad Sci USA 93, 10234-10239.

Koonce, M., Grissom, P., and McIntosh, J. (1992). Dynein from *Dictyostelium*: primary structure comparisons between a cytoplasmic motor enzyme and flagellar dynein. J. Cell Biol. *119*, 1597–1604.

Kornbluth, S., Dasso, M., and Newport, J. (1994). Evidence for a dual role for TC4 protein in regulating nuclear structure and cell cycle progression. J. Cell Biol. *125*, 705–719.

Kosik, K. S., Orecchio, L. D., Schnapp, B., Inouye, H., and Neve, R. L. (1990). The primary structure and analysis of the squid kinesin heavy chain. J. Biol. Chem. *265*, 3278–3283.

Kuang, J., Zhao, J. Y., Wright, D. A., Saunders, G. F., and Rao, P. N. (1989). Mitosis-specific monoclonal antibody MPM-2 inhibts *Xenopus* oocyte maturation and depletes maturation-promoting activity. Proc. Natl Acad. Sci. USA *86*, 4982–4986.

Kuang, J., Ashorn, C. L., Gonzalez-Kuyvenhoven, M., and Penkala, J. E. (1994). cdc25 is one of the MPM-2 antigens involved in the activation of maturation-promoting factor. Mol. Biol. Cell *5*, 135–145.

Kubiak, J. Z., Weber, M., Geraud, G., and Maro, B. (1992). Cell cycle modification during the transitions between meiotic M-phases in mouse oocytes. J. Cell Sci. *102*, 457–467.

Kuriyama, R., Rao, P. N., and Borisy, G. G. (1990). Immunocytochemical evidence for centrosomal phosphoproteins in mitotic sea urchin eggs. Cell Struct. Function *15*, 13–20.

Labbe, J., Lee, M., Nurse, P., Picard, A., and Doree, M. (1988). Activation at M-phase of a protein kinase encoded by a starfish homologue of the cell cycle control gene $cdc2^+$. Nature *335*, 251–254.

Laemmli, U., Cheng, S., Adolph, K., Paulson, J., Brown, J., and Baumbach, W. (1978). Metaphase chromosome structure: the role of nonhistone proteins. Cold Spring Harbor Symp. Quant. Biol. *423*, 351–360.

Lammers, J., Offenberg, H., VanAalderen, M., Vink, A., Dietrich, A., and Heyting, C. (1994). The gene encoding a major component of the lateral elements of synaptonemal complexes of the rat is related to X-linked lymphocyte-regulated genes. Mol. Cell. Biol. *14*, 1137–1146.

Landschultz, W. H., Johnson, P. F., and McKnight, S. L. (1988). The leucine zipper: a hypothetical structure common to a new class of DNA binding proteins. Science *240*, 1759–1763.

Lanford, R. E. and Butel, J. S. (1984). Construction and characterization of an SV40 mutant defective in nuclear transport of T antigen. Cell *37*, 801–813.

Lanini, L. and McKeon, F. (1995). Domains required for CENP-C assembly at the kinetochore. Mol. Biol. Cell *6*, 1049–1059.

Larin, Z., Fricker, M. D., and Tyler-Smith, C. (1994). *De novo* formation of several features of a centromere following introduction of a Y alphoid YAC into mammalian cells. Hum. Mol. Genet. *3*, 689–695.

Larsen, P. M. and Wolniak, S. M. (1993). Asynchronous entry into anaphase induced by okadaic acid: spindle organization and microtubule/kinetochore attachments. Protoplasma *177*, 53–65.

Lee, A., Tam, R., Belhumeur, P., DiPaolo, T., and Clark, M. W. (1993). Prp20, the *Saccharomyces cerevisiae* homolog of the regulator of chromosome condensation, RCC1, interacts with double-stranded DNA through a multi-component complex containing GTP-binding proteins. J. Cell Sci. *106*, 287–298.

Lee, M. G. and Nurse, P. (1987). Complementation used to clone a human homologue of the fission yeast cell cycle control gene *cdc2*. Nature *327*, 31–35.

Lee, M. G., Norbury, C. J., Spurr, N. K., and Nurse, P. (1988). Regulated expression and phosphorylation of a possible mammalian cell-cycle control protein. Nature *333*, 676–679.

Lee-Eiford, A., Ow, R., and Gibbons, I. (1986). Specific cleavage of dynein heavy chains by ultraviolet irradiation in the presence of ATP and vanadate. J. Biol. Chem. *261*, 2337–2342.

Lees-Miller, J. P., Helfman, D. M., and Schroer, T. A. (1992). A vertebrate actin-related protein is a component of a multisubunit complex involved in microtubule-based vesicle motility. Nature *359*, 244–246.

LeGuellec, R., Paris, J., Couturier, A., Roghi, C., and Philippe, M. (1991). Cloning by differential screening of a *Xenopus* cDNA that encodes a kinesin-related protein. Mol. Cell. Biol. *11*, 3395–98.

Lehner, C. F. and O'Farrell, P. H. (1990). The roles of *Drosophila* cyclins A and B in mitotic control. Cell *61*, 535–547.

Levinger, L. and Varshavsky, A. (1982(a)). Selective arrangement of ubiquitinated and D1 protein-containing nucleosomes with the *Drosophila* genome. Cell *28*, 375–385.

Levinger, L. and Varshavsky, A. (1982(b)). Protein D1 preferentially binds $(A + T)$-rich satellite DNA. Proc. Natl Acad. Sci. USA *79*, 7152–7156.

Li, R. and Murray, A. W. (1991). Feedback control of mitosis in budding yeast. Cell *66*, 519–531.

Li, Y., Yeh, E., Hays, T., and Bloom, K. (1993). Disruption of mitotic spindle orientation in a yeast dynein mutant. Proc. Natl Acad. Sci. USA *90*, 10096–10100.

Liao, H., Li, G., and Yen, T. J. (1994). Mitotic regulation of microtubule cross-linking activity of CENP-E kinetochore protein. Science *265*, 394–398.

Liao, H., Winkfein, R. J., Mack, G., Rattner, J. B., and Yen, T. J. (1995). CENP-F is a protein of the nuclear matrix that assembles onto kinetochores at late G2 and is rapidly degraded after mitosis. J. Cell Biol. *130*, 507–518.

Lima-de-Faria, A. (1955). The division cycle of the kinetochore. Heriditas *41*, 238–240.

Lin, H. and Church, K. (1982). Meiosis in *Drosophila melanogaster*, III. The effect of orientation disruptor (ord) on gonial mitotic and the meiotic divisions in males. Genetics *102*, 751–770.

Linder, P., Lasko, P. F., Ashburner, M., Leroy, L., Nielsen, P. J., Nishi, K. *et al.* (1989). Birth of the D-E-A-D box. Nature (Lond.) *337*, 121–122.

Lohka, M. J., Kyes, J. L., and Maller, J. L. (1987). Metaphase protein phosphorylation in *Xenopus laevis*. Mol. Cell. Biol. *7*, 760–768.

Lombillo, V. A., Nislow, C., Yen, T. J., Gelfand, V. I., and McIntosh, J. R. (1995). Antibodies to the kinesin motor domain and CENP-E inhibit microtubule depolymerization-dependent motion of chromosomes *in vitro*. J. Cell Biol. *128*, 107–115.

Lorca, T., Devault, A., Colas, P., Loon, A. V., Fesquet, D., Lazaro, J. *et al.* (1992). Cyclin A-Cys41 does not undergo cell cycle dependent degradation in *Xenopus* extracts. FEBS Lett. *306*, 90–93.

Luca, F. C. and Ruderman, J. V. (1989). Control of programmed cyclin destruction in a cell-free system. J. Cell Biol. *109*, 1895–1909.

Ludlow, J. W., Glendening, C. L., Livingston, D. M., and DeCarprio, J. A. (1993). Specific enzymatic dephosphorylation of the retinoblastoma protein. Mol. Cell. Biol. *13*, 367–372.

Lupas, A., Dyke, M. V., and Stock, J. (1991). Predicting coiled coils from protein sequences. Science (Wash.) *252*, 1162–1164.

Lydersen, B. K., and Pettijohn, D. E. (1980). Human-specific nuclear protein that associates with the polar region of the mitotic apparatus: distribution in a human/hamster hybrid cell. Cell *22*, 489–499.

Lye, R., Porter, M., Scholey, J., and McIntosh, J. (1987). Identification of a microtubule-based cytoplasmic motor in the nematode *C. elegans*. Cell *51*, 309–318.

Mackay, A. M. and Earnshaw, W. C. (1993). The INCENPs: structural and functional analysis of a family of chromosome passenger proteins. Cold Spring Harbor Symp. Quant. Biol. *LVIII*, 697–706.

Mackay, A. M., Eckley, D. M., Chue, C., and Earnshaw, W. C. (1993). Molecular analysis of the INCENPs (inner centromere proteins): separate domains are required for association with microtubules during interphase and with the central spindle during anaphase. J. Cell Biol. *123*, 373–385.

Maekawa, T., Leslie, R., and Kuriyama, R. (1991). Identification of a minus end-specific microtubule-associated protein located at the mitotic poles in cultured mammalian cells. Eur. J. Cell Biol. *54*, 255–267.

Mandelkow, E. and Mandelkow, E. M. (1995). Microtubules and microtubule-associated proteins. Curr. Opin. Cell. Biol. *7*, 72–81.

Mann, R. K. and Grunstein, M. (1992). Histone H3 N-terminal mutations allow hyperactivation of the yeast GAL1 gene *in vivo*. EMBO J. *11*, 3297–3306.

Mason, J. (1976). Orientation disruptor (ord): a recombination-defective and disjunction-defective meiotic mutant in *Drosophila melanogaster*. Genetics *84*, 545–572.

Masuda, H., Sevik, M., and Cande, W. Z. (1992). *In vitro* microtubule-nucleating activity of spindle pole bodies in fission yeast *Schizosaccharmyces pombe* cell cycle-dependent activation in *Xenopus* cell-free extracts. J. Cell Biol. *117*, 1055–1066.

Masumoto, H., Sugimoto, K., and Okazaki, T. (1989). Alphoid satellite DNA is tightly associated with centromere antigens in human chromosomes throughout the cell cycle. Exp. Cell Res. *181*, 181–196.

Matsumoto, T. and Beach, D. (1991). Premature initiation of mitosis in yeast lacking *RCC1* or an interacting GTPase. Cell *66*, 347–360.

Mayer-Jaekel, R. E., Ohkura, H., Gomes, R., Sunkel, C. E., Baumgartner, S., Hemmings, B. A. *et al*. (1993). The 55kd regulatory subunit of *Drosophila* protein phosphatase 2A is required for anaphase. Cell *72*, 621–633.

McDonald, H. B. and Goldstein, L. S. B. (1990). Identification and characterization of a gene encoding a kinesin-like protein in *Drosophila*. Cell *61*, 991–1000.

McDonald, H. B., Stewart, R. J., and Goldstein, L. S. (1990). The kinesin-like ncd protein of Drosophila is a minus end-directed microtubule motor. Cell *63*, 1159–1165.

McKay, S., Thomson, E., and Cooke, H. (1994). Sequence homologies and linkage group conservation of the human and mouse CENP-C genes. Genomics *22*, 36–40.

McKee, B. and Karpen, G. (1990). *Drosophila* ribosomal RNA genes function as an X-Y pairing site during male meiosis. Cell *61*, 61–72.

McNeilage, L. J., Whittingham, S., McHugh, N., and Barnett, A. J. (1986). A highly conserved 72 000 dalton centromeric antigen reactive with autoantibodies from patients with progressive systemic sclerosis. J. Immunol. *137*, 2541–2547.

Melchior, F., Paschal, B., Evans, J., and Gerace, L. (1993). Inhibition of nuclear protein import by nonhydrolyzable analogues of GTP and identification of the small GTPase Ran/TC4 as an essential transport factor. J. Cell Biol. *123*, 1649–1659.

Meluh, P. and Koshland, D. (1995). Evidence that the *MIF2* gene of *Saccharomyces cerevisiae* encodes a centromere protein with homology to the mammalian centromere protein CENP-C. Mol. Biol. Cell *6*, 793–807.

Middleton, K. and Carbon, J. (1994). *KAR3*-encoded kinesin is a minus-end-directed motor that functions with centromere binding proteins (CBF3) on an *in vitro* yeast kinetochore. Proc. Natl Acad. Sci. USA *91*, 7212–7216.

Mikami, A., Paschal, B., Mazumdar, M., and Vallee, R. (1993). Molecular cloning of the retrograde transport motor cytoplasmic dynein (MAP 1C). Neuron *10*, 787–796.

Minshull, J., Pines, J., Golsteyn, R., Standart, N., Mackie, S., Colman, A. *et al.* (1989). The control of cyclin synthesis modification and destruction in the control of cell division. J. Cell Sci. Suppl. *12*, 77–97.

Minshull, J., Golsteyn, R., Hill, C. S., and Hunt, T. (1990). The A- and B-type cyclin-associated cdc2 kinases in *Xenopus* turn on and off at different times in the cell cycle. EMBO J. *9*, 2865–2875.

Mirzabekov, A. D., Shik, V. V., Belyavsky, A. V., and Bavykin, S. G. (1978). Primary organization of the nucleosome core particle of chromatin: sequences of histone DNA arrangement along the DNA. Proc. Natl Acad. Sci. USA *75*, 4184–4188.

Mitchell, D. R. and Kang, Y. (1991). Identification of *oda6* as a *Chlamydomonas* dynein mutant by rescue with the wild-type gene. J. Cell Biol. *113*, 835–842.

Mitchison, T. and Kirschner, M. (1985). Properties of the kinetochore *in vitro* I. Microtubule nucleation and tubulin binding. J. Cell Biol. *101*, 755–765.

Miyabashira, J., Sekiguchi, T., and Nishimoto, T. (1994). Mammalian cells have two functional RCC1 proteins produced by alternative splicing. J. Cell Sci. *107*, 2203–2208.

Miyazaki, W. Y. and Orr-Weaver, T. L. (1992). Sister-chromatid misbehavior in *Drosophila ord* mutants. Genetics *132*, 1047–61.

Miyazaki, W. Y. and Orr-Weaver, T. L. (1994). Sister-chromatid cohesion in mitosis and meiosis. Annu. Rev. Genet. *28*, 167–187.

Moens, P. B. and Pearlman, R. E. (1989). Satellite DNAI in chromatin loops of rat pachytene chromosomes and in spermatids. Chromosoma *98*, 287–294.

Moens, P. and Spyropoulos, B. (1995). Immunocytology of chiasmata and chromosomal disjunction at mouse meiosis. Chromosoma *104*, 175–182.

Mole-Bajer, J., Bajer, A., Zinkowski, R., Balczon, R., and Brinkley, B. (1990). Autoantibodies from a patient with scleroderma CREST recognized kinetochores of the higher plant *Haemanthus*. Proc. Natl Acad. Sci. USA *87*, 3599–3603.

Moore, M. S. and Blobel, G. (1993). The GTP-binding protein Ran/TC4 is required for protein import into the nucleus. Nature *365*, 661–663.

Moore, M. S. and Blobel, G. (1994). A G protein involved in nucleocytoplasmic transport: the role of Ran. Trends Biochem. Sci. *19*, 211–216.

Moroi, Y., Peebles, C., Fritzler, M., Steigerwald, J., and Tan, E. (1980). Autoantibody to centromere (kinetochore) in scleroderma sera. Proc. Natl Acad. Sci. USA *77*, 1627–1631.

Moroi, Y., Hartman, A. L., Nakane, P. K., and Tan, E. M. (1981). Distribution of kinetochore (centromere) antigen in mammalian cell nuclei. J. Cell Biol. *90*, 254–259.

Moss, S. B., Challoner, P. B., and Groudine, M. (1989). Expression of a novel histone 2B during mouse spermiogenesis. Dev. Biol. *133*, 83–92.

242   *Centromere proteins of higher eukaryotes*

Mueller, P. R., Coleman, T. R., and Dunphy, W. G. (1995). Cell cycle regulation of a *Xenopus* Wee1-like kinase. Mol. Biol. Cell *6*, 119–134.

Mullner, E. W. and Kuhn, L. C. (1988). A stem-loop in the 3' untranslated region mediates iron-dependent regulation of transferrin receptor mRNA stability in the cytoplasm. Cell *53*, 815–825.

Murakami, Y., Huberman, J., and Hurwitz, J. (1996). Identification, purification, and molecular cloning of autonomously replicating sequence-binding protein 1 from fission yeast *Schizosaccharomyces pombe*. Proc. Natl Acad. Sci. USA *93*, 502–507.

Muro, Y., Masumoto, H., Yoda, K., Nozaki, N., Ohashi, M., and Okazaki, T. (1992). Centromere protein B assembles human centromeric alpha-satellite DNA at the 17-bp sequence, CENP-B Box. J. Cell Biol. *116*, 585–596.

Murphy, T. D. and Karpen, G. H. (1995). Interaction between the *nod*[+] kinesin-like gene and extracentromeric sequences are required for transmission of a *Drosophila* minichromosome. Cell *81*, 139–148.

Murray, A. W., Solomon, M. J., and Kirschner, M. W. (1989). The role of cyclin synthesis and degradation in the control of maturation promoting factor activity. Nature (Lond.) *339*, 280–286.

Murray, A. W. (1994). Cell cycle checkpoints. Curr. Opin. Cell Biol. *6*, 872–876.

Nasmyth, K. A. and Reed, S. I. (1980). Isolation of genes by complementation in yeast: molecular cloning of a cell-cycle gene. Proc. Natl Acad. Sci. USA *77*, 2119–2123.

Navone, F., Niclas, J., Hom-Booher, N., Sparks, L., Bernstein, H., McCaffrey, G. et al. (1992). Cloning and expression of a human kinesin heavy chain gene: interaction of the COOH-terminal domain with cytoplasmic microtubles in transfected CV-1 cells. J. Cell Biol. *117*, 1263–1275.

Nelson, W., Liu, L., and Coffey, D. (1986). Newly replicated DNA is associated with DNA topoisomerase II in cultured rat prostatic adenocarcinoma cells. Nature *322*, 187.

Nicklas, R. B. (1971). Mitosis. Adv. Cell Biol. *2*, 225–297.

Nicklas, R. B. (1988). The forces that move chromosomes in mitosis. Annu. Rev. Biophys. Chem. *12*, 431–449.

Nicklas, R. B., Ward, S. C., and Gorbsky, G. J. (1995). Kinetochore chemistry is sensitive to tension and may link mitotic forces to a cell cycle checkpoint. J. Cell Biol. *130*, 929–939.

Nicklas, T. B. (1974). Chromosome segregation mechanisms. Genetics *78*, 205–213.

Nigg, E. A. (1993). Cellular substrates of p34[cdc2] and its companion cyclin-dependent kinases. Trends Cell. Biol. *3*, 296–301.

Nishimoto, T., Eilen, E., and Basilico, C. (1987a). Premature chromosome condensation in a ts DNA mutant of BHK cells. Cell *15*, 475–483.

Nishimoto, T., Ajiro, K., Davis, F. M., Yamashita, K., Kai, R., Rao, P. N., and Sekiguchi, M. (1987b). Mitosis-specific phosphorylation associated with premature chromosome condensation in a ts cell cycle mutant. In: Molecular Regulation of Nuclear Events in Mitosis and Meiosis, eds. Schlegel, R.A., Halleck, M.S. and Rao, P.N. Academic Press, New York, 259–293.

Nishitani, H., Kobayashi, H., Ohtsubo, M., and Nishmoto, T. (1990). Cloning of *Xenopus RCC1* cDNA, a homolog of the human *RCC1* gene: complementation of tsBN2 mutation and identification of the product. J. Biochem. *107*, 228–235.

Nishitani, H., Ohtsubo, M., Yamashita, K., Iida, H., Pines, J., Yasuda, H. et al. (1991). Loss of RCC1, a nuclear DNA-binding protein, uncouples the completion of DNA replication from the activation of cdc2 protein kinase and mitosis. EMBO J. *10*, 1555–1564.

Norbury, C. and Nurse, P. (1992). Animal cell cycles and their control. Annu. Rev. Biochem. *61*, 441–470.

Nurse, P., Thuriaux, P., and Nasmyth, K. (1976). Genetic control of the cell division cycle in the fission yeast *Schizosaccharomyces pombe*. Mol. Gen. Genet. *146*, 167–176.

Nurse, P. (1990). Universal control mechanism regulating onset of M-Phase. Nature (Lond.) *344*, 503–508.

Nurse, C. N. P. (1992). Animal cell cycles and their control. Annu. Rev. Biochem. *61*, 441–470.

O'Brien, D. A. and Bellvé, A. R. (1980). Protein constituents of the mouse spermatozoan. Dev. Biol. *75*, 386–404.

O'Brien, S. J., Womack, J. E., Lyons, L. A., Moore, K. J., Jenkins, N. A., and Copeland, N. G. (1993). Anchored reference loci for comparative genomic mapping in mammals. Nature Genet. *3*, 103–112.

O'Connell, M. J., Meluh, P. B., Rose, M. D., and Morris, N. R. (1993). Suppression of the *bim*C4 mitotic spindle defect by deletion of *klp*A, a gene encoding a *KAR3*-related kinesin-like protein in *Aspergillus nidulans*. J. Cell Biol. *120*, 153–162.

Ohkura, H. and Yanagida, M. (1991). *S. pombe* gene sds22+ essential for a midmitotic transition encodes a leucine-rich repeat protein that positively modulates protein phosphatase-1. Cell *64*, 149–157.

Ohtsubo, M., Kai, R., Furuno, N., Sekiguchi, T., Sekiguchi, M., Hayashida, H. *et al.* (1987). Isolation and characterization of the active cDNA of the human cell cycle gene (RCC1) involved in the regulation of onset of chromosome condensation. Genes Dev. *1*, 585–593.

Ohtsubo, M., Okazaki, H., and Nishimoto, T. (1989). The RCC1 protein, a regulator for the onset of chromosome condensation locates in the nucleus and binds to DNA. J. Cell Biol. 1389–1397.

Ohtsubo, M., Yoshida, T., Seino, H., Nishitani, H., Clark, K., Sprague, G., Jr *et al.* (1991). Mutation of the hamster cell cycle gene RCC1 is complemented by the homologous genes of *Drosophila* and *S. cerevisiae*. EMBO J. *10*, 1265–1273.

Orr-Weaver, T. (1995). Meiosis in *Drosophila*: seeing is believing. Proc. Natl Acad. Sci. USA *92*, 10443–10449.

Otsuka, A. J., Jeyaprakash, A., Carcia-Anoveros, J., Tang, L. Z., Fisk, G., Hartshorne, T. *et al.* (1991). The *C. elegans unc-104* encodes a putative kinesin heavy chain-like protein. Neuron *6*, 113–22.

Ouspenski, I. I., Mueller, U. W., Matynia, A., Sazer, S., Elledge, S. J., and Brinkley, B. R. (1995). Ran-binding protein-1 is an essential component of the Ran/RCC1 molecular switch system in budding yeast. J. Biol. Chem. *270*, 1975–1978.

Pabo, C. O. and Sauer, R. T. (1992). Transcription factors: structural families and principles of DNA recognition. Annu. Rev. Biochem. *61*, 1053–1095.

Page, S. L., Earnshaw, W. C., Choo, K. H. A., and Shaffer, L. G. (1995). Further evidence that CENP-C is a necessary component of active centromeres: studies of a dic(X;15) with simultaneous immunofluorescence and FISH. Hum. Mol. Genet. *4*, 289–294.

Palevitz, B. (1990). Kinetochore behavior during generative cell division in *Tradescantia virginiana*. Protoplasma *157*, 120–127.

Palmer, D. K. and Margolis, R. L. (1985). Kinetochore components recognized by human autoantibodies are present on mononucleosomes. Mol. Cell. Biol. *5*, 173–186.

Palmer, D. K., O'Day, K., Wener, M. H., Andrews, B. S., and Margolis, R. L. (1987). A 17-kD centromere protein (CENP-A) copurifies with nucleosome core particles and with histones. J. Cell Biol. *104*, 805–815.

Palmer, D. K., O'Day, K., and Margolis, R. L. (1990). The centromere specific histone CENP-A is selectively retained in discrete foci in mammalian sperm nuclei. Chromosoma *100*, 32–36.

Palmer, D. K., O'Day, K., Trong, H. L., Charbonneau, H., and Margolis, R. L. (1991). Purification of the centromere-specific protein CENP-A and demonstration that it is a distinctive histone. J. Cell Biol. *88*, 3734–3738.

Pankov, R., Lemieux, M., and Hancock, R. (1990). An antigen located in the kinetochore region in metaphase and on polar microtubule ends in the midbody region in anaphase, characterised using a monoclonal antibody. Chromosoma *99*, 95–101.

Paschal, B. and Vallee, R. (1987). Retrograde transport by the microtubule-associated protein MAP1C. Nature *330*, 181–183.

Paschal, B., Shpetner, H., and Vallee, R. (1987). MAP 1C is a microtubule-activated ATPase which translocates microtubules *in vitro* and has dynein-like properites. J. Cell Biol. *105*, 1273–1282.

Paschal, B., Mikami, A., Pfister, K., and Vallee, R. (1992). Homology of the 74-kD cytoplasmic dynein subunit with a flagella dynein polypeptide suggests an intracellular targeting function. J. Cell Biol. *118*, 1133–1143.

Paschal, B., Holzbaur, E., Pfister, K., Clark, S., Meyer, D., and Vallee, R. (1993). Characterization of a 50-kDa polypeptide in cytoplasmic dynein preparations reveals a complex with p150^{Glued} and a novel actin. J. Biol. Chem. *268*, 15318–15323.

Paulson, J. and Laemmli, U. (1977). The structure of histone-depleted chromosomes. Cell *12*, 817–828.

Peterson, C. L. (1994). The SMC family: novel motor proteins for chromosome condensation? Cell *79*, 389–392.

Pfarr, C. M., Coue, M., Grissom, P. M., Hays, T. S., Porter, M. E., and McIntosh, J. R. (1990). Cytoplasmic dynein is localized to kinetochores during mitosis. Nature *345*, 263–265.

Philp, A. V., Axton, J. M., Saunders, R. D. C., and Glover, D. M. (1993). Mutations of the *Drosophila melanogaster* gene *three rows* permit aspects of mitosis to continue in the absence of chromatid segregation. J. Cell Sci. *106*, 87–98.

Pierre, P., Scheel, J., Rickard, J., and Kreis, T. (1992). CLIP-170 links endocytic vesicles to microtubules. Cell *70*, 887–900.

Pines, J. and Hunter, T. (1989). Isolation of a human cyclin cDNA: evidence for cyclin mRNA and protein regulation in the cell cycle and for interaction with p34^{cdc2}. Cell *58*, 833–846.

Pines, J. and Hunter, T. (1990). Human cyclin A is adenovirus E1A-associated protein p60 and behaves differently from cyclin B. Nature *346*, 760.

Pines, J. and Hunter, T. (1991). Human cyclins A and B1 are differentially located in the cell and undergo cell cycle-dependent nuclear transport. J. Cell Biol. *115*, 145–153.

Plamann, M., Minke, P., Tinsley, J., and Bruno, K. (1994). Cytoplasmic dynein and centractin are required for normal nuclear distribution in filamentous fungi. J. Cell Biol. *127*, 139–149.

Pluta, A., and Earnshaw, W. (1996). Specific interaction between human kinetochore protein CENP-C and a nucleolar transcriptional regulator. J Biol Chem 271, 18767-18774.

Pluta, A. F., Saitoh, N., Goldberg, I., and Earnshaw, W. C. (1992). Identification of a subdomain of CENP-B that is necessary and sufficient for localization to the human centromere. J. Cell Biol. *116*, 1081–1093.

Pollard, T., Doberstein, S., and Zot, H. (1991). Myosin-I. Annu. Rev. Physiol. *53*, 653–681.

Porter, M. E. and Johnson, K. A. (1989). Dynein structure and function. Annu. Rev. Cell Biol. *5*, 199–251.

Price, C. M. and Pettijohn, D. E. (1986). Redistribution of the nuclear mitotic apparatus protein (NuMA) during mitosis and nuclear assembly. Exp. Cell Res. *166*, 295–311.

Price, C. M., McCarty, G. A., and Pettijohn, D. E. (1984). NuMA protein is a human autoantigen. Arthritis Rheumatism *27*, 774–779.

Rappaport, R. (1986). Establishment of the mechanism of cytokinesis in animal cells. Int. Rev. Cytol. *105*, 245–81.

Rattner, J. B. (1986). Organization within the mammalian kinetochore. Chromosoma *93*, 515–520.

Rattner, J. B., Kingwell, B., and Fritzler, M. (1988). Detection of distinct structural domains within the primary constriction using autoantibodies. Chromosoma *96*, 360–367.

Rattner, J. B., Lew, J., and Wang, J. H. (1990). p34^{cdc2} kinase is localized to distinct domains within the mitotic apparatus. Cell Motil. Cytoskeleton *17*, 227–235.

Rattner, J. B., Wang, T., Mack, G., Fritzler, M., Martin, L., and Valencia, D. (1992). MSA-36: a chromosomal and mitotic spindle-associated protein. Chromosoma *101*, 625–633.

Rattner, J. B., Rao, A., Fritzler, M. J., Valencia, D. W., and Yen, T. J. (1993). CENP-F is a 400 kDa kinetochore protein that exhibits a cell-cycle dependent localization. Cell Motil. Cytoskeleton *26*, 214–226.

Rattner, J. B., Hendzel, M., Furbee, C., Muller, M., and Bazett-Jones, D. (1996). Topoisomerase IIα is associated with the mammalian centromere in a cell cycle- and species-specific manner and is required for proper centromere/kinetochore structure. J. Cell Biol. *134*, 1097-1107.

Rechsteiner, M. (1990). PEST sequences are signals for rapid intracellular proteolysis. Semin Cell Biol. *1*, 433–440.

Ren, M., Drivas, G., D'Eustachio, P., and Rush, M. G. (1993). Ran/TC4: a small nuclear GTP-binding protein that regulates DNA synthesis. J. Cell Biol. *120*, 313–323.

Ren, M., Coutavas, E., D'Eustachio, P., and Rush, M. G. (1994). Effects of mutant Ran/TC4 proteins on cell cycle progression. Mol. Cell. Biol. *14*, 4216–4224.

Riabowol, K., Draetta, G., Brizuela, L., Vandre, D., and Beach, D. (1989). The cdc2 kinase is a nuclear protein that is essential for mitosis in mammalian cells. Cell *57*, 393–401.

Richmond, T. J., Finch, J. T., Rushton, B., Rhodes, D., and Klug, A. (1984). Structure of the nucleosome core particle at 7Å resolution. Nature (Lond.) *311*, 532–537.

Rieder, C. L. (1982). The formation, structure, and composition of the mammalian kinetochore fiber. Intern. Rev. Cytol. *79*, 1–58.

Rieder, C. L. (1991). Mitosis: towards a molecular understanding of chromosome behavior. Curr. Opin. Cell Biol. *3*, 59–66.

Rieder, C. L. and Alexander, S. P. (1989). The attachment of chromosomes to the mitotic spindle and the production of aneuploidy in newt lung cells. In Mechanisms of Chromosome Distribution and Aneuploidy; eds. Resnick, M.A. and Vig, B.K. Alan R. Liss, New York, 185–194.

Rieder, C. L. and Alexander, S. P. (1990). Kinetochores are transported poleword along a single astral microtubule during chromosome attachment to the spindle in newt lung cells. J. Cell Biol. *110*, 81–95.

Rieder, C. L., Cole, R., Khodjakov, A., and Sluder, G. (1995). The checkpoint delaying anaphase in response to chromosome monoorientation is mediated by an inhibitory signal produced by unattached kinetochores. J. Cell Biol. *130*, 941–8.

Robbins, J., Dilworth, S. M., Laskey, R. A., and Dingwall, C. (1991). Two interdependent basic domains in nucleoplasmin nuclear targeting sequence. Cell *64*, 615–623.

Roca, J. (1995). The mechanisms of DNA topoisomerases. Trends Biochem. Sci. *20*, 156–160.

Rodriguez-Alfageme, C., Rudkin, G., and Cohen, L. (1980). Isolation, properties and cellular distribution of D1, a chromosomal protein of *Drosophila*. Chromosoma *78*, 1–31.

Rogers, S., Wells, R., and Rechsteiner, M. (1986). Amino acid sequences common to rapidly degraded proteins: the PEST hypothesis. Science (Wash.) *234*, 364–368.

Romanova, L., Deriagin, G., Mashkova, T., Tumeneva, I., Mushegian, A., Kisselev, L., and Alexandrov, I. (1996). Evidence for selection in evolution of alpha satellite DNA: The central role of CENP-B/pJα binding region. J. Mol. Biol. *261*, 334–340.

Roof, D. M., Meluh, P. B., and Rose, M. D. (1992). Kinesin-related proteins required for assembly of the mitotic spindle. J. Cell Biol. *118*, 95–108.

Rose, D., Thomas, W., and Holm, C. (1990). Segregation of recombined chromosomes in meiosis I requires DNA topoisomerase II. Cell *60*, 1009–1017.

Rothfield, N. F., Whitaker, D., Bordwell, B., Weiner, E., Senecal, J. L., and Earnshaw, W. C. (1987). Detection of anti-centromere antibodies using cloned CENP-B. Arthritis Rheumatism *30*, 1416–1419.

Saitoh, H. and Dasso, M. (1995). The RCC1 protein interacts with Ran, RanBP1, hsc70, and a 340-kDa protein in *Xenopus* extracts. J. Biol. Chem. *270*, 10658–10663.

Saitoh, H., Tomkiel, J., Cooke, C. A., H. R., Maurer, M., Rothfield, N. F., and Earnshaw, W. C. (1992). CENP-C, an autoantigen in scleroderma is a component of the human inner kinetochore plate. Cell *70*, 115–125.

Saitoh, N., Goldburg, I. G., Wood, E. R., and Earnshaw, W. C. (1994). ScII: an abundant chromosome scaffold protein is a member of a family of putative ATPases with an unusual predicted tertiary structure. J. Cell Biol. *127*, 303–318.

Saitoh, N., Goldberg, I. G., and Earnshaw, W. C. (1995). The SMC proteins and the coming of age of the chromosome scaffold hypothesis. BioEssays *17*, 759–766.

Sale, W., Goodenough, U., and Heuser, J. (1985). The substructure of isolated and *in situ* outer dynein arms of sea urchin flagella. J. Cell Biol. *101*, 1400–1412.

Sandler, L., Lindsley, D., Nicoletti, B., and Trippa, G. (1968). Mutants affecting meiosis in natural populations of *Drosophila melanogaster*. Genetics *60*, 525–558.

Saraste, M., Sibbald, P. R., and Wittinghofer, A. (1990). The P-loop—a common motif in ATP- and GTP-binding proteins. Trends Biochem. Sci. *15*, 430–434.

Saunders, W. S. and Hoyt, M. A. (1992). Kinesin-related proteins required for structural integrity of the mitotic spindle. Cell *70*, 451–458.

Saunders, W. S., Chue, C., Goebl, M., Craig, C., Clark, R., Powers, J. *et al.* (1993). Molecular cloning of a human homologue of *Drosophila* heterochromatin protein HP1 using anti-centromere autoantibodies with *anti-chromo* specificity. J. Cell Sci. *104*, 573–582.

Saunders, W. S., Koshland, D., Eshel, D., Gibbons, I. R., and Hoyt, M. A. (1995). *Saccharomyces cerevisiae* kinesin- and dynein-related proteins required for anaphase chromosome segregation. J. Cell Biol. *128*, 617–624.

Sawin, K. E., Mitchison, T. J., and Wordeman, L. G. (1992). Evidence for kinesin-related proteins in the mitotic apparatus using peptide antibodies. J. Cell Sci. *101*, 303–313.

Saxton, W. M., Hicks, J., Goldstein, L. S. B., and Raff, E. C. (1991). Kinesin heavy chain is essential for viability and neuromuscular functions in *Drosophila*, but mutants show no defects in mitosis. Cell *64*, 1093–1102.

Sazer, S. and Nurse, P. (1994). A fission yeast RCC1-related protein is required for the mitosis to interphase transition. EMBO J. *13*, 606–615.

Scholey, J. M., Heuser, J., Yang, J. T., and Goldstein, L. S. B. (1989). Identification of globular mechanochemical heads of kinesin. Nature (Lond.) *338*, 355–357.

Schrader, T. E. and Crothers, D. M. (1990). Effects of DNA sequence and histone–histone interactions on nucleosome placement. J. Mol. Biol. *216*, 69–84.

Schröder, H., Trolltsch, D., Friese, U., Bachmann, M., and Müller, W. (1987). Mature mRNA is selectively released from the nuclear matrix by an ATP/dATP-dependent mechanism sensitive to topoisomerase inhibitors. J. Biol. Chem. *262*, 8917–8925.

Schroer, T. (1994). Structure, function and regulation of cytoplasmic dynein. Curr. Opin. Cell Biol. *6*, 69–73.

Schroer, T. A. and Sheetz, M. P. (1991). Two activators of microtubule-based vesicle transport. J. Cell Biol. *115*, 1309–1318.

Schroer, T., Steuer, E., and Sheetz, M. (1989). Cytoplasmic dynein is a minus end-directed motor for membranous organelles. Cell *56*, 937–946.

Seino, H., Hisamoto, N., Uzawa, S., Sekiguchi, T., and Nishimoto, T. (1992). DNA-binding domain of RCC1 protein is not essential for coupling mitosis with DNA replication. J. Cell Sci. *102*, 393–400.

Seki, N., Saito, T., Kitagawa, K., Masumoto, H., Okazaki, T., and Horit, T. A. (1994). Mapping of the human centromere protein B gene (CENP-B) to chromosome 20p13 by fluorescence *in situ* hybridization. Genomics *24*, 187–188.

Shamu, C. and Murray, A. (1992). Sister chromatid separation in frog egg extracts requires DNA topoisomerase II activity during anaphase. J. Cell Biol. *117*, 921–934.

Shan, B., Zhu, X., Chen, P., Durfee, T., Yang, Y., Sharp, D., and Lee, W. (1992). Molecular cloning of cellular genes encoding retinoblastoma-associated proteins: identification of a gene with properties of the transcription factor E2F. Mol. Cell. Biol. 12, 5620-5631.

Simanis, V. and Nurse, P. (1986). The cell cycle control gene $cdc2^+$ of fission yeast encodes a protein kinase potentially regulated by phosphorylation. Cell *45*, 261–268.

Simerly, C., Balczon, R., Brinkley, B. R., and Schatten, G. (1990). Microinjected kinetochore antibodies interfere with chromosome movement in meiotic and mitotic mouse embryos. J. Cell Biol. *111*, 1491–1504.

Skibbens, R. V., Skeen, V. P., and Salmon, E. D. (1993). Directional instability of kinetochore motility during chromosome congression and segregation in mitotic newt lung cells; a push-pull mechanism. J. Cell Biol. *122*, 859–875.

Smit, A., and Riggs, A. (1996). Tiggers and other DNA transposon fossils in the human genome. Proc Natl Acad Sci USA 93, 1443-1448.

Smith, P. (1990). DNA topoisomerase dysfunction: a new goal for antitumour chemotherapy. Bioessays *12*, 167–172.

Smith, D., Baker, B., and Gatti, B. (1985). Mutations in genes controlling essential mitotic functions in Drosophila melanogaster. Genetics 110, 647-670.

Soloman, M., Strauss, F., and Varshavsky, A. (1986). A mammalian high mobility group protein recognizes any stretch of six A.T base pairs in duplex DNA. Proc. Natl Acad. Sci. USA *83*, 1276–1280.

Sparks, C. A., Bangs, P. L., McNeil, G. P., Lawrence, J. B., and Fey, E. G. (1993). Assignment of the nuclear mitotic apparatus protein NuMA gene to human chromosome 11q13. Genomics *17*, 222–224.

Standart, N., Minshull, J., Pines, J., and Hunt, T. (1987). Cyclin synthesis, modification and destruction during meiotic maturation of the starfish oocyte. Dev. Biol. *124*, 248–258.

Steuer, E., Wordeman, L., Schroer, T. A., and Sheetz, M. P. (1990). Localization of cytoplasmic dynein to mitotic spindles and kinetochores. Nature (Lond.) *345*, 266–268.

Stewart, R. J., Pesavento, P. A., Woerpel, D. N., and Goldstein, L. S. B. (1991). Identification and partial characterization of six members of the kinesin super-family in *Drosophila*. Proc. Natl Acad. Sci. USA *88*, 8470–8474.

Stewart, R. J., Thaler, J. P., and Goldstein, L. S. B. (1993). Direction of microtubule movement is an intrinsic property of the motor domains of kinesin heavy chain and *Drosophila* ncd protein. Proc. Natl Acad. Sci. USA *90*, 5209–5213.

Stoler, S., Keith, K. C., Curnick, K. E., and Fitzgerald-Hayes, M. (1995). A mutation in *CSE4*, an essential gene encoding a novel chromatin-associated protein in yeast, causes chromosome nondisjunction and cell cycle arrest at mitosis. Genes Dev. *9*, 573–586.

Stratmann, R. and Lehner, C. (1996). Separation of sister chromatids in mitosis requires the *Drosophila pimples* product, a protein degraded after the metaphase/ anaphase transition. Cell *84*, 25–35.

Strauss, F. and Varshavsky, A. (1984). A protein binds to a satellite DNA repeat at three specific sites that would be brought into mutual proximity by DNA folding in the nucleosome. Cell *37*, 889–901.

Strunnikov, A. V., Larionov, V. L., and Koshland, D. (1993). *SMC1*: an essential yeast gene encoding a putative head-rod-tail protein is required for nuclear division and defines a new ubiquitous protein family. J. Cell Biol. *123*, 1635–1648.

Sugimoto, K., Yata, H., and Himemo, M. (1993). Mapping of the human CENP-B gene to chromosome 20 and the CENP-C gene to chromosome 12 by a rapid cycle DNA amplification procedure. Genomics *17*, 240–242.

Sugimoto, K., Hagishita, Y., and Himeno, M. (1994a). Functional domain structure of human centromere protein B. Implication of the internal and C-terminal self-association domains in centromeric heterochromatin condensation. J. Biol. Chem. *269*, 24271–24276.

Sugimoto, K., Yata, H., Muro, Y., and Himeno, M. (1994b). Human centomere protein C (CENP-C) is a DNA-binding protein which possesses a novel DNA-binding motif. J. Biochem. *116*, 877–881.

Sullivan, B. and Schwartz, S. (1995). Identification of centromeric antigens in dicentric Robertsonian translocations: CENP-C and CENP-E are necessary components of functional centromeres. Hum. Mol. Genet. *4*, 2189–2197.

Sullivan, K. F. and Glass, C. A. (1991). CENP-B is a highly conserved mammalian centromere protein with homology to the helix-loop-helix family of proteins. Chromosoma *100*, 360–370.

Sullivan, K. F., Hechenberger, M., and Masri, K. (1994). Human CENP-A contains a histone H3 related histone fold domain that is required for targeting to the centromere. J. Cell Biol. *127*, 581–592.

Sumner, A. (1992). Inhibitors of topoisomerases do not block the passage of human lymphocyte chromosomes through mitosis. J. Cell Sci. *103*, 105–115.

Sumner, A. (1995). Inhibitors of topoisomerase II delay progress through mitosis and induce a doubling of the DNA content in CHO cells. Exp. Cell Res. *217*, 440–447.

Sumner, A. (1996). The distribution of topoisomerase II on mammalian chromosomes. Chromosome Res. *4*, 4–5.

Sundin, O. and Varshavsky, A. (1980). Terminal stages of SV40 DNA replication proceed via multiply intertwined catenated dimers. Cell *21*, 103–114.

Sundin, O. and Varshavsky, A. (1981). Arrest of segregation leads to accumulation of highly intertwined catenated dimers: dissection of the final stages of SV40 DNA replication. Cell *25*, 659–669.

Surana, U., Amon, A., Dowzer, C., McGrew, J., Byers, B., and Nasmyth, K. (1993). Destruction of the CDC228/CLB mitotic kinase is not required for the metaphase to anaphase transition in budding yeast. EMBO J. 1969–1978.

Swaroop, A., Swaroop, M., and Garen, A. (1987). Sequence analysis of the complete cDNA and encoded polypeptide for the *Glued* gene of *Drosophila melanogaster*. Proc. Natl Acad. Sci. USA *84*, 6501–6505.

Swedlow, J. R., Sedat, J., and Agard, D. (1993). Multiple chromosomal populations of topoisomerase II detected *in vivo* by time-lapse, three-dimensional wide-field microscopy. Cell *73*, 97–108.

Swenson, K. I., Farrell, K. M., and Ruderman, J. V. (1986). The clam embryo protein cyclin A induces entry into M phase and the resumption of meiosis in *Xenopus* oocytes. Cell *47*, 861–870.

Taagepera, S., Rao, P., Drake, F., and Gorbsky, G. (1993). DNA topoisomerase IIα is the major chromosome protein recognized by the mitotic phosphoprotein antibody MPM-2. Proc. Natl Acad. Sci. USA *90*, 8407–8411.

Taagepera, S., Dent, P., Her, J. H., Sturgill, T. W., and Gorbsky, G. J. (1994). The MPM-2 antibody inhibits mitogen-activated protein kinase activity by binding to an epitope containing phosphothreonine-183. Mol. Biol. Cell *5*, 1243–1251.

Taagepera, S., Campbell, M., and Gorbsky, G. (1995). Cell-cycle-regulated localization of tyrosine and threonine phosphoepitopes at the kinetochores of mitotic chromosomes. Exp. Cell Res. *221*, 249–260.

Tachibana, T., Imamoto, N., Seino, H., Nishimoto, T., and Yoneda, Y. (1994). Loss of RCC1 leads to suppression of nuclear protein import in living cells. J. Biol. Chem. *269*, 24542–24545.

Takai, Y., Kaibuchi, K., Kikuchi, A., and Kawata, M. (1992). Small GTP-binding proteins. Int. Rev. Cytol. *133*, 187–230.

Tamphaichitr, N., Sobhon, P., Taluppeth, N., and Chalermisarachai, P. (1978). Basic nuclear proteins in testicular cells and ejaculated spermatozoa in man. Exp. Cell Res. *117*, 347–356.

Tan, K., Mattern, M., Boyce, R., and Schein, P. (1987). Elevated DNA topoisomerase II activity in nitrogen mustard-resistant human cells. Proc. Natl Acad. Sci. USA *84*, 7668–7671.

Tang, T. K., Tang, C. C., Chen, Y. L., and Wu, C. W. (1993). Nuclear proteins of the bovine esophageal epithelium II. The NuMA gene gives rise to multiple mRNAs and gene products reactive with monoclonal antibody W1. J. Cell Sci. *104*, 249–260.

Tang, T. K., Tang, C. C., Chao, Y. J., and Wu, C. W. (1994). Nuclear mitotic apparatus protein (NuMA): spindle association, nuclear targeting and differential subcellular localization of various NuMA isoforms. J. Cell Sci. *107*, 1389–1402.

Testa, J. R., Zhou, J. Y., Bell, D. W., and Yen, T. J. (1994). Chromosomal localization of the genes encoding the kinetochore proteins CENP-E and CENP-F to human chromosomes 4q24-q25 and 1q32-q41, respectively, by fluorescence *in situ* hybridization. Genomics *23*, 691–693.

Tharappel, J. and Elgin, S. (1986). Identification of a nonhistone chromosomal protein associated with heterochromatin in *Drosophila melanogaster* and its gene. Mol. Cell. Biol. *6*, 3862–3872.

Theurkauf, W. and Hawley, R. (1992). Meiotic spindle assembly in *Drosophila* females: behavior of nonexchange chromosomes and the effects of mutations in the nod kinesin-like protein. J. Cell Biol. *116*, 1167–1180.

Thrower, D. A., Jordan, M. A., Schaar, B. T., Yen, T. J., and Wilson, L. (1995). Mitotic HeLa cells contain a CENP-E-associated minus end-directed microtubule motor. EMBO J. *14*, 918–926.

Tokai, N., Fujimoto-Nishiyama, A., Toyoshima, Y., Yonemura, S., Tsukia, S., Inoue, J. *et al.* (1996). Kid, a novel kinesin-like DNA binding protein, is localized to chromosomes and the mitotic spindle. EMBO J. *15*, 457–467.

Tokito, M., Lee, V., and Holzbaur, E. (1993). Characterization of the human cDNA encoding p150^{Glued} and its expression in the human NTera 2 cell line. Mol. Biol. Cell *4*, 162 (abstract).

Tombes, R. M., Peloquin, J. G., and Borisy, G. G. (1991). Specific association of an M-phase kinase with isolated mitotic spindles and identification of two of its substrates as MAP4 and MAP1B. Cell Regulat. *2*, 861–874.

Tomkiel, J., Cooke, C. A., Saitoh, H., Bernat, R. L., and Earnshaw, W. C. (1994). CENP-C is required for maintaining proper kinetochore size and for a timely transition to anaphase. J. Cell Biol. *125*, 531–545.

Toth, M., Grimsby, J., Buzsaki, G., and Donovan, G. (1995). Epileptic seizures caused by inactivation of a novel gene, jerky, related to centromere binding protein-B in transgenic mice. Nature Genet. *11*, 71–75.

Tousson, A., Zeng, C., Brinkley, B. R., and Valdivia, M. M. (1991). Centrophilin: a novel mitotic spindle protein involved in microtubule nucleation. J. Cell Biol. *112*, 427–440.

Trivinos-Lagos, L., Collins, C., and Chisholm, R. (1993). Cloning of dynein intermediate chain: multiple isoforms are expressed during *Dictyostelium* development. Mol. Biol. Cell *4*, 47 (Abstract).

Trowell, H. E., Nagy, A., Vissel, B., and Choo, K. H. A. (1993). Long-range analyses of the centromeric regions of human chromosomes 13, 14 and 21: identification of a narrow domain containing two key centromeric DNA elements. Hum. Mol. Genet. *2*, 1639–1649.

Tugendreich, S., Tomkiel, J., Earnshaw, W. C., and Hieter, P. (1995). CDC27Hs colocalizes with CDC16Hs to the centrosome and mitotic spindle and is essential for the metaphase to anaphase transition. Cell *81*, 261–268.

Uchida, S., Sekiguchi, T., Nishitani, H., Miyauchi, K., Ohtsubo, M., and Nishimoto, T. (1990). Premature chromosome condensation is induced by a point mutation in the hamster *RCC1* gene. Mol. Cell. Biol. *10*, 577–584.

Uemura, T. and Yanagida, M. (1984). Isolation of type I and II DNA topoisomerase mutants from fission yeast: single and double mutants show different phenotypes in cell growth and chromatin organization. EMBO J. *3*, 1737–1744.

Uemura, T. and Yanagida, M. (1986). Mitotic spindle pulls but fails to separate chromosomes in type II DNA topoisomerase mutants: uncoordinated mitosis. EMBO J. *5*, 1003–1010.

Uemura, T., Ohkura, H., Adachi, Y., Morino, K., Shiozaki, K., and Yanagida, M. (1987). DNA topoisomerase II is required for condensation and separation of mitotic chromosomes in *S. pombe*. Cell *50*, 917–925.

Uschewa, A., Auramova, Z., and Tsanev, R. (1982). Tightly bound somatic histones in mature ram sperm nuclei. FEBS Lett. *138*, 50–54.

Vaisberg, E., Koonce, M., and McIntosh, J. (1993). Cytoplasmic dynein plays a role in mammalian mitotic spindle formation. J. Cell Biol. *123*, 849–858.

Valdivia, M. and Brinkley, B. R. (1985). Fractionation and initial characterization of the kinetochore from mammalian metaphase chromosomes. J. Cell Biol. *101*, 1124–1134.

Vale, R. D. (1992). Microtubule motors: many new models off the assembly line. Trends Biochem. Sci. *17*, 300–304.

Vale, R. D., Reese, T. S., and Sheetz, M. P. (1985). Identification of a novel force-generating protein, kinesin, involved in microtubule-based motility. Cell *42*, 39–50.

Vallee, R. and Shpetner, H. (1990). Motor protein of cytoplasmic microtubules. Annu. Rev. Biochem. *59*, 909–932.

Vallee, R., Wall, J., Paschal, B., and Shpetner, H. (1988). Microtubule-associated protein 1C from brain is a two-headed cystolic dynein. Nature *332*, 561–563.

Vandre, D. D. and Borisy, G. G. (1989). Anaphase onset and dephosphorylation occur concomitantly. J. Cell Sci. *94*, 245–258.

Vandre, D. D. and Burry, R. W. (1992). Immunoelectron microscopic localization of phosphoproteins associated with the mitotic spindle. J. Histochem. Cytochem. *40*, 1837–1847.

Vaughan, K. and Vallee, R. (1993). Transfection of Cos-7 cells with cytoplasmic dynein intermediate chains. Mol. Biol. Cell *4*, 162 (Abstract).

Vandre, D. D., Davis, F. M., Rao, P. N., and Borisy, G. G. (1984). Phosphoproteins are components of mitotic microtubule organizing centers. Proc. Natl Acad. Sci. USA *81*, 4439–4443.

Vandre, D. D., Davis, F. M., and Borisy, G. G. (1986). Distribution of cytoskeletal proteins sharing a conserved phosphorylated epitope. Eur. J. Cell Biol. *41*, 72–81.

Vandre, D. D., Centonze, V. E., Peloquin, J., Tombes, R. M., and Borisy, G. G. (1991). Proteins of the mammalian mitotic spindle: phosphorylation/dephosphorylation of MAP-4 during mitosis. J. Cell Sci. *98*, 577–588.

Vaughan, K. and Vallee, R. (1995). Cytoplasmic dynein binds dynactin through a direct interaction between the intermediate chains and p150[glued]. J. Cell Biol. *131*, 1507–1516.

Vernos, I., Raats, J., Hirano, T., Heasman, J., Karsenti, E., and Wylie, C. (1995). Xklp1, a chromosomal *Xenopus* kinesin-like protein essential for spindle organization and chromosome positioning. Cell *81*, 117–127.

Vissel, B. and Choo, K. H. A. (1992). Evolutionary relationships of multiple alpha satellite subfamilies in the centromeres of human chromosomes 13, 14, and 21. J. Mol. Evol. *35*, 137–146.

Vissel, B., Nagy, A., and Choo, K. H. A. (1992). A satellite III sequence shared by human chromosomes 13, 14 and 21, that is contiguous with alpha satellite DNA. Cytogenet. Cell Genet. *61*, 81–86.

VonWettstein, D., Rasmussen, S., and Holm, P. (1984). The synaptonemal complex in genetic segregation. Annu. Rev. Genet. *18*, 331–431.

Walczak, C. E., Mitchison, T. J., and Desai, A. (1996). XKCM1: a *Xenopus* kinesin-related protein that regulates microtubule dynamics during mitotic spindle assembly. Cell *84*, 37–47.

Walker, J., Saraste, M., Runswick, M., and Gay, N. (1982). Distantly related sequences in the alpha- and beta-subunits of ATP synthase, myosin, kinases and other ATP-requiring enzymes and a common nucleotide binding fold. EMBO J. *1*, 945–951.

Walker, R. A., Salmon, E. D., and Endow, S. A. (1990). The *Drosophila* claret segregation protein is a minus-end directed motor molecule. Nature (Lond.) *347*, 780–782.

Wang, J. C. (1985). DNA topoisomerases. Annu. Rev. Biochem. *54*, 665–697.

Waterman-Storer, C., Karki, S., and Holzbaur, E. (1993). Analysis of the p150[Glued]–centractin complex reveals a microtubule-binding function *in vivo* and *in vitro*. Mol. Cell. Biol. *4*, 162 (Abstract).

Watt, P. and Hickson, I. (1994). Structure and function of type II topoisomerases. Biochem. J. *303*, 681–693.

Waye, J. S., England, S. B., and Willard, H. F. (1987). Genomic organization of alpha satellite DNA on human chromosome 7: evidence for two distinct alphoid domains on a single chromosome. Mol. Cell. Biol. *7*, 349–356.

Weiner, E. S., Hildebrandt, S., Senecal, J. L., Daniels, L., Noell, S., Joyal, F. *et al.* (1991). Prognostic significance of anticentromere antibodies and anti-topoisomerase I antibodies in Raynaud's disease. Arthritis Rheumatism *34*, 68–77.

Weith, A. and Traut, W. (1980). Synaptonemal complexes with associated chromatin in a moth *Ephestia kuehniella*. Chromosoma *78*, 275–291.

Westendorf, J. M., Rao, P. N., and Gerace, L. (1994). Cloning of cDNAs for M-phase phosphoproteins recognized by the MPM2 monoclonal antibody and determination of the phosphorylated epitope. Proc. Natl Acad. Sci. USA *91*, 714–718.

Wevrick, R. and Willard, H. F. (1989). Long-range organization of tandem arrays of alpha-satellite DNA at the centromeres of human chromosomes: high-frequency array-length polymorphism and meiotic stability. Proc. Natl Acad. Sci. USA *86*, 9394–9398.

Wevrick, R. and Willard, H. F. (1991). Physical map of the centromeric region of human chromosome 7: relationship between two distinct alpha satellite arrays. Nucleic Acids Res. *19*, 2295–2301.

Wevrick, R., Willard, V. P., and Willard, H. F. (1992). Structure of DNA near long tandem arrays of alpha satellite DNA at the centromeres of human chromosome 7. Genomics *14*, 912–923.

Wilkerson, C., King, S., and Wiman, G. (1994). Molecular analysis of the gamma heavy chain of *Chlamydomonas flagellar* outer-arm dynein. J. Cell Sci. *107*, 497–506.

Williams, B. C., Karr, T. L., Montgomery, J. M., and Goldberg, M. L. (1992). The *Drosophila L(1)Zw10* gene product, required for accurate mitotic chromosome segregation, is redistributed at anaphase onset. J. Cell Biol. *118*, 759–774.

Williams, B., Gatti, M., and Goldberg, M. (1996). Bipolar spindle attachments affect redistributions of ZW10, a Drosophila centromere/kinetochore component required for accurate chromosome segregation. J. Cell Biol. 134, 1127–1140.

Williams, B., and Goldberg, M. (1994). Determinants of Drosophila zw10 protein localization and function. J. Cell Sci. 107, 785–798.

Wilson, R. R., Anscough, K., Anderson, C., Baynes, M., Berks, J., Bonfield, J. *et al.* (1994). 2.2. Mb of contiguous nucleotide sequence from chromosome III of *C. elegans*. Nature (Lond.) *368*, 32–38.

Witt, P. L., Ris, H., and Borisy, G. G. (1980). Origin of kinetochore microtubules in Chinese hamster cells. Chromosoma *81*, 483–505.

Woessner, R. D., Chung, T., Hofmann, G., Mattern, M. R., Mirabelli, C. K., Drake, F. H. *et al.* (1990). Differences between nomal and *ras*-transformed NIH-3T3 cells in expression of the 170kD and 180kD forms of topoisomerase II. Cancer Res. *50*, 2901–2908.

Woessner, R. D., Mattern, M. R., Mirabelli, C. K., Johnson, R. K., and Drake, F. H. (1991). Proliferation- and cell cycle-dependent differences in expression of the 170 kilodalton and 180 kilodalton forms of topoisomerase II in NIH-3T3 cells. Cell Growth Differentiation *2*, 209–214.

Wood, E. and Earnshaw, W. C. (1990). Mitotic chromosome condensation *in vitro* using somatic cell extracts and nuclei with variable levels of endogenous topoisomerase II. J. Cell Biol. *111*, 2839–2850.

Wordeman, L. and Mitchison, T. J. (1995). Identification and partial characterization of mitotic centromere-associated kinesin, a kinesin-related protein that associates with centromeres during mitosis. J. Cell Biol. *128*, 95–105.

Wordeman, L., Steuer, E., Sheetz, M., and Mitchison, T. (1991). Chemical subdomains within the kinetochore domain of isolated CHO mitotic chromosomes. J. Cell Biol. *114*, 285–294.

Wright, B., Henson, J., Wedaman, K., Willy, P., Morand, J., and Scholey, J. (1991). Subcellular localization and sequence of sea urchin kinesin heavy chain: evidence for its association with membranes in the mitotic apparatus and interphase cytoplasm. J. Cell Biol. *113*, 817–833.

Wu, K., Strauss, F., and Varshavsky, A. (1983). Nucleosome arrangement in green monkey alpha-satellite chromatin. J. Mol. Biol. *170*, 93–117.

Xiang, X., Beckwith, S., and Morris, N. (1994). Cytoplasmic dynein is involved in nuclear migration in *Aspergillus nidulans*. Proc. Natl Acad. Sci. USA *91*, 2100–2104.

Yang, L., Word, M. S., Li, J. J., Kelly, T. J., and Liu, L. F. (1987). Roles of DNA topoisomerases in simian virus 40 DNA replication *in vitro*. Proc. Natl Acad. Sci. USA *84*, 950–954.

Yang, C. H., Lambie, E. J., and Snyder, M. (1992). NuMA: an unusually long coiled-coil related protein in the mammalian nucleus. J. Cell Biol. *116*, 1303–1317.

Yang, C H.., Tomkiel, J., Saitoh, H., Johnson, D., and Earnshaw, W. C. (1996). Identification of overlapping DNA-binding and centromere-targeting domains in the human kinetochore protein CENP-C. Mol. Cell. Biol. *16*, 3576–86.

Yang, J. T., Laymon, R. A., and Goldstein, L. S. B. (1989). A three domain structure of kinesin heavy chain revealed by DNA sequence and microtubule binding analyses. Cell *56*, 879–889.

Yang, J. T., Saxton, W. M., Stewart, R. J., Raff, E. C., and Goldstein, L. S. (1990). Evidence that the head of kinesin is sufficient for force generation and motility *in vitro*. Science *249*, 42–47.

Yeh, E., Skibbens, R. V., Cheng, J. W., Salmon, E. D., and Bloom, K. (1995). Spindle dynamics and cell cycle regulation of dynein in the budding yeast, *Saccharomyces cerevisiae*. J. Cell Biol. *130*, 687–700.

Yen, T. J., Compton, D. A., Wise, D., Zinkowski, R. P., Brinkley, B. R., Earnshaw, W. C. *et al.* (1991). CENP-E a novel human centromere-associated protein required for progression from metaphase to anaphase. EMBO J. *10*, 1245–1254.

Yen, T. J., Li, G., Schaar, B. T., Szilak, I., and Cleveland, D. W. (1992). CENP-E is a putative kinetochore motor that accumulates just before mitosis. Nature *359*, 536–539.

Yoda, K., Kitagawa, K., Masumoto, H., Muro, Y., and Okazaki, T. (1992). A human centromere protein, CENP-B, has a DNA binding domain containing four potential alpha helices at the NH_2 terminus, which is separable from dimerizing activity. J. Cell Biol. *119*, 1413–1427.

Yoda, K., Nakamura, T., Masumoto, H., Suzuki, N., Kitagawa, K., Nakano, M., Shinjo, A., and Okazaki, T. (1996). Centromere protein B of African green monkey cells: Gene structure, cellular expression, and centromeric localization. Mol. Cell. Biol. 16, 5169-5177.

Zeng, C., He, D., and Brinkley, B. (1994). Localization of NuMA protein isoforms in the nuclear matrix of mammalian cells. Cell Motil. Cytoskeleton *29*, 167–176.

Zhang, X., Fittler, F., and Horz, W. (1983). Eight different highly specific nucleosome phases on alpha-satellite DNA in African green monkey. Nucleic Acids Res. *11*, 4287–4305.

Zhang, P., Knowles, B. A., Goldstein, L. S., and Hawley, R. S. (1990). A kinesin-like protein required for distributive chromosome segregation in *Drosophila*. Cell *62*, 1053–1062.

Zhang, Z., Tanaka, Y., Noanaka, S., Aizawa, H., and Kawasaki, H. (1993). The primary structure of rat brain (cytoplasmic) dynein heavy chain, a cytoplasmic motor enzyme. Proc. Natl Acad. Sci. USA *90*, 7928–7932.

Zhu, X., Chang, K. H., He, D., Mancini, M. A., Brinkley, W. R., and Lee, W. H. (1995a). The C terminus of mitosin is essential for its nuclear localization, centromere/kinetochore targeting and dimerization. J. Biol. Chem. *270*, 19545–19550.

Zhu, X., Mancini, M. A., Chang, K. H., Liu, C. Y., Chen, C. F., Shan, B. *et al.* (1995b). Characterization of a novel 350-kilodalton nuclear phosphoprotein that is specifically involved in mitotic-phase progression. Mol. Cell. Biol. *15*, 5017–5029.

Zinkowski, R. P., Meyne, J., and Brinkley, B. R. (1991). The centromere–kinetochore complex: a repeat subunit model. J. Cell Biol. *113*, 1091–1110.

Zirkle, R. E. (1970). UV-microbeam irradiation of newt-cell cytoplasm: spindle destruction, false anaphase, and delay of true anaphase. Radiat. Res. *41*, 516–537.

7

Anomalies of the human centromere

Clinical abnormalities that can be directly attributed to defects of the centromere are rarely seen. This is understandable considering the important roles the centromere plays in mitosis and meiosis. Nonetheless, based on various cytogenetic and molecular criteria, some interesting anomalies of the human centromere have been described.

Premature centromere division (PCD)

The term premature centromere division (PCD) was introduced by Fitzgerald *et al.* (1975) to describe the premature division of the X chromosome centromere compared with the centromeres of other chromosomes in the same metaphase. The X chromosomes with divided centromeres appear as two entirely parallel (or 'railroad-track') chromatids due to the absence of a primary constriction (Fig. 7.1). The expression of PCD-X increases with age and it has been suggested that this causes X chromosome aneuploidy in older women (Fitzgerald and McEvan 1977; Izakovic and Vahancik 1984; Bajnóczky 1985). PCD of chromosome 21 has also been thought to be responsible for several cases of trisomy 21 (Fitzgerald *et al.* 1986; Bajnóczky and Méhes 1988).

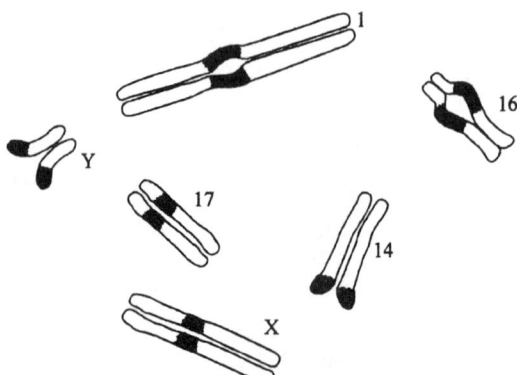

Fig. 7.1 Diagramatic representation of PCD of C-banded (dark area) chromosomes in Roberts' syndrome, showing puffing (chromosomes 1 and 16), splaying (chromosomes 14 and Y), and 'railroad-track' (chromosomes 17 and X) morphologies.

In addition to chromosomes X and 21 aneuploidy, the phenomenon of PCD has been shown to be associated with a number of other disorders, the best studied of which is Roberts' syndrome (discussed below). These disorders are summarized in Table 7.1. As can be seen, disturbance in the mechanics of centromere division may be manifested in at least some patients from a wide range of conditions. In the cases studied, PCD is generally observed in only a proportion of the metaphase cells, with the phenomenon often affecting all the chromosomes (but not in the same cell at once), although chromosomal preference has been noted for Roberts' syndrome (chromosomes 1, 9, 13–16, 21, 22, and Y), aneuploidy (chromosomes X, 18, and 21), tuberous sclerosis (chromosome 3), bladder cancer (chromosomes 7, 12–15, 17–18, and 20–22), and Fanconi's anaemia and ataxia teleangiectasia (chromosomes 13–15).

Table 7.1 Disorders with premature centromere division manifestation

Disorder	References
Roberts' syndrome (RS+ subgroup)	> 100 cases (see text)
Aneuploidy	Fitzerald *et al.* 1975, 1986; Méhes *et al.* 1978; Bajnóczky and Méhes 1988
Spontaneous abortions	Rudd *et al.* 1983; Murthy and Prabhakara 1990; Bajnóczky and Gardó 1993
Translocation carrier	Madan *et al.* 1987
Subfertility	Rudd *et al.* 1983; Gabarrón *et al.* 1986
Ambiguous genitalia	Miller *et al.* 1990
Pseudohermaphroditism	Rangnekar *et al.* 1990
Megaloblastic anemia	Heath 1966; Bamezai *et al.* 1986
Tuberous sclerosis	Scappaticci *et al.* 1988
Leukaemias	Sandberg 1980; Shiraishi *et al.* 1982; Gallo *et al.* 1984; Vig 1984; Littlefield *et al.* 1985
Burkitt's lymphoma	Zhang 1986
Nasopharingeal carcinoma	Zhang 1986
Bladder cancer	Berrozpe *et al.* 1990
Révész syndrome	Kajtár and Méhes 1994
Paracoccidioidomycosis	Freire-Maia *et al.* 1994
Fanconi's anaemia	Méhes and Bühler 1995
Ataxia teleangiectasia	Méhes and Bühler 1995
Alzheimer's	Moorhead and Heyman 1983; Kormann-Bortolotto *et al.* 1993

The genetics of PCD are not fully understood. While PCD-X appears to be an autosomal recessive trait (Fitzgerald *et al.* 1975), autosomal dominant inheritance of PCD involving all chromosomes has been reported in a number of families (Rudd *et al.* 1983; Gabarrón *et al.* 1986; Madan *et al.* 1987). On the other hand, the expression of PCD in some of the other disorders may represent epiphenomena related to events such as ageing (as in the case of Alzheimer's disease; Moorhead and Heyman 1983; Kormann-Bortolotto *et al.* 1993), malignancies, or chromosomal instability. At least in the case of PCD-X, loss of centromere function in aged women has been demonstrated by negative Cd banding (Nakagome *et al.* 1984).

Roberts' syndrome

Clinical and cytological phenotypes

Roberts' syndrome (RS) is a relatively rare, autosomal recessive disorder characterized by pre- and post-natal growth retardation, developmental delay, symmetrical reduction in the number of digits, and craniofacial anomalies. Affected persons have additional manifestations which may include renal defects, cardiac anomalies, mental retardation, and malignancies (reviewed by Van Den Berg and Francke 1993; Sinha *et al.* 1994). A number of syndromes expressing the hallmark craniofacial and bone malformations of RS have been referred to variably as Appelt–Gerken–Lenz syndrome (Appelt *et al.* 1966), hypomelia hypotrichosis facial hemangioma syndrome (Hall and Greenberg 1972), tetraphocomelia cleft palate syndrome (Kucheria *et al.* 1976), and pseudothalidomide or SC phocomelia syndrome (Herrmann *et al.* 1969). These syndromes may represent RS with variable manifestations or they may in fact be different syndromes.

Approximately 80% of RS patients (Van Den Berg and Francke 1993) display the characteristic cytogenetic phenomenon of PCD discussed above which, in this disorder, manifests as repulsion (or 'puffing') of heterochromatic regions near centromeres, particularly of chromosomes 1, 9, and 16, and 'splaying' of the short arms of the acrocentrics and of the distal Yq (Fig. 7.1) (Freeman *et al.* 1974; German 1979; Tomkins *et al.* 1979; Louie and German 1981). Several of the other chromosomes may display the 'railroad-track' appearance because of the absence of a constriction at the centromere. This cytogenetic phenomenon (known as the 'RS effect') is observed in cells of different tissue origin, with several chromosomes in each metaphase showing a visible abnormality. Every chromosome has been identified with this abnormal appearance, the most evident being the chromosomes containing the largest amount of heterochromatin.

A number of functional anomalies are associated with the cultured mitotic cells of patients demonstrating the RS effect. These include extended doubling time and metaphase duration, failure to enter mitosis or proceed past metaphase, anaphase lag, increased micronucleation, abnormal nuclear morphology, and aneuploidy (Louie and German 1983; Tomkins and Sisken 1984; Jabs *et al.* 1991). During anaphase, there is an increased incidence of outlying, lagging, or prematurely advancing chromosomes. Chromosomes that initiate but do not complete migration to the pole will come to lie just outside the major chromosomal mass. When these chromosomes fail to be completely incorporated into reforming nuclei, they most likely result in structures known as nuclear projections. Lagging chromosomes that show no evidence of chromosome movement are left at the spindle equator. This class of lagging chromosomes has been found to result in the formation of micronuclei. The micronuclei are passively distributed to the daughter cells and the daughter cells containing micronuclei do not undergo further division. Thus, the possibility arises that PCD leads to abnormal anaphase progression, which results in cells with micronuclei being removed from the cycling cell population, thus depleting developing tissues of valuable cells. Such a depletion, compounded by the abnormal cell cycle time and cell loss due to mitotic arrest, may contribute to the observed growth retardation and dysmorphic changes.

Genetic heterogeneity and evolutionary conservation

RS patients that are clinically indistinguishable may be divided into two groups on the basis of their cytogenetic profiles. One group, mentioned above, contains the chromosomal and cell culture growth abnormalities, and has been designated 'RS+' (Louie and German 1981). The second group, which constitutes about 20% of RS patients (Van Den Berg and Francke 1993), shows neither of these abnormalities and has been designated 'RS–' (Burns and Tomkins 1989). In addition, whereas fibroblasts and lympho-blastoid cells from RS+ patients show hypersensitivity to cell killing by DNA damaging agents such as mitomycin C, similar cell preparations from RS– patients do not. Complementation studies have indicated that when cells from unrelated RS+ patients are fused, no correction of abnormal phenotype is observed, suggesting that the patients belong to a single complementation group (Allingham-Hawkins and Tomkins 1995). Fusion between an RS+ and RS– cell lines, on the other hand, results in hybrids that demonstrate correction of both the chromosomal abnormality and mitomycin C hypersensitivity, suggesting that RS+ and RS– patients belong to different complementation groups and do not arise from mutation of the same gene.

Complementation has also been used to study the evolutionary conservation of the RS+ gene (Dev and Wertelecki 1984; Krassikoff *et al.* 1986; Gunby *et al.* 1987; Allingham-Hawkins and Tomkins 1991). The results demonstrate that fusion of RS+ cells with cell lines of rat, mouse, hamster, and human origin all lead to correction of the RS+ abnormalities in the fusion products. The accomplishment of such cross-species complementation with different mammalian cell lines suggests that the RS+ gene product is functionally conserved in mammals.

Centromere morphology and function

Although the cytogenetic manifestation of PCD is consistent with abnormalities of centromere structure and/or function, electron microscopy of lagging chromosomes from RS cells indicates a normal centromere morphology with a prominent kinetochore composed of a typical trilaminar organization (Jabs *et al.* 1991). Furthermore, the kinetochore appears to have the ability to capture microtubules and consequently to achieve normal orientation on the metaphase plate. It is, however, premature to rule out more subtle abnormalities in centromere structure or function that were not examined or detected by this study. It is conceivable, for example, that defects may reside at the centromere pairing domain and involve one or more of the sister chromatid cohesion proteins discussed in Chapter 6.

ICF (*i*mmunodeficiency, *c*entromeric instability and *f*acial anomalies) syndrome

Clinical and cytological phenotypes

ICF syndrome is a disorder that is probably transmitted in an autosomal recessive mode. The disorder is characterized by variable immunodeficiency, centromeric heterochromatin instability, and facial anomalies. The condition is extremely rare, with only 18 cases on record (17 cases reviewed by Brown *et al.* 1995; Franceschini *et al.* 1995; 1 new case by Sawyer *et al.* 1995). The immunodeficiency consists of the absence of or a severe reduction in at least two immunoglobulin isotypes (Hulten 1978; Tiepolo *et al.* 1979; Gimelli *et al.* 1993), and usually manifests as respiratory or gastrointestinal upset. Facial anomalies most often recorded are hypertelorism, a flat nasal bridge, epicanthic folds, protrusion of the tongue, and micrognathia. Mental retardation is variable, from severe neurodegeneration to special educational needs without any delay in motor function. There is usually growth retardation.

Cytogenetically, ICF patients show a characteristic instability in the centromeric heterochromatin of chromosomes 1, 9, 16, and (very rarely) chromosome 2. The instability is manifested as dramatic undercondensation (or 'stretching') of centromeric heterochromatin, centromeric breaks, somatic pairing and interchanges between homologous and non-homologous chromosomes, and formation of multibranched configurations with duplication or deletions of whole chromosome arms (Fig. 7.2). These chromosomal abnormalities are almost exclusively encountered in phytohemaglutinin-stimulated lymphocytes. In fibroblast cell lines, bone marrow cells, and Epstein–Barr virus-transformed lymphocytes, the cytogenetic manifestations of the syndrome are limited to occasional elongation of the heterochromatic regions.

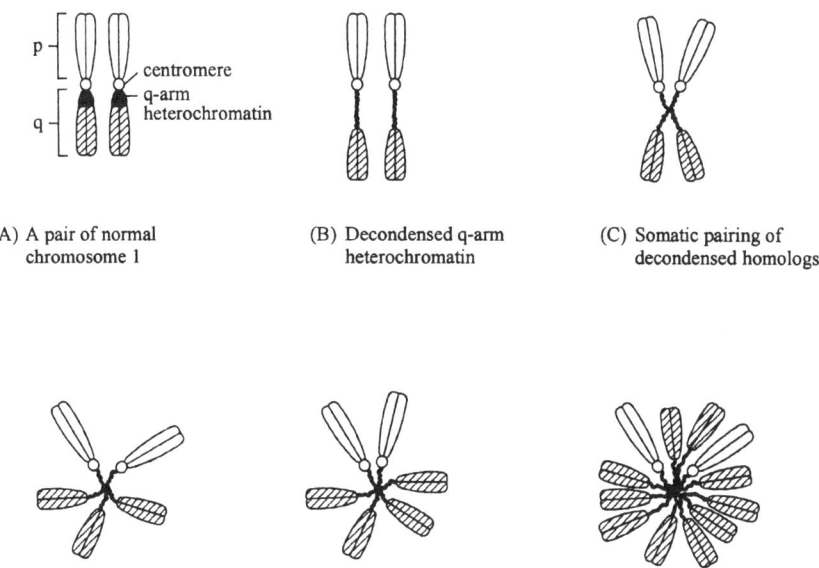

(A) A pair of normal chromosome 1

(B) Decondensed q-arm heterochromatin

(C) Somatic pairing of decondensed homologs

(D) Decondensed and multibranched chromosomes showing 2 copies of p-arms with 3, 4 or 10 copies of q-arms

Fig. 7.2 Representative cytogenetic aberrations in ICF syndrome. (A) A chromosome 1 pair that has undergone normal mitotic condensation. (B) Decondensed or 'stretched' proximal q arm heterochromatic region. (C) Somatic pairing of decondensed homologues. (D) Three variations of multibranched chromosomes. Note that stretching and pairing involve the q arm heterochromatin but not the centromere directly. Also, only the q arm, where the centromere is absent, is duplicated in the multibranched configuration.

The cytogenetic anomalies in ICF patients apparently prevent a complete migration of the affected chromosomes to the poles, resulting in nuclear abnormalities in interphase cells, as in Roberts' syndrome. In ICF patients, such nuclear abnormalities appear as nuclear projections and micronuclei which, by fluorescence *in situ* hybridization (or FISH) analysis, have been shown to contain the aberrant chromosomes (Fasth *et al.* 1990; Maraschio *et al.* 1992; Sawyer *et al.* 1995). As micronuclei are structures known to form from acentric chromosomes or chromosome fragments that are not incorporated into one or the other daughter nucleus during cell division, it is possible that decondensation and/or hypomethylation (see below) of juxtacentromeric heterochromatin may have resulted in kinetochore dysfunction, causing the stretched chromosomes to behave acentrically. The formation of micronuclei

can also be explained by the inability of the complex chromosomal configurations, such as somatically paired homologues or multibranched chromosomes, to undergo normal anaphase separation (Tawn and Holdsworth 1992).

Mechanisms for the observed cytogenetic anomalies

It was initially hypothesized that centric fission of chromosomes 1, 9, and 16
would lead to the formation of functional centromeres on the short or long
arms of these chromosomes and, through successive duplications, they gave
rise to the observed multibranched configurations (Fryns *et al.* 1988).
However, Cd-banding (which stains active centromeres; see Table 7.2)
and FISH analysis using chromosome-specific α-satellite DNA show that
the centromeres of these chromosomes are not split into two or more parts
(Gimelli *et al.* 1993) or in other ways involved in the chromosomal anomalies
(Maraschio *et al.* 1992). In fact, a single centromere is always present on the
short arm side and absent on the long arm side of the chromosomes, and only
the long arms of the chromosomes where the centromere is absent are shown
to be duplicated in the multibranched configuration (Tiepolo *et al.* 1979;
Gimelli *et al.* 1993). Scanning electron micrographs of multibranched
chromosomes also reveal genuine fusion at the heterochromatin in the long
arms, but not in the centromeres (Brown *et al.* 1995). It is clear from these
studies that centromeres are not directly involved in the origin of multibranched chromosomes but it is the juxtacentromeric heterochromatin,
especially that on chromosomes 1, 9, and 16, which plays the 'active'
role. Maraschio *et al.* (1989) postulated that mutation of a gene(s) that
controls the normal process of condensation of juxtacentromeric heterochromatin could be the cause of the chromosomal anomalies found in this
syndrome. Such a control mechanism, if it exists, may be tissue specific since
phytohemaglutinin-stimulated lymphocytes are the only tissue of a number
tested that show the chromosomal abnormalities.

Table 7.2 Characteristics of active and inactive centromeres in human
dicentric chromosomes

Characteristics	Active centromere	Inactive centromere
Primary constriction	+ ve	− ve
Heterochromatin/C-banding	+ ve	+ ve
Cd-banding	+ ve	− ve
Chromatid separation	− ve	+ ve
α-Satellite DNA	+ ve	+ ve
CENP-B protein	+ ve	+ ve
CENP-C protein	+ ve	− ve
CENP-E protein	+ ve	− ve

+ ve denotes presence, and −ve denotes absence, of a given characteristic.

Several lines of evidence suggest that the observed heterochromatin stretching, heterochromatin fusion and other chromosomal rearrangements may be associated with hypomethylation of pericentric repeated DNA. (i) The human classical satellite DNA sequences, which are found in the proximal long-arm pericentric heterochromatin of chromosomes 1, 9, and 16, but not the centromeric α-satellite sequences, are undermethylated in ICF patients (Jeanpierre *et al.* 1993; Miniou *et al.* 1994). (ii) Recombination is favoured by DNA hypomethylation (Engler *et al.* 1991; Kricker *et al.* 1992). (iii) Mammalian cells treated with 5-azacytidine, a DNA demethylating agent, are found to increase the frequency of sister chromatid exchange (Chambers and Taylor 1982; Hori 1983; Lavia *et al.* 1985). (iv) When normal lymphocyte cultures are treated with 5-azacytidine, not only is the heterochromatin extended but fusion of heterochromatin comparable with that seen in ICF patients also occurs (Schmid *et al.* 1983).

Heteromorphism of centromeric heterochromatin

By cytogenetic staining, the constitutive heterochromatin of the human centromeric region has long been recognized as a highly heteromorphic structure. This heteromorphism is considered harmless since it is not known to be associated with any overt clinical abnormalities. The cytogenetically observed heteromorphism has been supported by molecular studies employing cloned centromeric α-satellite DNA. For example, *in situ* hybridization analyses have demonstrated that the level of α-satellite DNA on apparently normal chromosomes may be amplified significantly to give rise to hugely enlarged heterochromatic regions that are several times those of the normal chromosomes (Jabs and Carpenter 1988; Dale *et al.* 1989), or the DNA may fall well below detection level (Verma and Luke 1992). Furthermore, by resolving genomic α-satellite DNA on pulsed field gels, a high degree of variation in the size of α-satellite DNA array has been found amongst homologous chromosomes (Tyler-Smith 1987; Mahtani and Willard 1990; Wevrick and Willard 1991; Ge *et al.* 1992). For example, in a series of 29 different X chromosomes, α-satellite array lengths have been shown to vary over a 3-fold range of between 1380 and 3730 kb (Mahtani and Willard 1990).

Dicentric chromosomes

Dicentric chromosomes are relatively common in the general human population and are occasionally found in patients with congenital anomalies (Dewald 1988). These chromosomes also occur in isolated cells in patients with chromosome breakage syndrome or in persons exposed to toxic agents such as chemotherapy or viruses (German 1974; Harnden 1974; Miller 1983). In addition, they are frequently found in cells of advanced malignant processes (Sandberg 1980). Dicentric chromosomes may consist of two nonhomologous chromosomes or they may involve two homologues (Therman 1986; Dewald 1988; Gardner and Sutherland 1989, 1996). Dicentrics between nonhomologous chromosomes often involve the exchange of acrocentric chromosomes through Robertsonian translocations, while dicentrics between homologous chromosomes are usually broken at identical points to form an isodicentric chromosome with a symmetrical band pattern. Most isodicentrics consist of two X chromosomes, but Y isodicentrics are also frequent.

Definition of active and inactive centromeres in dicentric chromosomes

The functional status of a centromere can be defined by a combination of cytogenetic, molecular, and biochemical criteria. By classical cytogenetic interpretation, an active centromere is one that forms a primary constriction and is positive for both C- (Therman 1986) and Cd-bandings (Daniel 1979; Maraschio *et al.* 1980; Lambiase *et al.* 1984; Benn and Perle 1987; Rivera *et al.* 1989) (Table 7.2). An inactive centromere, on the other hand, does not form a primary constriction, is positive for C-banding but negative for Cd-banding, and shows chromatid separation in metaphase chromosomes (Niebuhr 1972a; Therman *et al.* 1974; Daniel and Lam-Po-Tang 1976; Daniel 1979). By molecular analysis, both active and inactive centromeres are α-satellite DNA positive. Using antibodies raised against individual centromere proteins, active centromeres are demonstrated to contain CENP-B, -C, and -E, while inactive centromeres contain CENP-B, but not CENP-C and -E (Earnshaw *et al.* 1989; Page *et al.* 1995; Sullivan and Schwartz 1995).

Robertsonian translocations

Robertsonian translocations refer to whole arm exchange between two acrocentric chromosomes (Robertson 1916). In humans, these rearrangements are the most common structurally abnormal chromosomes and have a general population incidence of 1 in 1000 (Hamerton *et al.* 1975; Therman *et al.* 1989; Nielsen and Wohlert 1991). All 10 possible nonhomologous pairwise combinations of the five acrocentric chromosomes (13, 14, 15, 21, and 22) have been observed, but the distribution of the translocations is highly nonrandom, with rob(13q14q) and rob(14q21q) comprising 80–85% of all nonhomologous Robertsonian translocations (Hecht *et al.* 1968; Rowley and Pergament 1969; Hecht and Kimberling 1971; Therman *et al.* 1989). In comparison, the remaining eight types of Robertsonian translocations collectively account for only 15–20%. Clinically, Robertsonian translocations do not seem to significantly affect the phenotype of a balanced carrier, apart from being occasionally associated with aneuploid offspring, spontaneous abortion, male sterility, and an increased risk of uniparental disomy (reviewed by Therman 1986; Gardner and Sutherland 1989, 1996; Donnai 1993). In addition to nonhomologous Robertsonian translocations, cases involving homologous acrocentric long arms are also found. However, most of the homologous cases have been shown to be isochromosomes and not translocations, originating via mechanisms entirely different than those responsible for the nonhomologous events (Grasso *et al.* 1989; Antonarakis *et al.* 1990; Shaffer *et al.* 1991).

Breakpoints in Robertsonian translocations

For a period, it was assumed that Robertsonian translocations generally gave rise to monocentric chromosomes through the fusion of two centromeres (Therman 1986). However, more recent banding and molecular studies have indicated the contrary in that greater than 90% of these translocations involving nonhomologous chromosomes result in dicentric chromosomes (Niebuhr 1972a; Dittes *et al.* 1975; Daniel and Lam-Po-Tang 1976; Mattei *et al.* 1979; Gosden *et al.* 1981; Cheung *et al.* 1990; Earle *et al.* 1992; Gravholt *et al.* 1992; Wolff and Schwartz 1992; Kalitsis *et al.* 1993; Han *et al.* 1994; Shaffer *et al.* 1994; Sullivan *et al.* 1994; Page *et al.* 1996). In particular, FISH studies have shown that all rob(13q14q) and rob(14q21q) translocations are dicentric, suggesting that the chromosomal breakpoints in these two groups of translocations must be on the short arms of both participating chromosomes (Cheung *et al.* 1990; Earle *et al.* 1992; Gravholt *et al.* 1992; Wolff and Schwartz 1992; Kalitsis *et al.* 1993; Han *et al.* 1994; Luke *et al.* 1994; Shaffer *et al.* 1994; Page *et al.* 1996). In the remaining 10% of non-dicentric Robertsonian translocations, one breakpoint would occur in the proximal long arm region of one chromosome, while the other breakpoint would be in a region extending from any part of the short arm to within the centromere of the second chromosome (Cheung *et al.* 1990; Wolff and Schwartz 1992; Page *et al.* 1996). Interestingly, although theoretically possible, there is as yet no direct molecular evidence demonstrating centric Robertsonian fusions involving breakpoints within both the centromeres.

The high prevalence of the rob(13q14q) and rob(14q21q) translocations has led to the proposal of a mechanism of formation involving sequence-driven recombination between homologous DNA that exists on the short arms of chromosomes 13 and 14, or 14 and 21, and where such homologous sequences are present in opposite orientations (Fig. 7.3) so as to facilitate recombinations that would result in rob(13q14q) and rob(14q21q), but not rob(13q21q) (Ohno *et al.* 1961; Ferguson-Smith and Handmaker 1961, 1963; Ferguson-Smith 1967; Rowley and Pergament 1969; Hecht and Kimberling 1971; Choo *et al.* 1988; Therman *et al.* 1989). Support for this model has come from FISH studies using centromeric and acrocentric short arm probes, and the demonstration of a consistent breakpoint domain within the proximal short arm regions of chromosomes 13 and 14, or 14 and 21, in the rob(13q14q) and rob(14q21q) translocations, but highly variable break-point locations for the less frequent forms of Robertsonian translocations (Page *et al.* 1996).

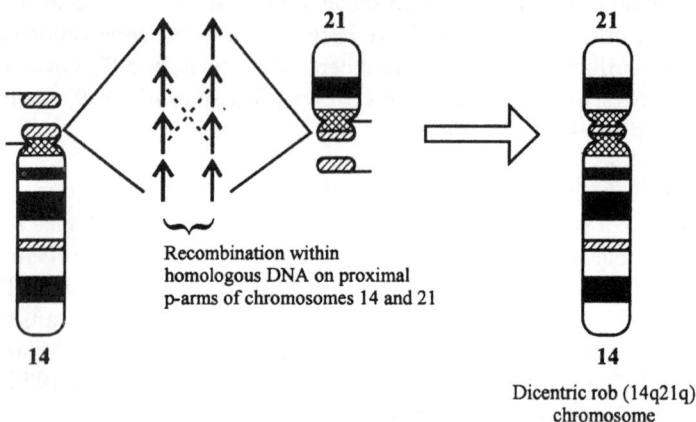

Fig. 7.3 Formation of a dicentric rob(14q21q) Robertsonian translocation chromosome. The orientation of the hypothesized homologous DNA region on the short arms is indicated by the arrows. A recombination (dotted lines) within this region results in the retention of both the centromeres on the long arms. It is presumed that rob(13q14q) occurs through a similar mechanism.

Centromere inactivation

It is generally assumed that true dicentric chromosomes (i.e. chromosomes with two active centromeres, as distinct from pseudodicentrics where there is only one active centromere) are unstable and prone to anaphase bridge formation and breakage due to forces generated by two active centromeres pulling the chromosomes toward opposite spindle poles. The continued existence of a dicentric chromosome therefore requires one of three conditions to be met. First, if the two centromeres are very close or physically fused together, they may function as a monocentric structure. The second condition requires that one of the centromeres be physically eliminated, e.g. by chromosomal deletion (Dewald *et al.* 1979; Vianna-Morgante and Rosenberg 1986). The third, and probably most common, condition involves the functional inactivation or suppression of one of the two centromeres (Niebuhr 1972a; Therman *et al.* 1974; Hsu *et al.* 1975; Daniel and Lam-Po-Tang 1976; Nakagome *et al.* 1976; Daniel 1979; Dewald *et al.* 1979; Zuffardi *et al.* 1980; Ing and Smith 1983; Earnshaw and Migeon 1985; Merry *et al.* 1985; Therman *et al.* 1986). In rare instances, it has been demonstrated that conditions 2 and 3 can separately operate in different cells of the same individual (Vianna-Morgante and Rosenberg 1986).

Is centromere inactivation random and meiotically stable?

Cytogenetic banding studies have indicated both random as well as apparently nonrandom centromere inactivation in dicentric translocations (Sears and Camara 1952; Angell *et al.* 1970; Warburton *et al.* 1973; Daniel and Lam-Po-Tang 1976; Nakagome *et al.* 1976; Roberts *et al.* 1977; Daniel 1979; Dewald *et al.* 1979; Wandall 1989). However, two recent molecular studies of Robertsonian translocations have provided evidence in support of the existence of a functional hierarchy of the centromeres, at least in the acrocentric group of chromosomes, in that there is an intrinsic preference for a particular centromere to be inactivated over the second centromere of a specific pairwise combination of a nonhomologous translocation (Gravholt *et al.* 1992; Sullivan *et al.* 1994). The results of these studies indicate that the centromere of chromosome 14 is most often the active centromere in the acrocentric hierarchy, while those of chromosomes 13, 21, and 22 are also more often active than the centromere of chromosome 15. In addition, these studies have demonstrated that the pattern of preferential centromere inactivation is meiotically stable.

Reversibility of inactivation

In a number of reported cases of human pseudodicentrics in which the origin of the centromere could be determined, the inactive centromere has been found to be the same in all the cells of an individual (Warburton *et al.* 1973; Pallister *et al.* 1974; Nakagome *et al.* 1976; Roberts *et al.* 1977; Daniel *et al.* 1979). However, cases of pseudodicentric chromosomes with either one or the other centromere inactivated in different cells of the same individual have also been described (Niebuhr 1972b; Dewald *et al.* 1979; Ing and Smith 1983; Daniel *et al.* 1985; Rivera *et al.* 1989; Wang and Li 1992). It is unclear in these latter cases whether such functional mosaicism results from successive inactivation–reversion processes, or from the persistence of a functionally dicentric chromosome (see below) long enough to be acted upon by random centromere inactivation to produce the distinct cell lines.

True functional dicentrics

Since most dicentric chromosomes are functionally monocentric, true functional dicentrics can be regarded as exceptions. Nonetheless, subjected to the reliability of the tests used to define centromere functions, these have been observed to co-exist, generally as a very rare component, with their sister pseudodicentric chromosomes in different cells of the same individual (Niebuhr 1972b; Dewald *et al.* 1979; Daniel *et al.* 1985; Wandall 1989, 1994; Sullivan and Schwartz 1995). The presence of abnormal chromosomes seen in variant karyotypes associated with various types of dicentric chromosomes has also been taken as evidence for the existence of functionally dicentric chromosomes, since these are expected to be unstable and associated with increased frequency of abnormal segregation (DeLaChapelle and Stenstrand 1974; Schmid *et al.* 1974; Hsu *et al.* 1978; Daniel *et al.* 1980; Ing and Smith 1983; Schwartz *et al.* 1983; Peretti *et al.* 1986; Haaf and Schmid 1990; Sharp *et al.* 1990; Sullivan *et al.* 1994). Again, whether these dicentric chromosomes originate from the reversion of an inactivated centromere, or are due to the persistence of a functionally dicentric chromosome, is not clear. Furthermore, it is possible that in these dicentric cases, one of the centromeres may appear to possess the recognizable features of an active centromere by the rather limited known criteria used, but may nonetheless be functionally impaired in a way that poses only a subtle effect on chromosomal segregation and, as such, is tolerated.

Marker chromosomes

Marker chromosomes with no detectable α-satellite DNA

A marker chromosome is defined as one that is morphologically different from any normal chromosome, with its origin generally not being readily discernible by classical cytogenetic techniques. The frequency of *de novo* constitutional marker chromosomes has been estimated to be 1 in 2500 (Warburton 1991). FISH analysis using chromosome-specific centromeric α-satellite DNA as probes indicates that the centromeres of the majority of marker chromosomes display this DNA (reviewed by Callen 1994). However, marker chromosomes that contain no detectable α-satellite DNA within their centromeres have also been described. Table 7.3 shows the centromere properties of five such α-satellite-negative marker chromosomes. In each of these cases, the chromosome has lost its normal centromere and has developed a new one in an entirely different, euchromatic region of the chromosome. Three of these markers have originated *de novo* from chromosomes 10 (Voullaire *et al.* 1993), 8 (Ohashi *et al.* 1994), and 14 (Magnani *et al.* 1993), and demonstrate full mitotic centromere function, as evident from the 100% stability of the chromosomes through many cell divisions. The remaining two marker chromosomes are derived from chromosome 15 (Blennow *et al.* 1994) and contain a functionally incomplete centromere, as indicated by the presence of these chromosomes in only 70–80% of the patients' lymphocytes, and their total loss in prolonged culture in rapidly dividing cell lines.

In addition to these five marker chromosomes, four other cases of less well-characterized marker chromosomes that contain no detectable α-satellite DNA at their primary constrictions have been described. One of these appears to be 100% stable (Crolla *et al.* 1992), another is 64% stable (Rauch *et al.* 1992), while the stability of the third marker (a ring chromosome) is not known (Callen *et al.* 1992). The fourth 'marker' chromosome is a morphologically normal Y chromosome that is present in 5% of the patient's lymphocytes. This chromosome has the added interest that it is a dicentric chromosome in which the normal centromere has become inactive and been replaced by a new, α-satellite DNA-negative centromere formed within the heterochromatic q arm region (Bukvic *et al.* 1996). The detection of such a marker chromosome suggests that the formation of neocentromeres in hitherto noncentromeric domains is not confined to euchromatic regions of chromosomes, and that the activity of the centromere so formed can, in an unknown way, prevail over that of a pre-existing centromere.

Table 7.3 Formation of active centromeres within apparently euchromatic regions of chromosomes

Property	Case 1	Case 2	Case 3	Case 4	Case 5
Chromosome of origin	10	8	14	15	15
Centromere position	10q25.2	8p23.1-pter	14q32	15q23	15q24
Primary constriction	+ ve	− ve	+ ve	+ ve	+ ve
C-banding	− ve	− ve	− ve	− ve	− ve
α-Satellite DNA	− ve	− ve	− ve	− ve	− ve
Satellite 3 DNA	− ve	n.a.	n.a.	n.a.	n.a.
CENP-A protein	+ ve*	n.a.	n.a.	n.a.	n.a.
CENP-B protein	− ve	− ve	n.a.	n.a.	n.a.
CENP-C protein	+ ve*	n.a.	n.a.	n.a.	n.a.
Mitotic stability	100%	100%	100%	70–80%	70–80%
Reference	Voullaire *et al.* 1993	Ohashi *et al.* 1994	Magnani *et al.* 1993	Blennow *et al.* 1994	Blennow *et al.* 1994

n.a. = data not available.
+ ve denotes presence and −ve denotes absence of a given property.
* Earle, E., duSart, D., and Choo, K.H.A., unpublished data.

Two possible mechanisms (Voullaire *et al.* 1993) have been proposed for the origin of these unusual neocentromeres of the marker chromosomes. The first requires the breakage of a part of a normal centromere, followed by its transposition into the marker chromosome. However, because of the lack of any evidence for either the disruption of a normal centromere or the presence of α-satellite DNA that is associated with a normal centromere, this possibility appears unlikely. The second, and probably more favourable, mechanism involves the *in situ* formation of the neocentromeres through the activation of pre-existing latent centromeres (discussed below).

Latent centromeres

The existence of 'latent intercalary centromeres' has been postulated as a possible mechanism for chromosome evolution (Dutrillaux 1979). Studies on primate karyotypes have indicated that a centromere can occasionally appear at a new position during speciation, and that centromere activation may be particularly important during the evolution of the group of Old World monkeys *Cercopithecidae*, where the evolution of the group is accompanied by a gradual increase in the number of chromosomes (Dutrillaux 1979). Similarly, comparative studies of mouse and human karyotypes have demonstrated two regions of conserved synteny which are interrupted by a centromere in humans, but not in mice (Searle *et al.* 1989). The detection of low levels of α-satellite DNA at noncentromeric regions (2q21 and 9q13) has also been interpreted as evidence of ancestral, and possibly latent, centromeres (Aleixandre *et al.* 1987; Baldini *et al.* 1993).

Voullaire *et al.* (1993) postulated that activation of a latent intercalary centromere is the most probable mechanism for the origin of the neocentromere in their marker chromosome (Case 1 in Table 7.3). These workers proposed that the activation could occur through one of two possible routes. The first would involve the mutational modification of a latent centromeric DNA in such a way that could lead to the emergence of new centromere functions. The second route involves an epigenetic effect, which is an effect that confers a heritable phenotypic difference on two identical DNA sequences without any specific modification of the nucleotide sequences (Brown and Tyler-Smith 1995). The concept of an epigenetic inheritance of the centromere is consistent with the observation that, although the centromeres of different mammalian species are stable in interspecific somatic cell hybrids, the centromeric DNA itself shows little or no conservation of the sequence that can be detected. Recently, the role of epigenetic factors in centromere function has been described in *S. pombe* (Steiner and Clarke 1994) (Chapter 3). These workers demonstrated that epigenetic factors can result in the conversion of a nonfunctional centromere to a functional one without changes in the content, structural arrangement, or chemical modification state of a centromere-containing plasmid DNA, and proposed that the conversion may involve the assembly of a functional DNA–protein complex through specific folding of the centromeric DNA elements. Whether the origin of the human marker centromeres described by Voullaire *et al.* (1993) and other workers (Table 3) is due to a similar epigenetic effect can now be directly tested by cloning the centromeric DNA from the marker chromosomes and comparing it with the DNA sequences of the corresponding regions of normal chromosomes.

Molecular characterization of an α-satellite-negative marker centromere

Using positional cloning techniques, du Sart *et al.* (1997) investigated the DNA of the neocentromere found in their chromosome 10-derived marker chromosome (Case 1, Table 7.3). These workers employed centromere-specific anti-CENP-A and anti-CENP-C antibodies in the localization of the centromeric DNA containing the antibody-binding domain. Detailed restriction mapping and partial sequence analyses of this DNA revealed no detectable difference between the marker centromere and the correponding region of the normal chromosome 10, suggesting a closely similar, if not identical, genomic organization of the DNA in the two regions. This result therefore does not support the transposition mechanism for the origin of the neocentromere. Instead, it favours a mechanism whereby epigenetic factors are responsible for the activation of a latent centromeric DNA, akin to that reported in *S. pombe* (Steiner and Clark 1994). Confirmation of this mechanism will, however, await full sequence comparison between the normal and marker DNA, as well as demonstration that there is no differential chemical modification in the two DNA sequences. It is possible that this phenomenon of latent centromere activation may account for the origin of other neocentromeres, such as those described in Table 7.3, and may constitute an important mechanism to compensate for the loss of normal centromeres due to the chromosomal rearrangement events that precede the formation of the marker chromosomes.

Marker chromosomes that associate with the centromeres of other chromosomes

Although human marker chromosomes normally exist and undergo mitotic segregation as independent structures, rare exceptions have been observed. For example, two small supernumerary marker chromosomes that show a preferential association with the centromeres of normal chromosomes during metaphase have been described (Schmid *et al.* 1989; Rivera *et al.* 1993). One of these marker chromosomes (Schmid *et al.* 1989) is present in ~ 60% of metaphases and shows a strong association with the centromeres of other chromosomes in 25% of cases. The marker is C-band positive, shows no cytogenetically detectable euchromatin, consists mainly or entirely of α-satellite DNA, and appears to be a ring chromosome (Schmid *et al.* 1989; Haaf *et al.* 1992). Somatically, it has remained relatively stable for several decades. This observation, together with results of positive staining with CREST antibody, suggests that the marker chromosome contains an active centromere. However, the fact that the chromosome is present in less than 60% of the patient's cells, and that prolonged culture of an actively dividing cell line results in a rapid drop in the frequency of the marker chromosome, indicate that the centromere may not be fully functional and/or the ring structure may contribute to mitotic instability.

The second marker is found in about 50% of metaphase cells, gives pale C- and late replicating-bandings, is associated with the centromere of other chromosomes in about 58% of the cells, and has remained relatively stable for over a decade (Rivera *et al.* 1993). In addition to these two examples, centromere association in some cancer cell lines has been observed (Haaf and Schmid 1989). Furthermore, in non-human cases, the small Y chromosome in colchicine-treated bone marrow cells of a South American marsupial has been shown to associate with the centromeres of other chromosomes in 36% of metaphases (Schmid *et al.* 1989), and the small B chromosomes in *Tradescantia paludosa* are known to associate with the centromeres (and telomeres) of the A chromosomes (Swanson *et al.* 1981).

The mechanisms responsible for these centromeric associations, and the biological significance of such an association, remain to be determined. Rivera *et al.* (1993) speculated that this association may simply be a relic of the Rabl orientation, which is a configuration formed when condensed chromosomes move toward the pole with the centromeres leading the way and giving the appearance of clustering or 'association' of centromeres when viewed from the centromere end (see Chapter 4). If this is the case, then the non-associativeness of most, usually the larger, marker chromosomes implies that factors such as chromosome size and length of chromosome arms may prevent or mask the centromere association.

Change of centromere position due to pericentric inversions

In pericentric inversions, two breaks occur in opposite chromosome arms, followed by rejoining of the middle, centromere-bearing segment onto the arms in an inverted orientation (Therman 1986; Gardner and Sutherland 1989, 1996). These inversions are often discovered because they change the position of the centromere, and thus differ from paracentric inversions (reviewed by Pettenati *et al.* 1995), in which both breaks occur on the same chromosome arm and there is no alteration in centromere position. Although most or all human chromosomes are involved in pericentric inversions, different chromosomes are involved nonrandomly; for instance inv(9) accounts for about 40% of them (Kaiser 1980). The breakpoints are also nonrandom. For example, 7 out of 10 inversions involving chromosome 2 show breakpoints at p11 and q13 (Leonard *et al.* 1975; Kaiser 1980). The reorientation of the centromere in these pericentric inversions does not appear to have any effect on the activity of the centromere.

Centric fission

Centric fission occurs when one functional centromere of a metacentric or submetacentric chromosome is horizontally split into two functional parts, giving rise to two stable telocentric products. The heterozygous person has 47 chromosomes and a balanced complement of genetic material, and may thus be phenotypically normal. Centric fission is extremely rare, with less than 10 families on record, involving five chromosomes—4, 7, 9, 10, and 21 (Hansen 1975; Dallapiccola *et al.* 1976; Surana *et al.* 1976; Fryns *et al.* 1980; Janke 1982; Del Porto *et al.* 1984; Nucaro *et al.* 1988; Elakis *et al.* 1993; Bogart *et al.* 1995). Both products of a centric fission are C-band positive and, where examined by FISH, contain α-satellite DNA (Bogart *et al.* 1995). Quite often, the two products have an asymmetrical amount of the C-band-positive constitutive heterochromatin. However, despite this asymmetrical distribution of heterochromatin, the two resulting centromeres have been shown to be functional not only in mitosis, but also in meiosis (Fryns *et al.* 1980). At meiosis in the heterozygotes, the centric fission products presumably form a trivalent with the intact homologue, essentially as in the Robertsonian translocation carriers. Alternate segregation produces normal and balanced gametes, while adjacent segregation results in gametes disomic or nullisomic for either of the fission products. Monosomy would probably be associated with occult abortion, and trisomy with miscarriage or, in exceptional cases, with the live birth of an abnormal child (Gardner and Sutherland 1989, 1996).

Examples of centric fission have also been reported in other species, including grasshoppers (Southern 1969), birds (Takagi and Sasaki 1974), root voles (Fredga and Bergstrom 1970), and primates (Egozcue 1971). As in humans, such fission events occur only rarely in these species.

Centric fusion

Centric fusion has been postulated as a significant driving force in mammalian karyotype evolution (Elder and Hsu 1988). Such fusions most often involve acrocentric chromosomes and, by classical cytogenetic staining, appear to occur at the centromeres. However, whether the fusion breakpoints are truly within the centromeres and not in some juxtacentromeric domains, may need to be considered in the light of recent findings on human Robertsonian translocations (see above) in which detailed molecular analysis of a relatively large number of cases has indicated that all have breakpoints involving at least one chromosome arm, with as yet no direct evidence for the existence of centromere-to-centromere fusion.

Conclusion

It can be seen from the above discussion that the centromere, like all genetic loci, is subjected to polymorphic variations (more so indeed than most other loci due to the high prevalence of tandemly repetitive DNA), and to the occasional mutational change that directly affect centromere morphology and function. The analysis of these polymorphic and mutational events has unravelled a number of interesting phenomena, such as premature centromere division, centric fission, centromere inactivation, dicentricity, centromere latency, and epigenetic effects. The study of these phenomena is important since they directly impinge upon the basic mechanisms that underlie the regulation as well as the evolution of the centromere. Some of the unusual cytogenetic cases presented in this chapter should provide invaluable experimental materials for such a study.

References

Aleixandre, C., Miller, D. A., Mitchell, A. R., Warburton, D. A., Gersen, S. L., Disteche, C. *et al.* (1987). p82H identifies sequences at every human centromere. Hum. Genet. *77*, 46–50.

Allingham-Hawkins, D. and Tomkins, D. (1991). Somatic cell hybridization of Roberts' syndrome and normal lymphoblasts resulting in correction of both the cytogenetic and mutagen hypersensitivity cellular phenotypes. Somat. Cell Mol. Genet. *17*, 455–462.

Allingham-Hawkins, D. and Tomkins, D. (1995). Heterogeneity in Roberts' syndrome. Am. J. Med. Genet. *55*, 188–194.

Angell, R., Cianelli, F., and Polani, P. (1970). Three dicentric Y chromosomes. Annls. Hum. Genet. *34*, 39–50.

Antonarakis, S., Adelsberger, P., Petersen, M., Binkert, F., and Schinzel, A. (1990). Analysis of DNA polymorphisms suggests that most *de novo* dup(21q) chromosomes in patients with Down's syndrome are isochromosomes and not translocations. Am. J. Hum. Genet. *47*, 968–972.

Appelt, H., Gerken, H., and Lenz, H. (1966). Tetraphokomelie mit lippen-kiefergaumenspalte und clitorishypertrophie-ein syndrome. Paediat. Paedol. *2*, 119–124.

Bajnóczky, K. (1985). Centromere separation sequence in aged women and men. Acta Biol. Hung. *36*, 313–318.

Bajnóczky, K. and Gardo, S. (1993). 'Premature anaphase' in a couple with recurrent miscarriages. Hum. Genet. *92*, 388–390.

Bajnóczky, K. and Méhes, K. (1988). Parental centromere separation sequence and aneuploidy in the offspring. Hum. Genet. *78*, 286–288.

Baldini, A., Ried, T., Shridhar, V., Ogura, K., D'Aiuto, L., Rocchi, M. *et al.* (1993). An alphoid DNA sequence conserved in all human and great ape chromosomes: evidence for ancient centromeric sequences at human chromosomal regions 2q21 and 9q13. Hum. Genet. *90*, 577–583.

Bamezai, R., Shiraishi, Y., and Taguchi, H. (1986). Centromere spreading in a case of megaloblastic anemia 'cured' under TC 199 culture condition. Cancer Genet. Cytogenet. *20*, 341–343.

Benn, P. A. and Perle, M. A. (1987). Chromosome staining and banding techniques. In Human Cytogenetics—A Practical Approach, eds. Rooney, D.E. and Czepulkowski B.H. IRL Press. Oxford. 57–84.

Berg, D. V. D. and Francke, U. (1993). Roberts' syndrome: a review of 100 cases and a new rating system for severity. Am. J. Med. Genet. *47*, 1104–1123.

Berrozpe, G., Caballin, M., Miro, R., Gelabert, A., and Egozcue, J. (1990). Centromere splitting in bladder cancer. Hum. Genet. *85*, 184–186.

Blennow, E., Telenius, H., Vos, D. d., Larsson, C., Henriksson, P., Johansson, O. *et al.* (1994). Tetrasomy 15q: two marker chromosomes with no detectable alphasatellite DNA. Am. J. Hum. Genet. *54*, 877–883.

Bogart, M., Fujita, N., Serles, L., and Hsia, Y. (1995). Prenatal diagnosis of a stable *de novo* centric fission: a case report. Am. J. Med. Genet. *59*, 36–37.

Brown, D., Grace, E., Sumner, A., Edmunds, A., and Ellis, P. (1995). ICF syndrome (immunodeficiency, centromeric instability and facial anomalies): investigation of heterochromatin abnormalities and review of clinical outcome. Hum. Genet. *96*, 411–416.

Brown, W. and Tyler-Smith, C. (1995). Centromere activation. Trends Genet. *11*, 337–339.

Bukvic, N., Susca, F., Gentile, M., Tangari, E., Ianniruberto, A., and Guanti, G. (1996). An unusual dicentric Y chromosome with a functional centromere with no detectable alpha-satellite. Hum. Genet. *97*, 453–456.

Burns, M. and Tomkins, D. (1989). Hypersensitivity to mitomycin C cell killing in Roberts' syndrome fibroblasts with, but not without, the heterochromatin abnormality. Mutat. Res. *216*, 2434–2439.

Callen, D. F., Eyre, H., Yip, M., Freemantle, J., and Haan, E. A. (1992). Molecular cytogenetic and clinical studies of 42 patients with marker chromosomes. Am. J. Med. Genet. *43*, 709–715.

Callen, D. F. (1994). Characterization of constitutive marker chromosomes in humans. In *In Situ* Hybridization Protocols, ed. Choo, K.H.A. Humana Press, Totowa, 439–457.

Chambers, J. C. and Taylor, J. (1982). Induction of sister chromatid exchanges by 5-fluorodeoxycytidine: correlation with DNA methylation. Chromosoma *85*, 603–609.

Cheung, S., Sun, L., and Featherstone, T. (1990). Molecular cytogenetic evidence to characterize breakpoint regions in Robertsonian translocations. Cytogenet. Cell Genet. *54*, 97–102.

Choo, K. H., Vissel, B., Brown, R., Filby, R. G., and Earle, E. (1988). Homologous alpha satellite sequences on human acrocentric chromosomes with selectivity for chromosomes 13, 14 and 21: implications for recombination between nonhomologues and Robertsonian translocations. Nucleic Acids Res. *16*, 1273–1284.

Crolla, J. A., Dennis, N. R., and Jacobs, P. A. (1992). A non-isotopic *in situ* hybridisation study of the chromosomal origin of 15 supernumerary marker chromosomes in man. J. Med. Genet. *29*, 699–703.

Dale, S., Earle, E., Voullaire, L., Rogers, J., and Choo, K. H. (1989). Centromeric alpha satellite DNA amplification and translocation in an unusually large chromosome 14p+ variant. Hum. Genet. *82*, 154–158.

Dallapiccola, B., Mastroiacco, P., and Gandini, E. (1976). Centric fission of chromosome No. 4 in the mother of two patients with trisomy 4p. Hum. Genet. *31*, 121–125.

Daniel, A. (1979). Single Cd band in dicentric translocations with one suppressed centromere. Hum. Genet. *48*, 85–92

Daniel, A., and Lam-Po-Tang, P. (1976). Structure and inheritance of some heterozygous Robertsonian translocations in man. J. Med. Genet. *13*, 381–388.

Daniel, A., Perel, I., Clarke, A., and Saville, T. (1979). Familial dicentric translocation, t(13;18)(p13;p11.2), ascertained by recurrent miscarriages. J. Med. Genet. *16*, 73–75.

Daniel, A., Lyons, N., Casey, J., and Gras, L. (1980). Two dicentric Y iso-chromosomes, one without the Yqh heterochromatic segment. Hum. Genet. *54*, 31–39.

Daniel, A., Ekblom, L., Phillips, S., FitzGerald, J., and Opitz, J. (1985). NOR activity and centromere suppression related in a *de novo* fusion tdic(9;13)(p22;p13) chromosome in a child with del(9p) syndrome. Am. J. Med. Genet. *22*, 577–584.

DeLaChapelle, A., and Stenstrand, K. (1974). Dicentric human X chromosomes. Hereditas *76*, 259–268.

DelPorto, G., Fusco, D. D., Baldi, M., Grammatico, P., and D'Alessandro, E. (1984). Familial centric fission of chromosome 4. J. Med. Genet. *21*, 388–391.

Dev, V. and Wertelecki, W. (1984). Elimination of abnormal centromere chromatid apposition (ACCA) in selected human–mouse cell hybrids. Am. J. Hum. Genet. *36*, 90S.

Dewald, G., Boros, S., Conroy, M., Dahl, R., Spurbeck, J., and Vitek, H. (1979). A tdic(5;15)(p13;p11) chromosome showing variation for constriction in the centromeric region in a patient with Cri-du-Chat syndrome. Cytogenet. Cell Genet. *24*, 15–26.

Dewald, G. (1988). Translocation dicentric chromosomes among human autosomes and centromeric suppression. In The Cytogenetics of Mammalian Autosomal Rearrangements, ed. Daniel, A. Alan R. Liss, New York 533–550.

Dittes, H., Drone, W., Bross, K., Schmid, M., and Jogel, W. (1975). Biochemical and cytogenetic studies on the nucleolus organizing regions (NOR) of man. II. A family with the 15/21 translocation. Humangenetik *26*, 47–59.

Donnai, D. (1993). NICHD conference: Robertsonian translocation: clues to imprinting. Am. J. Med. Genet. *46*, 681–682.

duSart, D., Cancilla, M., Earle, E., Mao, J., Saffery, R., Tainton, K. R., Kalitsis, P. M., Martyn, J., Barry, A. E., and Choo, K. H. A. A functionally critical human centromere DNA region consisting of a unique sequence formed through the activation of a latent centromere. Nature Genet. (in press).

Dutrillaux, B. (1979). Chromosomal evolution in primates: tentative phylogeny from *Microcebus murinus* (Prosimian) to man. Hum. Genet. *48*, 251–314.

Earle, E., Shaffer, L. G., Kalitsis, P., McQuillan, C., Dale, S., and Choo, K. (1992). Identification of DNA sequences flanking the breakpoint of human t(14q21q) Robertsonian translocations. Am. J. Hum. Genet. *50*, 717–724.

Earnshaw, W. C. and Migeon, B. (1985). Three related centromere proteins are absent from the inactive centromere of a stable isodicentric chromosome. Chromosoma *92*, 290–296.

Earnshaw, W. C., Ratsie III, H., and Stetten, G. (1989). Visualization of centromere proteins CENP-B and CENP-C on a stable dicentric chromosome in cytological spreads. Chromosoma (Berl.) *98*, 1–12.

Egozcue, J. (1971). A possible case of centric fission in a primate. Experientia (Basel) *27*, 969–970.

Elakis, G., Moriarty, H., Saville, T., Purvis-Smith, S., Robertson, R., and Roach, T. (1993). Familial transmission of a rare centromeric fission of chromosome 10. Bull. Hum. Genet. Soc. Australasia *6*, 40.

Elder, F., and Hsu, T. (1988). Tandem fusion in the evolution of mammalian chromosomes. In The Cytogenetics of Mammalian Autosomal Rearrangements, Eds Daniel, A. 481–506. Alan R. Liss, New York.

Engler, P., Haasch, D., Pinkert, C., Doglio, L., Glymour, M., Brinster, R., *et al.* (1991). A strain-specific modifier on mouse chromosome 4 controls the methylation of independent transgene loci. Cell *65*, 939–947.

Fasth, A., Forestier, E., Holmberg, E., Holmgren, G., Nordenson, I., Soderstrom, T., *et al.* (1990). Fragility of the centromeric region of chromosome 1 associated with combined immunodefiency in siblings. A recessively inherited entity? Acta Paediat. Scand. *79*, 605–612.

Ferguson-Smith, M. (1967). Chromosomal satellite association. Lancet *i*, 1156–1157.

Ferguson-Smith, M. and Handmaker, S. (1961). Observations on the satellited human chromosomes. Lancet *i*, 638–640.

Ferguson-Smith, M. and Handmaker, S. (1963). The association of satellited chromosomes with specific chromosomal regions in cultured human somatic cells. Annls. Hum. Genet. *27*, 143–156.

Fitzgerald, P. and McEvan, C. (1977). Total aneuploidy and age-related sex chromosome aneuploidy in cultured lymphocytes of normal men and women. Hum. Genet. *39*, 329–337.

Fitzgerald, P., Pickering, A., Mercer, J., and Miethke, P. (1975). Premature centromere division: a mechanism of non-disjunction causing X chromosome aneuploidy in somatic cells of man. Annls. Hum. Genet. *38*, 417–428.

Fitzgerald, P., Archer, S., and Morris, C. (1986). Evidence for the repeated primary non-disjunction of chromosome 21 as a result of premature centromere division (PCD). Hum. Genet. *72*, 58–62.

Franceschini, P., Martino, S., Ciocchini, M., Ciuti, E., Vardeu, M., Guala, A. *et al.* (1995). Variability of clinical and immunological phenotype in immunodeficiency-centromeric instability-facial anomalies syndrome. Eur. J. Pediatr. *154*, 840–846.

Fredga, K. and Bergstrom, U. (1970). Chromosome polymorphism in the root vole (*Microtus oeconomus*). Hereditas (Lund) *66*, 145–152.

Freeman, M., Williams, D., Schimke, R., Temtamy, S., Vachier, E., and German, J. (1974). The Roberts' syndrome. Clin. Genet. *5*, 1–16.

Freire-Maia, D., Garcia, M., Mendes, R., and Marques, S. (1994). Chromosome aberrations in lymphocyte cultures from paracoccidioidomycosis patients. J. Med. Vet. Mycol. *32*, 199–203.

Fryns, J. P., Buckle, J., Hens, L., and VandenBerghe, H. (1980). Balanced transmission of centromeric fission products in man. Hum. Genet. *54*, 127–128.

Fryns, J., Azou, M., Jaeken, J., Eggermont, E., and Pedersen, J. (1988). Centromeric instability of chromosomes 1, 9, and 16 associated with combined immunodeficiency. Hum. Genet. *57*, 108–110.

Gabarron, I., Jimenez, A., and Glover, G. (1986). Premature centromere division dominantly inherited in a subfertile family. Cytogenet. Cell Genet. *43*, 69–71.

Gallo, J., Misawa, S., and Testa, J. (1984). Centromere spreading in acute non-lymphocytic leukemia. Cancer Genet. Cytogenet. *12*, 105–109.

Gardner, R. J. M. and Sutherland, G. R. (1989). 1st Edn. Chromosome Abnormalities and Genetic Counseling, Oxford University Press, Oxford.

Gardner, R. J. M. and Sutherland, G. R. (1996). Ibid. 2nd Edn.

Ge, Y., Wagner, M. J., Siciliano, M., and Wells, D. E. (1992). Sequence, higher order repeat structure, and long-range organization of alpha satellite DNA specific to human chromosome 8. Genomics *13*, 585–593.

German, J. (1974). Chromosomes and Cancer. John Wiley & Son, New York, 601–617.

German, J. (1979). Roberts' syndrome. I. Cytological evidence for a disturbance in chromatid pairing. Clin. Genet. *16*, 441–447.

Gimelli, G., Varone, P., Pezzolo, A., Lerone, M., and Pistoia, V. (1993). ICF syndrome with variable expression in sibs. J. Med. Genet. *30*, 429–432.

Gosden, J. R., Lawrie, S. S., and Gosden, C. (1981). Satellite DNA sequences in the human acrocentric chromosomes: information from translocations and heteromorphisms. Am. J. Hum. Genet. *33*, 243–251.

Grasso, M., Uzielli, M. G., Tavellini, F., Perroni, L., and Bricarelli, F. D. (1989). Isochromsome not translocation in trisomy 21q21q. Hum. Genet. *84*, 63–65.

Gravholt, C., Friedrich, U., Caprani, M., and Jorgensen, A. (1992). Breakpoints in Robertsonian translocations are localized to satellite III DNA by fluorescence *in situ* hybridization. Genomics *14*, 924–930.

Gunby, J., Tomkins, D., and Chang, P. (1987). Somatic cell hybridization of Roberts' syndrome and normal human fibroblasts transfected with plasmids carrying dominant selection markers. Somat. Cell Mol. Genet. *13*, 245–252.

Haaf, T. and Schmid, M. (1989). Centromeric association and non-random distribution of centromeres in human tumour cells. Hum. Genet. *81*, 137–143.

Haaf, T. and Schmid, M. (1990). Y isochromosome associated with a mosaic karyotype and inactivation of the centromere. Hum. Genet. *85*, 486–490.

Haaf, T., Summer, A., Kohler, J., Willard, H. F., and Schmid, M. (1992). A microchromosome derived from chromosome 11 in a patient with the CREST syndrome of scleroderma. Cytogenet. Cell Genet. *60*, 12–17.

Hall, B. and Greenberg, M. (1972). Hypomelia-hypotrichosis-facial hemangioma syndrome. Am. J. Dis. Child *123*, 602–604.

Hamerton, J., Canning, N., Ray, M., and Smith, J. (1975). A cytogenetic survey of 14,069 newborn infants. I. Incidence of chromosome abnormalities. Clin. Genet. *84*, 63–65.

Han, J. Y., Choo, K. H. A., and Shaffer, L. G. (1994). Molecular cytogenetic characterization of 17 rob(13q14q) Robertsonian translocation by FISH, narrowing the region containing the breakpoints. Am. J. Hum. Genet. *55*, 960–967.

Hansen, S. (1975). A case of centric fission in man. Humangenetik *26*, 257–259.

Harnden, D. G. (1974). Viruses, chromosomes, and tumors: the interaction between viruses and chromosomes. In Chromosomes and Cancer, ed. German, J. John Wiley & Sons, New York, 154–190.

Heath, C. (1966). Cytogenetic observation in vitamin B12 and folate deficiency. Blood *27*, 800–815.

Hecht, F., Case, M., Lovrien, E., Higgins, J., Thuline, H., and Melnyk, J. (1968). Nonrandomness of translocations in man:preferential entry of chromosomes into 13–15/21 translocations. Science *161*, 371–372.

Hecht, F. and Kimberling, W. (1971). Patterns of D chromosome involvement in human (DqDq) Robertsonian rearrangements. Am. J. Hum. Genet. *23*, 361–367.

Herrmann, J., Feingold, M., Tuffli, G., and Opitz, J. (1969). A familial dysmorphogenetic syndrome of limb deformities, characteristic facial appearance and associated anomalies: the 'pseudothalidomide' or 'SC-syndrome'. Birth Defects: Original Article Series *5*, 81–89.

Hori, T. (1983). Induction of chromosome decondensation, sister chromatid exchanges and endoreduplication by 5-azacytidine, an inhibitor of DNA methylation. Mutat. Res. *121*, 47–52.

Hsu, C., Pathnak, S., and Chen, T. R. (1975). The possibility of latent centromeres and a proposed nomenclature system for total chromosome and whole arm translocations. Cytogenet. Cell Genet. *15*, 41–49.

Hsu, L., Paciuc, S., David, K., Christian, S., Moloshok, R., and Hirschorn, K. (1978). Number of C-bands of human isochromosome Xqi and relation to 45,X mosaicism. J. Med. Genet. *15*, 222–226.

Hulten, M. (1978). Selective somatic pairing and fragility at 1q12 in a boy with common variable immunodeficiency. Clin. Genet. *14*, 294.

Ing, P. S. and Smith, S. D. (1983). Cytogenetic studies of a patient with mosaicism of isochromosome 13q and a dicentric (Y;13) translocation showing differential centromeric activity. Clin. Genet. *24*, 194–199.

Izakovic, V. and Vahancik, A. (1984). Karyotypes in lymphocytes of peripheral blood and bone marrow cells in 80-year-old and older subjects (in Slovakian). Vnitr Lek *30*, 974–983.

Jabs, E. and Carpenter, N. (1988). Molecular cytogenetic evidence for amplification of chromosome-specific alphoid sequences at enlarged C-bands on chromosome 6. Am. J. Hum. Genet. *43*, 69–74.

Jabs, E., Tuck-Muller, C., Cusano, R., and Rattner, J. (1991). Studies of mitotic and centromeric abnormalities in Roberts' syndrome: implications for a defect in the mitotic mechanism. Chromosoma *100*, 251–261.

Janke, D. (1982). Centric fission of chromosome no. 7 in three generations. Hum. Genet. *60*, 200–201.

Jeanpierre, M., Turleau, C., Aurias, A., Prieur, M., Ledeist, F., Fischer, A. *et al.*

(1993). An embryonic-like methylation pattern of classical satellite DNA is observed in ICF syndrome. Hum. Mol. Genet. *2*, 731–735.

Kaiser, P. (1980). Pericentrische inversionen menschlicher chromosomen. Thieme, Stuttgart.

Kajtar, P. and Méhes, K. (1994). Bilateral coats retinopathy associated with aplastic anaemia and mild dyskeratotic signs. Am. J. Med. Genet. *49*, 374–377.

Kalitsis, P., Earle, E., Vissel, B., Shaffer, L. G., and Choo, K. H. (1993). A chromosome 13-specific human satellite I DNA subfamily with minor presence on chromosome 21: further studies on Robertsonian translocations. Genomics *16*, 104–112.

Kormann-Bortolotto, M., Smith, M. d. A. C., and Neto, J. T. (1993). Alzheimer's disease and ageing: a chromosomal approach. Gerontology *39*, 1–6.

Krassikoff, N., Cowan, J., Parry, D., and Francke, U. (1986). Chromatid repulsion associated with Roberts'/SC phocomelia syndrome is reduced in malignant cells and not expressed in interspecies somatic-cell hybrids. Am. J. Hum. Genet. *39*, 618–630.

Kricker, M., Drake, J., and Radman, M. (1992). Duplication-targeted DNA methylation and mutagenesis in the evolution of eukaryotic chromosomes. Proc. Natl Acad. Sci. USA *89*, 1075–1079.

Kucheria, K., Bhargava, S., Bamezai, R., and Bhutani, P. (1976). A familial tetraphocomelia syndrome involving limb deformities, cleft lip, cleft palate, and associated anomalies—a new syndrome. Hum. Genet. *33*, 323–326.

Lambiase, S., Maraschio, P., and Zuffardi, O. (1984). The Cd technique identifies a specific structure related to centromeric function. Hum. Genet. *67*, 214–215.

Lavia, P., Ferraro, M., Micheli, A., and Olivieri, G. (1985). Effect of 5-azacytidine (5-azaC) on the induction of chromatid aberrations (CA) and sister-chromatid exchanges (SCE). Mutat. Res. *149*, 463–467.

Leonard, C., Hazael-Massieux, P., and Bocquet, L. (1975). Inversion pericentrique inv(2)(p11q13) dans des families non apparentees. Humangenetik *28*, 121–128.

Littlefield, L., Joiner, E., and Sayer, A. (1985). Premature separation of centromeres in marrow chromosomes from an untreated pateint with acute myelogenous leukemia. Cancer Genet. Cytogenet. *16*, 109–116.

Louie, E. and German, J. (1981). Roberts' syndrome. II. Aberrant Y-chromosome behavior. Clin. Genet. *19*, 71–74.

Louie, E. and German, J. (1983). Morphological evidence of disturbed chromosome separation in Roberts' syndrome: a human mitotic mutant. J. Cell Biol. *97*, 192A.

Luke, S., Aggarwal, G., Stetka, D., and Verma, R. (1994). Alphoid DNA diversity of a so-called monocentric Robertsonian fusion. Chromosome Res. *2*, 73–75.

Madan, K., Lindhout, D., and Palan, A. (1987). Premature centromere division (PCD): a dominantly inherited cytogenetic anomaly. Hum. Genet. *77*, 193–196.

Magnani, I., Sacchi, N., Darfler, M., Nisson, P., Tornaghi, R., and Fuhrman-Conti, A. (1993). Identification of the chromosome 14 origin of a C-negative marker associated with a 14q32 deletion by chromosome painting. Clin. Genet. *43*, 180–185.

Mahtani, M. M., and Willard, H. F. (1990). Pulsed-field gel analysis of alpha satellite DNA at the human X chromosome centromere: high-frequency polymorphisms and array size estimate. Genomics *7*, 607–613.

Maraschio, P., Zuffardi, O., and LoCurte, F. (1980). Cd bands and centromeric function in dicentric chromosomes. Hum. Genet. *54*, 265–267.

Maraschio, P., Tupler, R., Dainotti, E., Piantanida, M., Cazzola, G., and Tiepolo, L. (1989). Differential expression of the ICF (immunodeficiency, centromeric heterochromatin, facial anomalies) mutation in lymphocytes and fibroblasts. J. Med. Genet. *26*, 452–456.

Maraschio, P., Cortinovis, M., Dainotti, E., Typler, R., and Tiepolo, L. (1992). Interphase cytogenetics of the ICF syndrome. Annls. Hum. Genet. *56*, 273–278.

Mattei, G., Mattei, J., Ayme, S., and Giraud, F. (1979). Dicentric Robertsonian translocations in man: 17 cases studied by R, C, and N banding. Hum. Genet. *50*, 33–38.

Méhes, K. (1978). Non-random centromere division: a mechanism of non-disjunction causing aneuploidy? Hum. Hered. *28*, 255–260.

Méhes, K. and Buhler, E. (1995). Premature centromere division: a possible manifestation of chromosome instability. Am. J. Med. Genet. *56*, 76–79.

Merry, D., Pathak, S., Hsu, T., and Brinkley, B. (1985). Antikinetochore antibodies: use as probes for inactive centromeres. Am. J. Hum. Genet. *37*, 425–430.

Miller, O. J. (1983). Dicentric chromosomes. In Principles and Practice of Medical Genetics, eds. Emery, A.E.H. and Rimion, D.L. Churchill Livingstone, New York, 57.

Miller, K., Muller, W., Winkler, L., Hadam, M., Ehrich, J., and Flatt, S. (1990). Mitotic disturbances associated with mosaic aneuploidies. Hum. Genet. *84*, 361–364.

Miniou, P., Jeanpierre, M., Blanquet, V., Sibella, V., Bonneau, D., Herbelin, C. *et al.* (1994). Abnormal methylation pattern in constitutive and facultative (X inactive chromosome) heterochromatin of ICF patients. Hum. Mol. Genet. *3*, 2093–2102.

Moorhead, P. and Heyman, A. (1983). Chromosome studies of patients with Alzheimer disease. Am. J. Med. Genet. *14*, 545–556.

Murthy, S. and Prabhakara, K. (1990). Mitotic disturbances associated with inversion 9qh. A case report. Annls. Genet. *33*, 169–172.

Nakagome, Y., Teramura, F., Kataoka, K., and Hosono, F. (1976). Mental retardation, malformation syndrome and partial 7p monosomy [45,XX,tdic(7;15)(p21;p11)]. Clin. Genet. *9*, 621–624.

Nakagome, Y., Abe, T., Misawa, S., Takeshita, T., and Iinuma, K. (1984). The 'loss' of centromeres from chromosomes of aged women. Am. J. Hum. Genet. *36*, 398–404.

Niebuhr, E. (1972a). Dicentric and monocentric Robertsonian translocations in man. Humangenetik *16*, 217–226.

Niebuhr, E. (1972b). A 45,XX,5-,13-,dic + karyotype in a case of cri-du-chat syndrome. Cytogenetics *11*, 165–177.

Nielsen, J. and Wohlert, M. (1991). Chromosome abnormalities found among 34910 newborn children: results from a 13-year incidence study in Arhus, Denmark. Hum. Genet. *87*, 81–83.

Nucaro, A., Falchi, A., Monni, G., and Cao, A. (1988). Pseudomosaic centric fission of chromosome 4 in amniotic cells. Prenatal Diagnosis *8*, 629–631.

Ohashi, H., Wakui, K., Ogawa, K., Okano, T., Niikawa, N., and Fukushima, Y. (1994). A stable acentric marker chromosome: possible existence of an intercalary ancient centromere at distal 8p. Am. J. Hum. Genet. *55*, 1202–1208.

Ohno, S., Trujillo, J., Kaplan, W., and Kinosita, R. (1961). Nucleolus-organisers in the causation of chromosomal anomalies in man. Lancet *ii*, 123–126.

Page, S. L., Shin, J.-C., Han, J.-Y., Choo, K. H. A., and Shaffer, L. G. (1996). Breakpoint diversity illustrates distinct mechanisms for Robertsonian translocation formation. Hum. Mol. Genet. *5*, 1279–1288.

Pallister, P., Patau, K., Inhorn, S., and Opitz, J. (1974). A woman with congenital abnormalities, mental retardation and mosaicism for an unusual translocation chromosome, t(6;19). Clin. Genet. *5*, 188–195.

Peretti, D., Maraschio, P., Lambiase, S., Curto, F. L., and Zuffardi, O. (1986). Indirect immunofluorescence of inactive centromeres as indicator of centromeric function. Hum. Genet. *73*, 12–16.

Pettenati, M., Rao, P., Phelan, M., Grass, F., Rao, K., Cosper, P. *et al.* (1995). Paracentric inversions in humans: a review of 446 paracentric inversions with presentation of 120 new cases. Am. J. Med. Genet. *55*, 171–187.

Rangnekar, G., Loya, B., Gowwami, L., and Sengupta, L. (1990). Premature centromeric division and prominent telomeres in a patient with persistent Mullerian duct syndrome. Clin. Genet. *37*, 69–73.

Rauch, A., Pfeiffer, R. A., Trautmann, U., Liehr, T., Rott, H. D., and Ulmer, R. A. (1992). A study of ten small supernumerary (marker) chromosomes identified by fluorescence *in situ* hybridization (FISH). Clin. Genet. *42*, 84–90.

Rivera, H., Zuffardi, O., Maraschio, P., Caiulo, A., Anichini, C., Scarinci, R. *et al.* (1989). Alternate centromere inactivation in a pseudodicentric (15;20)(pter;pter) associated with a progressive neurological disorder. J. Med. Genet. *6*, 626–630.

Rivera, H., Dominguez, M., Vasquez, A., Ramos, A., and Fragoso, R. (1993). Centromeric association of a microchromosome in a Turner syndrome patient with a pseudodicentric Y. Hum. Genet. *92*, 522–524.

Roberts, S., Howell, R., Laurence, K., and Heathcote, M. (1977). Stable dicentric autosome, tdic(8;22)(p23;p13), in a mentally retarded girl. J. Med. Genet. *14*, 66–68.

Robertson, W. (1916). Chromosome studies. I. Taxonomic relationships shown in the chromosomes of Tettigidae and Acrididae: V-shaped chromosomes and their significance in Acrididae, Locustidae, and Gryllidae: chromosomes and variation. J. Morphol. *27*, 179–181.

Rowley, J. and Pergament, E. (1969). Possible non random selection of D group chromosomes involved in centric-fusion translocations. Annls. Genet. *12*, 177–183.

Rudd, N., Teshima, I., Martin, R., Sisken, J., and Weksberg, R. (1983). A dominantly inherited cytogenetic anomaly: a possible cell division mutant. Hum. Genet. *65*, 117–121.

Sandberg, A. A. (1980). The chromosomes in Human Cancer and Leukemia. Elsevier North Holland, New York.

Sawyer, J., Swanson, C., Wheeler, G., and Cunniff, C. (1995). Chromosome instability in ICF syndrome: formation of micronuclei from multibranched chromosomes 1 demonstrated by fluorescence *in situ* hybridization. Am. Med. Genet. *56*, 203–209.

Scappaticci, S., Cerimele, D., Tondi, M., Vivarelli, R., Fois, A., and Fraccaro, M. (1988). Chromosome abnormalities in tuberous sclerosis. Hum. Genet. *79*, 151–156.

Schmid, W., Naef, E., Murset, G., and Prader, A. (1974). Cytogenetic findings in 89 cases of Turner's syndrome with abnormal karyotypes. Humangenetik *24*, 93–104.

Schmid, M., Grunert, D., Haaf, T., and Engel, W. (1983). A direct demonstration of somatically paired heterochromatin in human chromosomes. Cytogenet. Cell Genet. *36*, 554–561.

Schmid, M., Haaf, T., Schindler, D., and Meurer, M. (1989). Centromeric association of a microchromosome. Hum. Genet. *81*, 127–136.

Schwartz, S., Palmer, C., Weaver, D., and Priest, J. (1983). Dicentric chromosome 13 and centromere inactivation. Hum. Genet. *63*, 332–337.

Searle, A. G., Peters, J., Lyon, M. F., Hall, J. G., Evans, E. P., Edwards, J. H. *et al.* (1989). Chromosome maps of man and mouse. IV. Annls. Hum. Genet. *53*, 89–140.

Sears, E. and Camara, A. (1952). A transmissible dicentric chromosome. Genetics *37*, 125–135.

Shaffer, L. G., Jackson-Cook, C., Meyer, J., Brown, J., and Spence, J. (1991). A molecular genetic approach to the identification of isochromosomes of chromosome 21. Hum. Genet. *86*, 375–382.

Shaffer, L. G., McCaskill, C., Han, J.-Y., Choo, K. H. A., Cutillo, D. M., Donnenfeld, A. E. *et* al. (1994). Molecular characterization of de novo secondary trisomy 13. Am. J. Hum. Genet. *55*, 968–974.

Sharp, C., Bedford, H., and Willard, H. F. (1990). Pericentromeric structure of human X 'isochromosomes': evidence for molecular heterogeneity. Hum. Genet. *85*, 330–336.

Shiraishi, Y., Taguchi, H., Niiya, K., Shiomi, F., Kikukawa, K., Kubonishi, S., *et* al. (1982). Diagnostic and prognostic signficance of chromosome abnormalities in marrow and mitogen response of lymphocytes of acute nonlymphocytic leukemia. Cancer Genet. Cytogenet. *5*, 1–24.

Sinha, A., Verma, R., and Mani, V. (1994). Clinical heterogeneity of skeletal dysplasia in Roberts' syndrome: a review. Hum. Hered. *44*, 121–126.

Southern, D. (1969). Stable telocentric chromosomes produced following centric misdivision in *Myrmeleotettix maculatus* (Thunb.). Chromosoma (Berl.) *26*, 140–147.

Steiner, N. C. and Clarke, L. (1994). A novel epigenetic effect can alter centromere function in fission yeast. Cell *79*, 865–874.

Sullivan, B. and Schwartz, S. (1995). Identification of centromeric antigens in dicentric Robertsonian translocations: CENP-C and CENP-E are necessary components of functional centromeres. Hum. Mol. Genet. *4*, 2189–2197.

Sullivan, B., Wolff, D., and Schwartz, S. (1994). Analysis of centromeric activity in Robertsonian translocations: implications for a functional acrocentric hierarchy. Chromosoma *103*, 459–467.

Surana, R., Stevens, L., Gardner, H., and Bailey, J. (1976). Partial trisomy 9. Exerpta Med. (Amst.) *397*, 157.

Swanson, C., Merz, T., and Young, W. (1981). Cytogenetics. The Chromosome in Division, Inheritance and Evolution, 2nd Edn. Prentice-Hall, Englewood Cliffs, 170–175.

Takagi, N. and Sasaki, M. (1974). A phylogenetic study of bird karyotypes. Chromosoma (Berl.) *46*, 91–120.

Tawn, E. and Holdsworth, D. (1992). Mutagen-induced chromosome damage in human lymphocytes. In Human Cytogenetics: A Practical Approach, Vol II, eds. Rooney, D.E. and Czepulkowski, B.H. IRL Press, Oxford.

Therman, E. (1986). Human Chromosomes Structure, Behavior, Effects, 2nd Ed. Springer-Verlag New York.

Therman, E., Sarto, G., and Patau, K. (1974). Apparently isodicentric but functionally monocentric X chromosome in man. Am. J. Hum. Genet. *26*, 83–92.

Therman, E., Trunca, C., Kuhn, E., and Sarto, G. (1986). Dicentric chromosomes and the inactivation of the centromere. Hum. Genet. *72*, 191–195.

Therman, E., Susman, B., and Denniston, C. (1989). The nonrandom participation of human acrocentric chromosomes in Robertsonian translocations. Annls. Hum. Genet. *53*, 49–65.

Tiepolo, L., Maraschio, P., Gimelli, G., Cuoco, C., Gargani, G., and Romano, C. (1979). Multibranched chromosomes 1, 9 and 16 in a patient with combined IgA and IgE deficiency. Hum. Genet. *51*, 127–137.

Tomkins, D. and Sisken, J. (1984). Abnormalities in the cell-division cycle in Roberts' syndrome fibroblasts: a cellular basis for the phenotypic characteristics? Am. J. Hum. Genet. *36*, 1332–1340.

Tomkins, D., Hunter, A., and Roberts, M. (1979). Cytogenetic findings in Roberts'-SC phocomelia syndrome(s). Am. J. Med. Genet. *4*, 17–26.

Tyler-Smith, C. (1987). Structure of repeated sequences in the centromeric region of the human Y chromosome. Development *101*, 93–100.

Van Den Berg, D. and Francke, U. (1993). Roberts' syndrome: a review of 100 cases and a new rating system for severity. Am. J. Med. Genet. *47*, 1104–1123.

Verma, R. A. and Luke, S. (1992). Variations in alphoid DNA sequences escape detection of aneuploidy at interphase by FISH technique. Genomics *14*, 113–116.

Vianna-Morgante, A. and Rosenberg, C. (1986). Deletion of the centromere as a mechanism for achieving stability of a dicentric chromosome. Cytogenet. Cell Genet. *42*, 119–122.

Vig, K. (1984). Sequence of centromere separation: another mechanism for the origin of nondisjunction. Hum. Genet. *66*, 239–243.

Voullaire, L. E., Slater, H. R., Petrovic, V., and Choo, K. H. A. (1993). A functional marker centromere with no detectable alpha-satellite, satellite III, or CENP-B protein: activation of a latent centromere? Am. J. Hum. Genet. *52*, 1153–1163.

Wandall, A. (1989). Kinetochore development in two dicentric chromosomes in man. Hum. Genet. *82*, 137–141.

Wandall, A. (1994). A stable dicentric chromosome: both centromeres develop kinetochores and attach to the spindle in monocentric and dicentric configuration. Chromosoma *103*, 56–62.

Wang, F. and Li, Y. (1992). A new stable human dicentric chromosome, tdic(4;21)(p16;q22), in a woman with first trimester abortion. J. Med. Genet. *30*, 696.

Warburton, D., Henderson, A., Shapiro, L., and Hsu, L. (1973). A stable human dicentric chromosome, tdic (12;14)(p13;p13) including an intercalary satellite region between centromeres. Am. J. Hum. Genet. *25*, 439–445.

Warburton, D. (1991). *De novo* balanced chromosome rearrangements and extra marker chromosomes identified at prenatal diagnosis: clinical significance and distribution of breakpoints. Am. J. Hum. Genet. *49*, 995–1013.

Wevrick, R. and Willard, H. F. (1991). Physical map of the centromeric region of human chromosome 7: relationship between two distinct alpha satellite arrays. Nucleic Acids Res. *19*, 2295–2301.

Wolff, D. and Schwartz, S. (1992). Characterization of Robertsonian translocations by using fluorescence *in situ* hybridization. Am. J. Hum. Genet. *50*, 174–181.

Zhang, S. (1986). Centromere spreading and out-of-phase chromatid separation in Burkitt's lymphoma and nasopharyngeal carcinoma. Cancer Genet. Cytogenet. *23*, 211–217.

Zuffardi, O., Danesino, C., Poloni, L., Pavesi, F., Bianchi, C., and Gargantini, L. (1980). Ring chromosome 12 and latent centromeres. Cytogenet. Cell Genet. *28*, 151–157.

8

Practical applications

Apart from its intrinsic biological importance and interest, the study of the centromere offers a number of immediate and potential practical applications. Two specific applications, namely the detection of chromosomal abnormalities and the construction of mammalian artificial chromosomes, are discussed below.

Detection of chromosomal abnormalities

The identification of numerical and structural chromosomal abnormalities is an essential component in the diagnosis and treatment of various diseases, including both congenital and acquired disorders. Using standard cytogenetic techniques, one can achieve a typical visual resolution of a single band to within 5–10 megabases. These techniques, although highly precise and now well-standardized for routine cytogenetics laboratories, are labour-intensive, time-consuming, and require fresh viable specimens. Furthermore, the identification of the origins of marker chromosomes and the determination of complex rearranged chromosomes, especially in cancer cytogenetic cases, present particularly frustrating challenges. Fluorescence *in situ* hybridization (FISH) technology overcomes most of these difficulties in an increasing array of applications for chromosome analysis. In certain situations, it increases the sensitivity and resolution levels of diagnosis that are previously unattainable by conventional cytogenetics methods, shortens the reporting time, and often bypasses the culturing process to accommodate less viable specimens. Through the development of multi-colour probes, epifluorescence filter sets, and advanced computer softwares, it is now possible to simultaneously detect and discriminate as many as 27 different DNA probes hybridized simultaneously to the chromosomes (Beau 1996; Speicher *et al.* 1996). The commercial availability of many different types of FISH probes has also made these probes more accessible to the investigators and has accelerated their rapid acceptance by clinical laboratories as an important adjunct study.

The DNA probes that are currently available for FISH studies can be grouped into four classes (Yung 1996). These are single-copy probes, whole- or part-chromosome painting probes, telomeric probes, and centromeric probes. The limitations and strengths of each of these classes of probes are discussed below.

Single-copy probes

This class of FISH probes consists of locus-specific DNA segments and is suitable for hybridizing both metaphase and interphase cells. The DNA segments may be as small as 1–2 kb and may contain only unique sequences (e.g. cDNA or selected genomic DNA segments that have been screened for the absence of repetitive DNA), or they may be as large as 40–45 kb (e.g. cosmid probes), 100–300 kb (e.g. BAC or PAC probes), or up to 1,500 kb (e.g. YAC probes) (see below for definitions of BAC, PAC and YAC). These larger probes almost invariably contain one or more types of interspersed repetitive DNA, such as Alu repeats, SINES, and LINES, and their use in FISH generally requires an additional experimental step involving competition with a vast excess of unlabelled total human genomic DNA or human "cot-1 DNA" (i.e. DNA enriched for moderately to highly repeated DNA of the human genome) to suppress the hybridization activity of the repetitive sequences. Since this class of probes is locus-specific, it is ideal for detecting microdeletion or microduplication of a specific chromosome region that is unrecognizable by routine cytogenetic methods. These probes are also useful for identifying known recurring translocations found in certain types of leukaemia patients, such as t(9;22) in CML patients, and for detecting aneuploidy (numerical chromosomal abnormality).

Chromosome painting probes

These probes are made up of a mixture of many DNA segments, each complementary to a specific site along the length of a single chromosome. The DNA segments may be derived from an entire chromosome, one arm of a chromosome, or only a portion of a chromosome arm. The combined use of such a mixture of DNA segments in a single hybridization results in the corresponding region of the chromosome lighting up or being "painted". This class of probes is especially useful for the accurate diagnosis of small marker chromosomes. For effective application of these probes in clinical diagnosis, it is important to know the precise regions in which the DNA mixture can paint. The problem of choosing the right chromosome-paint is increasingly being overcome with the development of multicolour-labelling and detection systems which allow simultaneous analysis of multiple or even all human chromosomes (Beau 1996; Speicher *et al.* 1996). A potential limitation for these painting probes is that they may not completely cover the whole designated chromosome region. Furthermore, they are not useful for structural changes within the same chromosome, such as inversions.

Telomeric probes

The telomere is the terminal cap structure of each chromosome. These structures are essential for cell survival because intact telomeres are important for replication and stability of chromosomes. A number of studies suggest that telomeres may be involved in breakage and healing events that can result in terminal deletion, gene amplification and cryptic translocation (Brown *et al.* 1990; Smith *et al.* 1992). Several reports have described the use of telomeric probes in the identification of clinical cases which result from cryptic translocation that would otherwise have gone undetected by standard cytogenetic analysis, indicating the utility of these probes as a diagnostic tool for such translocations (Lamb *et al.* 1989; Overhauser *et al.* 1989; Altherr *et al.* 1991; Kuwano *et al.* 1991). The recent publication by the National Institute of Health and Institute of Molecular Medicine Collaboration (1996) of the isolation of a complete set of specific FISH probes for each human telomere should increase the usefulness of these probes. These new probes will be especially suitable for rapid prenatal determination of fetal karyotype, where one parent is a known reciprocal translocation carrier. Other applications will include detection of cryptic translocations and deletions in families with unexplained mental retardation and/or dysmorphic features, and in couples with numerous spontaneous miscarriages.

A complete collection of telomeric probes can also be used as a screening tool to diagnose congenital anomalies and acquired aberrations presented in cancer patients. Routine cytogenetic study of chromosome preparations from bone marrow and solid tumours is often very testing, since these preparations are known for their poor morphology and limited resolution, and yet they often carry the most complex and unpredictable translocations and structural rearrangements. Use of telomeric probes to pre-screen the terminal integrity of all chromosomes will save labour, time, and material. With so many telomeric probes now available, an important future development is to incorporate the multi-colour labelling and detection system now established for whole-chromosome painting to improve the efficiency and cost-effectiveness of the procedure.

Centromeric probes

Human centromeric DNA contains repetitive α-satellite sequences that are found in great abundance on each centromere. As described in Chapter 5, α-satellite DNA probes specific for individual or a small subgroup of human chromosomes are now available for all the human chromosomes, making these probes very useful for chromosome identification. Because of the highly repetitive nature of these probes, a short hybridization time is required to produce strong signals. Another advantage of these probes is that both condensed metaphase chromosomes and interphase nuclei can be studied, since the centromeric DNA remains condensed at these cell cycle phases and will generate compact and well-resolved signals. These advantages make centromeric probes especially useful for prenatal and neonatal cases which require rapid diagnostic information for aneuploidy (reviewed by Kearns and Pearson 1994). Other applications include rapid assessment of the outcome of a sex-mismatched bone marrow transplant and determination of the percentage of abnormal cells in patients with mosaic numerical abnormality. This class of probes is also suitable for detecting numerical chromosome abnormality in interphase nuclei from neoplastic or product-of-conception specimens, where the process of culturing may present a problem, such as distortion of the natural distribution of aneuploidic cells in the original specimen. In addition, the centromeric probes have been shown to be good reagents for the classification of constitutive marker chromosomes (reviewed by Callen 1994).

A limitation of the centromeric probes is that, without prior cytogenetic or other FISH studies, a structurally rearranged chromosome cannot be distinguished from a normal one. Because of this limitation, these probes are best used for rapid preliminary diagnosis of numerical abnormality followed by confirmation using cytogenetic analysis. Another drawback when using the α-satellite probes is the high degree of quantitative heteromorphism associated with this DNA on different centromeres (Chapter 7). This means that some centromeres could occasionally escape detection due to extremely low levels or absence of this DNA, which could result in a false negative diagnosis (Verma and Luke 1992).

Mammalian artificial chromosomes (MACs)

An artificial chromosome is a defined structure that carries all the necessary functional elements for its long-term survival, replication, and segregation in a cell. Yeast artificial chromosomes (YACs) have been created and used in genetic studies in budding and fission yeasts for some time now and should serve as an excellent model system for the higher eukaryotes (Murray and Szostak 1983; Hahnenberger *et al.* 1989). Artificial minichromosomes have also been created in other lower organisms, including the fruit fly *Drosophila melanogaster* (Murphy and Karpen 1995) (Chapter 5), and the parasitic protozoan *Trypanosoma brucei* (Patnaik *et al.* 1996).

The construction of MACs has recently received a good deal of attention. The main reasons for this are that these artificial chromosomes could potentially offer a highly tractable system to study the structure and function of the mammalian chromosomes, provide a much needed functional assay for specific chromosomal components (including the centromere, telomere, and origin of replication), allow the cloning and analysis of large genes or clusters of genes in mammalian cells, used as an extrachromosomal vector for the production of transgenic animals, and provide an alternative strategy for somatic gene therapy (Huxley 1994; Monaco and Larin 1994).

Construction of MACs

Extrapolating from the YAC system, a functional MAC has to have at least three components for structural stability, DNA replication, and proper mitotic segregation. These components are: two telomeres, one or more origins of DNA replication, and a centromere (Tyler-Smith and Willard 1993). At present, the telomeres are the only component that has been cloned and understood sufficiently to be reliably used in the construction of a MAC. Since replication origins are thought to be abundant, with one located every 30–200 kb in the mammalian genome, it may be assumed, without actually undertaking the specific steps to identify or localize these origins, that they are likely to be present on any large piece of DNA in the hundreds of kilobase range that is used for the construction of a MAC. At present, the insufficiently defined nature of the mammalian centromere DNA appears to be the major stumbling block in the construction of MACs. It is around this stumbling block that the two currently used construction strategies have evolved.

In vitro *construction of a MAC*

This strategy involves the *in vitro* assembly of a MAC from cloned DNA sequences predicted to provide functions for the three minimal MAC components prior to introduction of the MAC into mammalian cells. In this approach, a DNA of 100–1000 kb, containing a candidate mammalian centromere sequence and, it is hoped, one or more origins of replication, is cloned into a yeast vector system carrying mammalian telomere ends. (The use of natural yeast telomere ends is not a viable option since these have never been reported to function in mammalian cells.) Following this, the construct is transformed into yeast cells as a linear molecule. A major limitation of this strategy is that yeast telomerase will cap mammalian chromosome ends with DNA characteristic of yeast telomeres (Cross *et al.* 1990), and it has been shown that the functional efficiency of the telomeres of a YAC retrofitted with cloned human telomeric DNA and propagated in yeast is dramatically reduced (Taylor *et al.* 1994). One possible solution to this problem is shown in Fig. 8.1A and involves the preparation of constructs in which the yeast and mammalian telomeric DNA sequences are arranged in tandem, with the yeast telomeres placed at the ends of the linear constructs in such a way that, following propagation in yeast and DNA isolation, the yeast telomeric ends can be removed specifically to expose the mammalian telomeres before the introduction of the MACs into mammalian cells.

Another potential limitation of this strategy is the technical difficulty associated with the cloning of large genomic DNA fragments into YAC vectors and the handling of yeast cells. This problem may in part be overcome using bacterial cloning systems involving bacterial

(A)

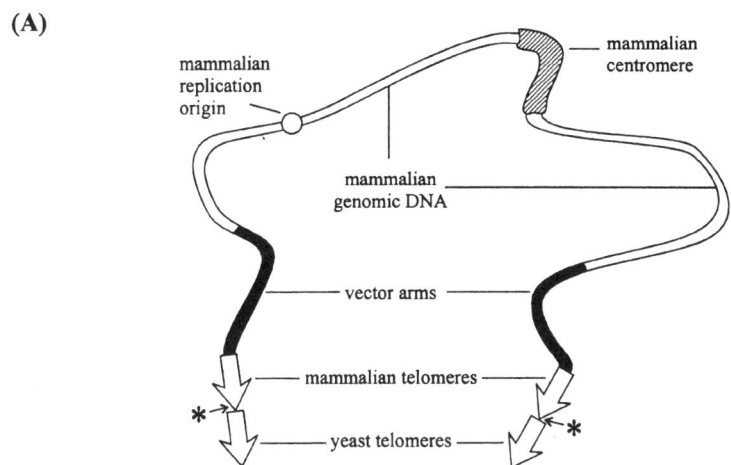

(B)

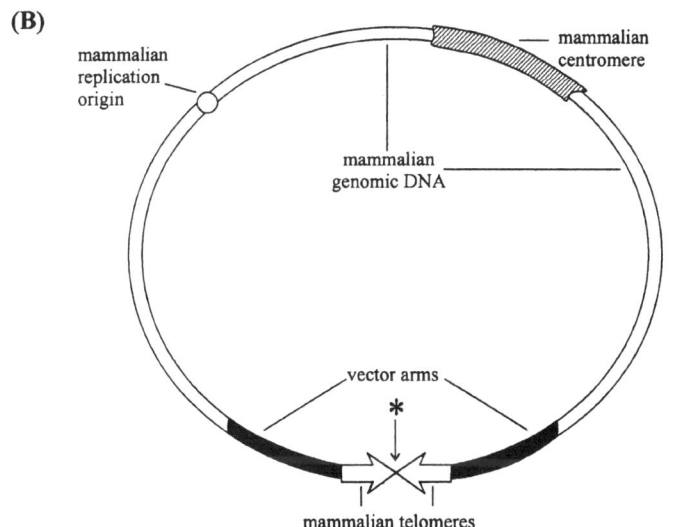

Fig. 8.1 Minimal structural components of a linear (A) or circular (B) MAC. The asterisk indicates a unique restriction enzyme site (such as I-*Sce*I, which is not present in the mammalian genome) that has been included between the mammalian and yeast telomeres (for A) or between two mammalian telomeres (for B) to allow the mammalian telomeres to be exposed prior to introduction of the MAC DNA into mammalian cells.

artificial chromosome (BAC; Shizuya *et al.* 1992) or P1-derived artificial chromosome(PAC; Ioannou *et al.* 1994) vectors (Monaco *et al.* 1994). The basic structure for these artificial chromosomes is similar to that described for the yeast constructs except that the chromosomes will initially be prepared as circular molecules as shown in Fig. 8.1B. However, whilst the BAC and PAC cloning systems offer a number of advantages over the yeast system, such as higher transformation efficiencies and easier handling, they are suitable only for the cloning of genomic sequences that are below 300 kb.

The *in vitro* MAC construction approach has the benefit that it gives good control over what goes into the MAC, including the incorporation of functional genes for gene expression and gene therapy studies. It allows genetic manipulation, such as *in vitro* mutagenesis of the mammalian centromere or any other chromosomal structures or expressed sequences to define critical elements. However, this approach suffers a number of drawbacks. (i) There is as yet no mammalian centromere DNA that has been clearly shown to work in this assay; various workers who are attempting to use human α-satellite or mouse minor satellite DNA have so far not reported any success. (ii) Cloning instability is a major problem, especially when the highly repetitive DNA sequences of the normal mammalian centromeres are involved. One possible solution is to use a repetitive DNA-free mammalian centromere, such as those described for the stable human marker chromosomes (Table 7.3). (iii) There may be a size limit for an artificial chromosome below which the structure will not be stable in mammalian cells. If this limit is substantially greater than 1 Mb, then preparing and handling such mega-size MACs will be difficult using current cloning technology.

MAC production by chromosome fragmentation

The second approach to MAC construction entails the progressive fragmentation of an existing chromosome *in situ* down to the size of a suitable minichromosome (Fig. 8.2). Several workers have reported that short stretches (several hundred base pairs) of the hexameric repeat $(TTAGGG)_n$ can function to produce *de novo* telomeres at previously interstitial chromosomal sites when introduced into mammalian cells (Farr *et al.* 1991; Barnett *et al.* 1993; Hanish *et al.* 1994). By combining this approach with the application of biochemical selection, it is possible to isolate fragmentation events involving specific chromosomes (Farr *et al.* 1992, 1995). Furthermore, by incorporating genomic DNA segments into the fragmentation construct to facilitate homologous recombination events in the nucleus of the mammalian cells, it is possible to achieve targetted truncation at designated chromosomal sites (Itzhaki *et al.* 1992; Brown *et al.* 1994). For example, using a construct containing α-satellite DNA, Brown *et al.* (1994) fragmented the human Y chromosome within its centromeric α-satellite array and created two derivative artificial chromosomes, one consisting of the long arm and 480 kb of α-satellite DNA, and the other consisting of the short arm and 140 kb of α-satellite DNA.

Two groups have specifically used telomere-associated chromosome fragmentation to produce human-derived minichromosomes. The first group used two different selectable markers in two rounds of telomere-associated chromosome breakage to generate a single X-derived minichromosome (Farr *et al.* 1995). The first break was within the centromeric α-satellite DNA while the second break was in the proximal short arm, thus deleting the entire long arm and generating a short arm acrocentric derivative of the chromosome. The resulting minichromosome is ∼7-8 Mb in size and contains ∼2.5 Mb of α-satellite DNA array. Cytogenetic analysis indicates that the chromosome is stably retained for 60 cell divisions in the absence of selection pressure. The second group, employing a three-selectable-marker system, has achieved three rounds of telomere-associated chromosome fragmentation and generated a set of eight minichromosome derivatives of the Y chromosome (Heller *et al.* 1996; Brown *et al.* 1996). These minichromosomes, which include representatives of both the short and long arms of the chromosome, range between 2.5 Mb and ∼8 Mb in size. The single minichromosome derivative of the short arm has a structure that is consistent with it being a simple truncation of the short arm, whereas all of the long-arm derivatives are rearranged by more than a simple truncation, including rearrangement in the structure of the α-satellite DNA array with respect to the starting cell line. Except for the smallest, 2.5 Mb-minichromosome, which shows an elevated and variable copy number, the other minichromosomes with sizes 4 Mb or larger all appear to be mitotically stable. The aberration seen with the 2.5 Mb-minichromosome may reflect the possibility that the minichromosome is approaching the lower size limit for stability in the host cells or because critical cis-acting functional DNA elements have been deleted from the minichromosome.

The telomere-associated chromosome fragmentation approach has several advantages over the *in vitro* MAC assembly strategy. First, it does not involve the technically difficult handling and cloning of large genomic DNA fragments, and the subsequent introduction of the large constructs into mammalian cells. Second, it utilizes the endogenous centromere structure and therefore avoids uncertainties associated with the choice of centromere sequences in the *in vitro* cloning strategy. Third, by controlling the chromosomal sites for fragmentation, the size of the artificial chromosome can be increased to significantly beyond the 1–2 Mb constraint imposed by existing cloning methods. However, a major disadvantage of this approach is that it does not allow easy genetic manipulation of the MAC, such as the introduction of large therapeutic genes, or extensive mutagenesis for functional dissection. This limitation may be alleviated to some extent if the MAC produced can be isolated and transformed into yeast, but the loss of human telomeric ends in yeast could pose a problem during subsequent re-introduction into mammalian cells and could require re-fitting of such ends.

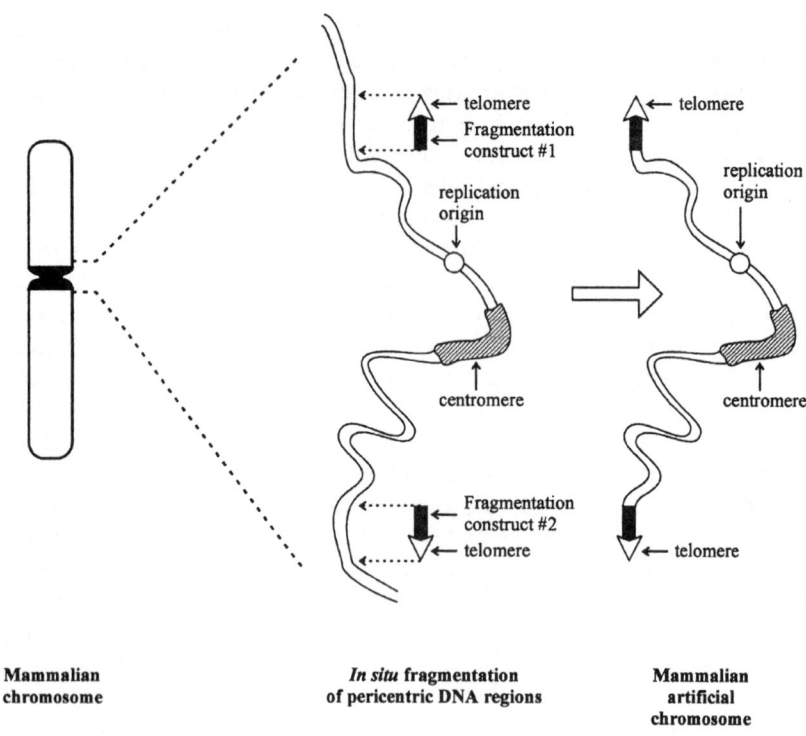

Mammalian chromosome

In situ **fragmentation of pericentric DNA regions**

Mammalian artificial chromosome

Fig. 8.2 *In situ* construction of mammalian artificial chromosome by successive rounds of random or targeted telomere-associated fragmentation of an existing chromosome using two constructs (1 and 2) on opposite sides of the centromere.

Future efforts in MAC construction

Future efforts in MAC production would clearly need to focus on defining the minimum DNA elements that are required for centromere and chromosome replication functions. Whether a stable chromosome can be generated that consists of only these minimum DNA elements flanked by mammalian telomeres also needs to be established, since there may be interference of activities of the different DNA elements when they are juxtaposed. It is also important to determine if a variety of genes can be stably maintained in the MACs and be properly regulated (see Position effect variegation, p. 89) when introduced into mammalian cells, although encouraging results have indicated that sandwiching a gene within 12 kb of a functional telomere and α-satellite DNA does not result in the silencing of gene activity (Bayne *et al.* 1994). Experience with yeast has indicated that YACs that are too small (< 100 kb) are generally not stable. Such a size constraint may similarly be true for MACs, but this needs to be empirically determined. It is possible that 'stuffer' DNA may have to be added to the MACs both to moderate potential interference between functional DNA elements and to increase the size of a MAC to provide stability. In addition to defining these parameters, it is important to establish conditions that would allow MACs that have been introduced or formed inside mammalian cells to be retrieved into bacteria or yeast to allow manipulation. Finally, efficient methods for delivering large DNA into mammalian cells need to be developed. Such methods should satisfy not only delivery into cells in culture, but also into specific tissues in a whole animal for gene therapy purposes.

Use of MACs in gene therapy

Gene therapy is increasingly being heralded as a promising new approach for the treatment of a variety of diseases. Such diseases include metabolic and neurological genetic disorders (Friedmann 1994; Kay and Woo 1994), infectious diseases such as AIDS (Gilboa and Smith 1994), and cancer (Culver and Blaese 1994; Hargest and Williamson 1996). Most current gene therapy protocols are based on viral vectors that utilize the infectivity of viruses, such as retrovirus, adenovirus, adeno-associated virus, and herpes virus, to deliver genes into body cells. These viral vectors have limitations of small insert size capacity, narrow target tissue range, integration of DNA into host genome, and risk of viral infection and immune reactions. Alternative nonviral strategies involving the use of naked DNA encapsulated in cationic liposomes, direct injection of plasmid DNA, and air-gun injection of a DNA-coated pellet, are also currently being developed (see Marshall 1995). DNA carrying large genes can potentially be delivered using these physical methods, but the introduced DNA needs to be integrated into the host genome to achieve stability. Any integration event is potentially undesirable since it will carry the risk of mutagenizing the host genome and of exerting position effects (reviewed by Milot *et al.* 1996) that could alter gene expression.

Compared with the above strategies, the use of MACs as an alternative approach for human gene therapy offers several distinct advantages. Firstly, since MACs do not involve viruses, they are not associated with the limitations seen with the viral vectors. Secondly, unlike both the viral and naked DNA approaches, a properly constructed MAC does not need to integrate into the host genome, but will remain extrachromosomally as a stable 47th human chromosome. Thirdly, the large insert size capacity of a MAC means that big genes or gene complexes can be used in the context of most or all of the necessary *cis*-acting regulatory elements.

Conclusion

In this chapter, we have described two specific practical applications of centromere study. A third, and potentially highly significant, application is to understand the aetiology of human anuploidies sufficiently to allow possible diagnosis and prevention in some situations. Human aneuploidies, as a group, constitute the single most common cause of genetic abnormality, but their aetiological mechanisms are largely unknown. Although many factors are expected to contribute to these mechanisms, the centromere, being a key component of the chromosomal segregation and cell division processes, is likely to play a major part. However, despite recent major advances in the centromere field, the level of our understanding remains insufficient to permit a systematic investigation of such mechanisms. This difficulty, together with those described for MAC construction, clearly underlie the need to significantly increase our present knowledge of the centromere.

References

Altherr, M., Bengtsson, U., Elder, F., Ledbetter, D., Wasmuth, J., McDonald, M., Gusella, J., and Greenberg, F. (1991). Molecular confirmation of Wolf-Hirschhorn syndrome with a subtle translocation of chromosome 4. Am. J. Hum. Genet. 49, 1235-1242.

Barnett, M. A., Buckle, V. J., Evans, E. P., Porter, A. C. G., Rout, D., Smith, A. G., et al. (1993). Telomere directed fragmentation of mammalian chromosomes. Nucleic Acids Res. 21, 27–36.

Bayne, R. A., Broccoli, D., Taggart, M. H., Thomson, E. J., Farr, C. J., and Cooke, H. J. (1994). Sandwiching of a gene within 12 kb of a functional telomere and alpha satellite does not result in silencing. Hum. Mol. Genet. 3, 539–546.

Beau, M. M. L. (1996). One FISH, two FISH, red FISH, blue FISH. Nature Genet. 12, 341–344.

Brown, K. E., Barnett, M. A., Burgtorf, C., Shaw, P., Buckle, V. J., and Brown, W. R. A. (1994). Dissecting the centromere of the human Y chromosome with cloned telomeric DNA. Hum. Mol. Genet. 3, 1227–1237.

Brown, W., Heller, R., Loupart, M., Shen, M., and Chand, A. (1996). Mammalian artificial chromosomes. Curr. Opin. Genet. Develop. 6, 281–288.

Brown, W., MacKinnon, P., Villasante, A., Spurr, N., Buckle, V., and Dobson, M. (1990). Structure and polymorphism of human telomere-associated DNA. Cell 63, 119–132.

Callen, D. F. (1994). Characterization of constitutive marker chromosomes in humans. In *In Situ* Hybridization Protocols, ed. Choo, K.H.A. Humana Press, Totowa, 439–457.

Cross, S., Lindsey, J., Fantes, J., McKay, S. J., McGill, N. I., and Cooke, H. J. (1990). The structure of a subtelomeric repeated sequence present on many human chromosomes. Nucleic Acids Res. 18, 6649–6657.

Culver, K. and Blaese, M. (1994). Gene therapy for cancer. Trends Genet. 10, 174–178.

Farr, C. J., Fantes, J., Goodfellow, P., and Cooke, H. J. (1991). Functional reintroduction of human telomeres into mammalian cells. Proc. Natl Acad. Sci. USA 88, 7006–7010.

Farr, C. J., Stevanovic, M., Thomson, E. J., Goodfellow, P. N., and Cooke, H. J. (1992). Telomere-associated chromosome fragmentation: applications in genome manipulation and analysis. Nature Genet. 2, 275–282.

Farr, C. J., Bayne, R. A. L., Kipling, D., Mills, W., Critcher, R., and Cooke, H. J. (1995). Generation of a human X-derived minichromosome using telomere-associated chromosome fragmentation. EMBO J. 14, 5444–5454.

Friedmann, T. (1994). Gene therapy for neurological disorders. Trends Genet. *10*, 210–214.

Gilboa, E. and Smith, C. (1994). Gene therapy for infectious diseases: the AIDS model. Trends Genet. *10*, 139–144.

Hahnenberger, K. M., Baum, M. P., Polizzi, C. M., Carbon, J., and Clarke, L. (1989). Construction of functional artificial minichromosomes in the fission yeast *Schizosaccharomyces pombe*. Proc. Natl Acad. Sci. USA *86*, 577–581.

Hanish, J. P., Yanowitz, J. L., and Lange, T. d. (1994). Stringent sequence requirements for the formation of human telomeres. Proc. Natl Acad. Sci. USA *91*, 8861–8865.

Hargest, R. and Williamson, R. (1996). Prophylactic gene therapy for cancer. Gene Therapy *3*, 97–102.

Heller, R., Brown, K., Burgtorf, C., and Brown, W. (1996). Mini-chromosomes derived from the Y chromosome by telomere directed chromosome breakage. Proc Natl Acad Sci USA 93, 7125–7130.

Huxley, C. (1994). Mammalian artificial chromosomes: a new tool for gene therapy. Gene Therapy *1*, 7–12.

Ioannou, P., Amemiya, C., Garnes, J., Kroisel, P., Shizuya, H., Chen, C. *et al.* (1994). A new bacteriophage P1-derived vector for the propagation of large human DNA fragments. Nature Genet. *6*, 84–89.

Itzhaki, J. E., Barnett, M. A., MacCarthy, A. B., Buckle, V. J., Brown, W. R. A., and Porter, A. C. G. (1992). Targeted breakage of a human chromosome mediated by cloned human telomeric DNA. Nature Genet. *2*, 283–287.

Kay, M. and Woo, S. (1994). Gene therapy for metabolic disorders. Trends Genet. *10*, 253–257.

Kearns, W. G. and Pearson, P. L. (1994). Detection of chromosomal aberrations in interphase and metaphase cells in prenatal and postnatal studies. In *In Situ* Hybridization Protocols, ed. Choo, K.H.A. Humana Press, Totowa, 459–476.

Kuwano, A., Ledbetter, S., Dobyns, W., Emanuel, B., and Ledbetter, D. (1991). Detection of deletions and cryptic translocations in Miller-Dieker syndrome by in situ hybridization. Am. J. Hum. Genet. 49, 707–714.

Lamb, J., Wilkie, A., Harris, P., Buckle, V., Lindenbaum, R., Barton, N., Reeders, S., Weatherall, D., and Higgs, D. (1989). Detection of breakpoints in submicroscopic chromosomal translocation, illustrating an important mechanism for genetic disease. Lancet ii, 819-824.

Marshall, E. (1995). Gene therapy's growing pains. Science *269*, 1050–1055.

Milot, E., Fraser, P., and Grosveld, F. (1996). Position effects and genetic disease. Trends Genet. *12*, 123–126.

Monaco, A. P. and Larin, Z. (1994). YACs, BACs, PACs and MACs: artificial chromosomes as research tools. Trends Biotechnol. *12*, 280–286.

Murphy, T. D. and Karpen, G. H. (1995). Localization of centromere function in a *Drosophila* minichromosome. Cell *82*, 599–609.

Murray, A. W. and Szostak, J. W. (1983). Construction of artificial chromosomes in yeast. Nature *305*, 189–193.

National Institute of Health and Institute of Molecular Medicine Collaboration. (1996). A complete set of human telomeric probes and their clinical application. Nature Genet 14, 86–89.

Overhauser, J., Bengtsson, U., McMahon, J., Ulm, J., Butler, M., Santiago, L., and Wasmuth, J. (1989). Prenatal diagnosis and carrier detection of a cryptic translocation by using DNA markers from the short arm of chromosome 5. Am. J. Hum. Genet. 45, 296–303.

Patnaik, P. K., Axelrod, N., Ploeg, L. H. T. V. d., and Cross, G. A. M. (1996).

Artificial linear mini-chromosomes for *Trypanosoma brucei*. Nucleic Acids Res. *24*, 668–675.

Shizuya, H., B, B., Kim, U., Mancino, V., Slepak, T., Tachiri, Y., and Simon, M. (1992). Cloning and stable maintenance of 300-kilobase-pair fragments of human DNA in *Escherichia coli* using an F-factor-based vector. Proc. Natl Acad. Sci. USA *89*, 8794–8797.

Smith, K., Stark, M., Gorman, P., and Stark, G. (1992). Fusion near telomeres occur very early in the amplification of CAD genes in Syrian hamster cells. Proc. Natl. Acad. Sci. USA 89, 5427–5431.

Speicher, M. R., Ballard, S. G., and Ward, D. C. (1996). Karyotyping human chromosomes by combinatorial multi-fluor FISH. Nature Genet. *12*, 368–375.

Taylor, S. S., Larin, Z., and Tyler-Smith, C. (1994). Addition of functional human telomeres to YACs. Hum. Mol. Genet. *3*, 1383–1386.

Tyler-Smith, C. and Willard, H. F. (1993). Mammalian chromosome structure. Curr. Biol. *3*, 390–397.

Verma, R. A. and Luke, S. (1992). Variations in alphoid DNA sequences escape detection of aneuploidy at interphase by FISH technique. Genomics *14*, 113–116.

Yung, J. (1996). New FISH probes—the end in sight. Nature Genet 14, 10–12.

Index